Martingale Limit Theory and Its Application

This is a volume in
PROBABILITY AND MATHEMATICAL STATISTICS

A Series of Monographs and Textbooks

Editors: Z. W. Birnbaum and E. Lukacs

A complete list of titles in this series can be obtained from the Publisher upon request.

Martingale Limit Theory and Its Application

P. Hall

Department of Statistics, SGS
Australian National University
Canberra, Australia

C. C. Heyde

CSIRO Division of Mathematics and Statistics
Canberra City, Australia

1980

ACADEMIC PRESS
A Subsidiary of Harcourt Brace Jovanovich, Publishers
New York London Toronto Sydney San Francisco

The history of probability (and of mathematics in general) shows a stimulating interplay of the theory and applications: theoretical progress opens new fields of applications, and in turn applications lead to new problems and fruitful research. The theory of probability is now applied to many diverse fields, and the flexibility of a general theory is required to provide appropriate tools for so great a variety of needs.

W. Feller

... the epistemological value of the theory of probability is revealed only by limit theorems. Moreover, without limit theorems it is impossible to understand the real content of the primary concept of all our sciences—the concept of probability.

B. V. Gnedenko and A. N. Kolmogorov

ACADEMIC PRESS, INC.
111 Fifth Avenue, New York, New York 10003

United Kingdom Edition published by
ACADEMIC PRESS, INC. (LONDON) LTD.
24/28 Oval Road, London NW1 7DX

Library of Congress Cataloging in Publication Data

Hall, P
Martingale limit theory and its application.

(Probability and mathematical statistics)
Bibliography: p.
Includes indexes.
1. Martingales (Mathematics) 2. Limit theorems (Probability theory) I. Heyde, C. C., joint author. II. Title.
QA274.5.H34 519.2'87 80–536
ISBN 0–12–319350–8

PRINTED IN THE UNITED STATES OF AMERICA

80 81 82 83 9 8 7 6 5 4 3 2 1

Contents

4 Invariance Principles in the Central Limit Theorem and Law of the Iterated Logarithm

5 Limit Theory for Stationary Processes via Corresponding Results for Approximating Martingales

6 Estimation of Parameters from Stochastic Processes

7 Miscellaneous Applications

Appendix

Preface

This book was commenced by one of the authors in late 1973 in response to a growing conviction that the asymptotic properties of martingales provide a key prototype of probabilistic behaviour, which is of wide applicability. The evidence in favor of such a proposition has been amassing rapidly over the intervening years—so rapidly indeed that the subject kept escaping from the confines of the text. The coauthor joined the project in late 1977.

The thesis of this book, that martingale limit theory occupies a central place in probability theory, may still be regarded as controversial. Certainly the story is far from complete on the theoretical side, and many interesting questions remain over such issues as the relationship between martingales and processes embeddable in or approximable by Brownian motion.[1] On the other hand, the picture is much clearer on the applied side. The vitality and principal source of inspiration of probability theory comes from its applications. The mathematical modeling of physical reality and the inherent nondeterminism of many systems provide an expanding domain of rich pickings in which martingale limit results are demonstrably of great usefulness.

The effective birth of probability as a subject took place in hardly more than a decade around 1650,[2] and it has been largely wedded to independence theory for some 300 years. For all the intrinsic importance and intuitive content of independence, it is not a vital requirement for the three key limit laws of probability—the strong law of large numbers, the central limit theorem, and the law of the iterated logarithm. As far as these results are concerned, the time has come to move to a more general and flexible framework in which suitable generalizations can be obtained. This is the story of the first part of the book, in which it is argued that martingale limit theory provides the most general contemporary setting for the key limit trio. The basic martingale tools, particularly the inequalities, have applications beyond

[1]For recent contributions, see Drogin (1973), Philipp and Stout (1975), and Monroe (1978). We have not directly concerned ourselves with these issues.

[2]See, for example, Heyde and Seneta (1977, Chapter 1).

the realm of limit theory. Moreover, extensions of the martingale concept offer the prospect of increased scope for the methodology.[3]

Historically, the first martingale limit theorems were motivated by a desire to extend the theory for sums of independent random variables. Very little attention was paid to possible applications, and it is only in much more recent times that applied probability and mathematical statistics have been a real force behind the development of martingale theory.[4] The independence theory has proved inadequate for handling contemporary developments in many fields. Independence-based results are not available for many stochastic systems, and in many more an underlying regenerative behaviour must be found in order to employ them. On the other hand, relevant martingales can almost always be constructed, for example by devices such as centering by subtracting conditional expectations given the past and then summing.

In this book we have chosen to confine our attention to discrete time. The basic martingale limit results presented here can be expected to have corresponding versions in continuous time, but the context has too many quite different ramifications and connotations[5] to be treated satisfactorily in parallel with the case of discrete time.

The word application rather than applications in the title of the book reflects the scope of the examples that are discussed. The rapidly burgeoning list of applications has rendered futile any attempt at an exhaustive or even comprehensive treatment within the confines of a single monograph. As a sample of the very recent diversity which we do not treat, we mention the coupon collectors problem [Sen (1979)], random walks with repulsion [Pakes (1980)], the assessment of epidemics [Watson (1980a, b)], the weak convergence of U-statistics [Loynes (1978), Hall (1979)] and of the empirical process [Loynes (1978)] and of the log-likelihood process [Hall and Loynes (1977)], and determining the order of an autoregression [Hannan and Quinn (1979)]. Martingale methods have also found application in many areas that are not usually associated with probability or statistics. For example, martingales have been used as a descriptive device in mathematical economics for over ten years, and more recently the limit theory has proved to be a powerful tool.[6] Our applications rather reflect the authors' interests, although it is hoped that they are diverse enough to establish beyond any doubt the usefulness of the methodology.

The book is intended for use as a reference work rather than as a textbook, although it should be suitable for certain advanced courses or seminars. The

[3]For example linear martingales [McQueen (1973)], weak martingales [Nelson (1970), Berman (1976)], and mixingales [McLeish (1975b, 1977)].

[4]This is a little ironic in view of the roots of the martingale concept in gambling theory.

[5]For example, through the association with stochastic integrals.

[6]See, for example, Foldes (1978), Plosser *et al.* (1979), and Pagan (1979).

prerequisite is a sound basic course in measure theoretic probability, and beyond that the treatment is essentially self-contained.

In bringing this book to its final form we have received advice from many people. Our grateful thanks are due particularly to G. K. Eagleson and also to D. Aldous, E. J. Hannan, M. P. Quine, G. E. H. Reuter, H. Rootzén, and D. J. Scott, who have suggested corrections and other improvements. Thanks are also due to the various typists who struggled with the manuscript and last but not least to our wives for their forbearance.

C. C. HEYDE
P. HALL

Canberra, Australia
November 1979

Notation

The following notation is used throughout the book.

a.s.	almost surely (that is, with probability one)
i.i.d.	independent and identically distributed
p.g.f.	probability generating function
r.v.	random variable
CLT	central limit theorem
LIL	law of the iterated logarithm
SLLN	strong law of large numbers
ML	maximum likelihood.

Almost sure convergence, convergence in probability, and convergence in distribution are denoted by $\xrightarrow{\text{a.s.}}$, $\xrightarrow{\text{p}}$, and $\xrightarrow{\text{d}}$, respectively.

For a random variable X, we use $\| X \|_p$ for $(E|X|^p)^{1/p}$, $p > 0$, while var X denotes the variance of X.

The metric space $C[0,1]$ is the space of continuous functions on the interval $[0,1]$ with the uniform metric ρ defined by

$$\rho(x,y) = \sup_{0 \leq t \leq 1} |x(t) - y(t)|.$$

The complement of an event E is denoted by E^c, and the indicator function of E by $I(E)$, where

$$I(E)\,(\omega) = \begin{cases} 1 & \text{if } \omega \in E \\ 0 & \text{otherwise.} \end{cases}$$

The normal distribution with mean μ and variance σ^2 is denoted by $N(\mu, \sigma^2)$.

The real and imaginary parts of a function f are denoted by $\mathbf{Re}\, f$ and $\mathbf{Im}\, f$, respectively.

For real numbers, $x+$ denotes max $(0, x)$, and sgn x is the sign of x, while $a \wedge b$ is $\min(a,b)$.

The transpose of a vector v is denoted by v', and the trace of a matrix A is written as tr A.

The square root of a nonnegative variable is taken to be nonnegative.

1
Introduction

1.1. General Definition

Let $(\Omega,\mathcal{F},P)$ be a probability space: Ω is a set, $\mathcal{F}$ a σ-field of subsets of Ω, and P a probability measure defined on $\mathcal{F}$. Let I be any interval of the form (a,b), $[a,b)$, $(a,b]$ or $[a,b]$ of the ordered set $\{-\infty,\ldots,-1,0,1,\ldots,\infty\}$. Let $\{\mathcal{F}_n, n \in I\}$ be an increasing sequence of σ-fields of $\mathcal{F}$ sets. Suppose that $\{Z_n, n \in I\}$ is a sequence of random variables on Ω satisfying

(i) Z_n is measurable with respect to $\mathcal{F}_n$,
(ii) $E|Z_n| < \infty$,
(iii) $E(Z_n|\mathcal{F}_m) = Z_m$ a.s. for all $m < n$, $m, n \in I$.

Then, the sequence $\{Z_n, n \in I\}$ is said to be a *martingale* with respect to $\{\mathcal{F}_n, n \in I\}$. We write that $\{Z_n,\mathcal{F}_n, n \in I\}$ is a martingale. If (i) and (ii) are retained and (iii) is replaced by the inequality $E(Z_n|\mathcal{F}_m) \geqslant Z_m$ a.s. ($E(Z_n|\mathcal{F}_m) \leqslant Z_m$ a.s.), then $\{Z_n,\mathcal{F}_n, n \in I\}$ is called a *submartingale* (*supermartingale*).

A *reverse martingale* or *backwards martingale* $\{Z_n, n \in I\}$ is defined with respect to a *decreasing* sequence of σ-fields $\{\mathcal{F}_n, n \in I\}$. It satisfies conditions (i) and (ii) above, and instead of (iii),

(iii′) $E(Z_n|\mathcal{F}_m) = Z_m$ a.s. for all $m > n$, $n, m \in I$.

Clearly $\{Z_i,\mathcal{F}_i, 1 \leqslant i \leqslant n\}$ is a reverse martingale if and only if $\{Z_{n-i+1}, \mathcal{F}_{n-i+1}, 1 \leqslant i \leqslant n\}$ is a martingale, and so the theory for finite reverse martingales is just a dual of the theory for finite (forward) martingales. The duality does not always extend so easily to limit theory.

1.2. Historical Interlude

The name *martingale* was introduced into the modern probabilistic literature by Ville (1939) and the subject brought to prominence through the work of Doob in the 1940s and early 1950s.

Martingale theory, like probability theory itself, has its origins partly in gambling theory, and the idea of a martingale expresses a concept of a fair game (Z_n can represent the fortune of the gambler after n games and $\mathscr{F}_n$ the information contained in the first n games). The term martingale has, in fact, a long history in a gambling context, where originally it meant a system for recouping losses by doubling the stake after each loss. The *Oxford English Dictionary* dates this usage back to 1815. The modern concept dates back at least to a passing reference in Bachelier (1900).

Work on martingale theory by Bernstein (1927, 1939, 1940, 1941) and Lévy (1935a,b, 1937) predates the use of the name martingale. These authors introduced the martingale in the form of consecutive sums with a view to generalizing limit results for sums of independent random variables. The subsequent work of Doob however, including the discovery of the celebrated martingale convergence theorem, completely changed the direction of the subject. His book (1953) has remained a major influence for nearly three decades. It is only comparatively recently that there has been a resurgence of real interest and activity in the area of martingale limit theory which deals with generalizations of results for sums of independent random variables. It is with this area that our book is primarily concerned.

1.3. The Martingale Convergence Theorem

This powerful result has provided much motivation for the continued study of martingales.

Theorem. *Let $\{Z_n, \mathscr{F}_n, n \geqslant 1\}$ be an L^1-bounded submartingale. Then there exists a random variable Z such that $\lim_{n\to\infty} Z_n = Z$ a.s. and $E|Z| \leqslant \liminf_{n\to\infty} E|Z_n| < \infty$. If the submartingale is uniformly integrable, then Z_n converges to Z in L^1, and if $\{Z_n, \mathscr{F}_n\}$ is an L^2-bounded martingale, then Z_n converges to Z in L^2.*

This is an existence theorem; it tells us nothing about the limit random variable save that it has a finite first or second moment. The theorem seems rather unexpected *a priori* and it is a powerful tool which has led to a number of interesting results for which it seems essentially a unique method of approach. Of course one is often still faced with finding the limit law, but that can usually be accomplished by other methods.

As a simple example of the power of the theorem, consider its application to show that if $S_n = \sum_{i=1}^n X_i$ is a sum of independent random variables with S_n converging in distribution as $n \to \infty$, then S_n converges a.s. This result is a straightforward consequence of the martingale convergence theorem when it is noted that $\{Z_n, n \geqslant 1\}$ defined by

$$Z_n = e^{itS_n}/E(e^{itS_n}) = e^{itS_n}\Big/\prod_{i=1}^{n} E(e^{itX_i})$$

is a martingale with respect to the sequence of σ-fields generated by the X_i, $1 \leqslant i \leqslant n$. The result is rather difficult to establish by other methods.

A proof of the martingale convergence theorem together with a discussion of related issues is provided in Chapter 2.

1.4. Comments on Classical Limit Theory and Its Analogs

It is our opinion that probability theory has its greatest impact through its limit theorems, and this book is concerned with the basic techniques for proving these theorems.

Take as a starting point the classical (Kolmogorov) form of the strong law of large numbers (SLLN). Let X_i, $i = 1, 2, \ldots$, be independent and identically distributed random variables (i.i.d. r.v.) and write $S_n = \sum_{i=1}^{n} X_i$, $n \geqslant 1$. Suppose that $E|X_1| < \infty$ and $EX_1 = \mu$. Then

$$n^{-1}S_n \xrightarrow{\text{a.s.}} \mu$$

as $n \to \infty$. This result embodies the idea of probability as a strong limit of relative frequencies [take $X_i = I_i(A)$, the indicator function of the set A at the ith trial; then $\mu = P(A)$] and on these grounds may be regarded as most basic in probability theory, in fact, the underpinning of the axiomatic theory as a physically realistic subject.

Of course, when we have an a.s. convergence result this is not the end of the story, and we may hope to say something about the rate of the convergence. As is well known, the imposition of the additional condition var $X_1 = \sigma^2 < \infty$ enables us to make two different kinds of rate statements. These are the central limit theorem (CLT),

$$\sigma^{-1}n^{1/2}(n^{-1}S_n - \mu) \xrightarrow{\text{d}} N(0,1)$$

($\xrightarrow{\text{d}}$ denotes convergence in distribution), and the law of the iterated logarithm (LIL),

$$n^{-1}S_n - \mu = \sigma\zeta(n)(2n^{-1}\log\log n)^{1/2},$$

where $\limsup_{n\to\infty} \zeta(n) = 1$ a.s. and $\liminf_{n\to\infty} \zeta(n) = -1$ a.s. These expressions may conveniently be interpreted as rate results about the SLLN. Their usual expressions suppress this basic relationship. The central limit theorem tells us just the right rate at which to magnify the difference $n^{-1}S_n - \mu$, which is tending a.s. to zero, in order to obtain convergence in distribution to a nondegenerate law. The law of the iterated logarithm gives, in a sense, the first term in the asymptotic expansion of $n^{-1}S_n - \mu$.

To highlight the delicacy of the LIL, we note from the CLT that

$$x_n^{-1} n^{1/2}(n^{-1}S_n - \mu) \xrightarrow{\text{p}} 0$$

($\xrightarrow{\text{p}}$ denotes convergence in probability) for any sequence of constants $\{x_n\}$ with $x_n \uparrow \infty$, while from the LIL,

$$x_n^{-1}\{n(\log\log n)^{-1}\}^{1/2}(n^{-1}S_n - \mu) \xrightarrow{\text{a.s.}} 0.$$

Hence the norming of the LIL provides a boundary between convergence in probability to zero and a.s. convergence to zero.

Next, we move to a discussion of limit results of a similar type in a more general setting. Suppose that $\{W_n\}$ is a stochastic process with $W_n \xrightarrow{\text{a.s.}} W$ for some W. It is useful to interpret this result as an SLLN analog and to ask whether analogs of the rate of convergence results hold. Thus, we seek a result of the kind "$B_n(W_n - W)$ converges in distribution to a proper limit law" for some sequence $\{B_n\}$ (possibily random) which increases to infinity (a CLT analog). Also, we seek a similar sequence $\{C_n\}$, again possibly random, such that

$$\limsup_{n\to\infty} C_n(W_n - W) = 1 \quad \text{a.s.}, \qquad \liminf_{n\to\infty} C_n(W_n - W) = -1 \quad \text{a.s.}$$

(an LIL analog). Examples in which these analogs exist abound in the literature; we shall quote some samples here and give further illustrations in later chapters. There are, in fact, many cases in which the use of the analogy leads to prospective research problems.

Example 1. (From renewal theory) Let $X_i, i = 1, 2, \ldots,$ be i.i.d. nonnegative random variables representing (say) lifetimes of mechanical components. Suppose that $0 < EX_1 = \mu < \infty$ and define for $x > 0$,

$$N(x) = \max[k | X_1 + \cdots + X_k < x],$$

the number of renewals before time x. Then, it is well known that [e.g., Chung (1974, p. 136)]

$$x^{-1}N(x) \xrightarrow{\text{a.s.}} \mu^{-1}$$

as $x \to \infty$ (SLLN analog). Furthermore, if $\operatorname{var} X_1 = \sigma^2 < \infty$, it is also well known that [e.g., Prabhu (1965, p. 159)]

$$\mu^{3/2}\sigma^{-1}x^{1/2}(x^{-1}N(x) - \mu^{-1}) \xrightarrow{\text{d}} N(0,1)$$

(CLT analog). More recently, it has been shown by Iglehart (1971) that

$$\limsup_{x\to\infty} \mu^{3/2}\sigma^{-1}[x(2\log\log x)^{-1}]^{1/2}(x^{-1}N(x) - \mu^{-1}) = 1 \quad \text{a.s.}$$

and correspondingly the lim inf of the same expression is -1 a.s. (LIL analog).

Example 2. (From extreme value theory) Let X_i, $i = 1, 2, \ldots$, be i.i.d. r.v. with a continuous distribution and set

$$V_1 = \min(n: X_n > X_1),$$
$$V_{r+1} = \min(n: n > V_r \text{ and } X_n > X_{V_r}), \qquad r \geqslant 1.$$

The V are the indices of successive record observations. Let $\Delta_1 = V_1$ and $\Delta_r = V_r - V_{r-1}$, $r > 1$, so that the Δ are the waiting times for the successive record observations. Then

$$r^{-1} \log \Delta_r \xrightarrow{\text{a.s.}} 1$$

[Holmes and Strawderman (1969)],

$$r^{1/2}(r^{-1} \log \Delta_r - 1) \xrightarrow{\text{d}} N(0,1)$$

[Neuts (1967)], and

$$\limsup_{r\to\infty} [r(2 \log\log r)^{-1}]^{1/2}(r^{-1} \log \Delta_r - 1) = 1 \quad \text{a.s.}$$

with the corresponding lim inf equal to -1 a.s. [Holmes and Strawderman (1970)]. These results provide analogs of the SLLN, CLT, LIL, respectively.

The results in Examples 1 and 2 retain a norming which is very similar to that of the standard case of the SLLN, CLT, and LIL. Such need not be the case, however, and norming by random variables rather than by constants may be appropriate, as in the next example.

Example 3. (From branching process theory) Let $Z_0 = 1, Z_1, Z_2, \ldots$, denote a supercritical Bienaymé–Galton–Watson process with nondegenerate offspring distribution, and let $1 < m = EZ_1 < \infty$. This process evolves in such a way that

$$Z_{n+1} = Z_n^{(1)} + Z_n^{(2)} + \cdots + Z_n^{(Z_n)},$$

where the $Z_n^{(i)}$ are i.i.d., each with the offspring distribution, and are independent of Z_n. It is well known that

$$m^{-n} Z_n \xrightarrow{\text{a.s.}} W$$

as $n \to \infty$ (a simple consequence of the martingale convergence theorem) [e.g., Harris (1963, p. 13)], where W is a random variable which turns out to be nondegenerate if and only if $EZ_1 \log^+ Z_1 < \infty$ [$\log^+ x = \max(0, \log x)$]. This result provides an SLLN analog. To obtain the corresponding CLT and LIL analogs we need to impose the additional condition $\operatorname{var} Z_1 = \sigma^2 < \infty$. Then, conditional on $Z_n > 0$,

$$\sigma^{-1}(m^2 - m)^{1/2} m^n Z_n^{-1/2}(m^{-n} Z_n - W) \xrightarrow{\text{d}} N(0,1)$$

(Heyde, 1971), and on the nonextinction set $\{W > 0\}$,

$$\limsup_{n\to\infty} \sigma^{-1}(m^2 - m)^{1/2} m^n (2Z_n \log n)^{-1/2} (m^{-n} Z_n - W) = 1 \quad \text{a.s.},$$

$$\liminf_{n\to\infty} \sigma^{-1}(m^2 - m)^{1/2} m^n (2Z_n \log n)^{-1/2} (m^{-n} Z_n - W) = -1 \quad \text{a.s.}$$

(Heyde and Leslie, 1971).

1.5. On the Repertoire of Available Limit Theory

Now let us take up the general question of establishing limit results of the type that we have just been discussing in Section 1.4. This raises the matter of tools and hence of what types of stochastic processes have been found amenable to general analysis along the lines of establishing limit results akin to the trio SLLN, CLT, and LIL. Broadly, such kinds of stochastic processes are in one of the following (overlapping) categories:

(1) Processes with independent increments.
(2) Markov processes.
(3) Gaussian processes.
(4) Stationary processes.
(5) Processes with increments satisfying asymptotic independence conditions.
(6) Processes whose increments form a multiplicative system.
(7) Martingales.

It can be claimed with reasonable justification that, at least as far as the SLLN, CLT, and LIL results are concerned, the limit theory for (7) essentially covers that for the other categories with the partial exception of (3). In the case of (3), special properties of the normal distribution can be important, but the power and usefulness of martingale methods has been amply illustrated in this context by Philipp and Stout (1975).

That (7) essentially covers (1) will be discussed in the next section in some detail (we are of course ignoring cases where means are not finite). The sufficiency parts of the key results for sums of independent random variables have been extended to the martingale case. This includes the Kolmogorov condition for the SLLN, the Lindeberg–Feller CLT, and the LIL for the stationary case. Generalizations will be given in later chapters.

Many Markov processes (such as those satisfying the so-called Doeblin condition) satisfy mixing conditions and hence can be classified under (5). However, the SLLN, CLT, and LIL for sums of functionals defined on a recurrent Markov chain may be obtained using the corresponding results for sums of independent random variables. The sum of interest is dissected into a sum of i.i.d. r.v. plus other (roundoff) terms which are asymptotically negligible under appropriate norming. See, for example, Chung (1967, Sections 14–16). The more recent work of Pakshirajan and Sreehari (1970) mimics the classical proof of the LIL for sums of independent random variables as given in Loève (1977). In a more general context, such as the work of Dobrushin (1956) on central limit behavior in the case of inhomogeneous Markov chains, the key tools are martingale limit results. Certain Markov chain central limit results have not to date been covered in the literature by the martingale methodology, for example, that of Keilson and Wishart (1964), for processes defined on a finite Markov chain, but it seems that martingale-based proofs can usually be found. Indeed, Resnick, in a private communication (1975) showed how to construct a martingale proof of the principal result of Keilson and Wishart (1964). This work of Keilson and Wishart has been used in Matthews (1970) to provide a central limit result for the case of absorbing Markov chains, conditional on absorption not having taken place.

In the case of limit results for processes with stationary increments, the ergodic theorem provides the basic SLLN for this context and this is a major exception to our thesis. The CLT and LIL results have been based, however, on various concepts of asymptotic independence (such as mixing conditions) and again make use of the classical limit theory for the independence case applied to appropriate blocks of summands. For recent discussions see, for example, Ibragimov and Linnik (1971, Chapter 18) and Oodaira and Yoshihara (1971a,b, 1972). A general martingale-based approach to some of these problems is discussed in detail in Chapter 5.

For processes having asymptotically independent increments which are not necessarily stationary, the limit theory for the independence case has again provided the basic tool with minor modification, for example, involving comparison of the dependent sequence with a certain independent one. For some recent discussions see Serfling (1968), Philipp (1969a,b), Bergström (1970), and Cocke (1972). The martingale approach has been discussed by Philipp and Stout (1975).

Processes whose increments form a multiplicative system and which have none of the other six properties arise comparatively infrequently in applications of stochastic processes. However, they are important in harmonic analysis and have algebraic applications in group theory. There is an extensive heritage of limit theory for such processes, dating back to the early CLT and LIL results of Salem and Zygmund (1947, 1948, 1950). Recent work has been done by Berkes (1973, 1975, 1976a,b), McLeish (1975a), Takahashi (1975), and Villalobos (1974). The proofs are often based on independence methods and make extensive use of orthogonality properties. However,

Philipp and Stout (1975) have used martingale techniques to re-prove and extend earlier results for lacunary trigonometric series.

The fact of the matter is, as we shall observe in the next section and in greater detail in later chapters, that the martingale limit apparatus (certainly as far as the sufficiency parts of the SLLN, CLT, and LIL are concerned) contains the independence-based apparatus and provides a more flexible tool for obtaining limit theorems. In the absence of a more inviting probabilistic structure, the best pieces of equipment in our armory of useful limit theorems are the martingale limit theorems. Furthermore, recent work of Gilat (1977), in which it is shown that every nonnegative submartingale can be represented as the absolute value of a martingale, offers the prospect of a significant extension of the scope of possible applications.

1.6. Martingale Limit Theorems Generalizing Those for Sums of Independent Random Variables

Sums of independent random variables centered at their expectations are martingales. The standard theory for sums of independent random variables has of course developed in its own right. Limit results such as the laws of large numbers, the central limit theorem, and the law of the iterated logarithm have been major achievements and have had a substantial impact. As mentioned above, however, this standard theory can for the most part be put into a more general martingale setting. From the point of view of applications there is considerable advantage in doing this; this is in fact the thesis of the book. We shall sketch the basic ideas here and give more details in the subsequent chapters.

Let $\{S_n, \mathscr{F}_n, n \geqslant 1\}$ be a zero-mean martingale and write $S_n = \sum_{i=1}^{n} X_i$. The first thing to note is that various results on sums of independent random variables in fact require only orthogonality of the increments ($E(X_i X_j) = 0$, $i \neq j$) and that this property holds for martingales whose increments have finite variance. As a simple example we see that the so-called Chebyshev inequality continues to hold for martingales. By Markov's inequality we have for any $\varepsilon > 0$,

$$P(|S_n| \geqslant \varepsilon) \leqslant \varepsilon^{-2} E S_n^2,$$

while

$$E S_n^2 = \sum_{j=1}^{n} E X_j^2 + 2 \sum_{i>j} E(X_i X_j) = \sum_{j=1}^{n} E X_j^2$$

since for $i > j$,

$$E(X_i X_j) = E(X_j E(X_i | \mathscr{F}_{i-1})) = E(X_j \{E(S_i | \mathscr{F}_{i-1}) - S_{i-1}\}) = 0.$$

This inequality for martingales yields immediately a *weak law of large numbers*. In fact, for square-integrable martingales,

$$P(|S_n| \geqslant n\varepsilon) \leqslant n^{-2}\varepsilon^{-2} \sum_{j=1}^{n} EX_j^2,$$

so that $n^{-1}S_n \xrightarrow{\text{p}} 0$ if $n^{-2}\sum_{j=1}^{n} EX_j^2 \to 0$.

In the general case of the weak law of large numbers, where the finite variance assumption is dropped, there is a well-known set of necessary and sufficient conditions for convergence in probability to zero of $n^{-1}S_n$, $S_n = \sum_{i=1}^{n} X_i$ denoting a sum of independent random variables. In fact [Loève (1977, p. 290)], $n^{-1}S_n \xrightarrow{\text{p}} 0$ if and only if

(i) $\sum_{j=1}^{n} P(|X_j| \geqslant n) \to 0$,
(ii) $n^{-1}\sum_{j=1}^{n} EX_{jn} \to 0$,
(iii) $n^{-2}\sum_{j=1}^{n} [EX_{jn}^2 - (EX_{jn})^2] \to 0$,

where X_{jn} is defined by $X_{jn} = X_j$ or 0 according as $|X_j| < n$ or $|X_j| \geqslant n$. An inspection of the proof given by Loève shows that the "if" part (which is after all the most important part) depends only on truncation and the use of the so-called Chebyshev inequality. A minor variant obtained in (ii) by replacing EX_{jn} by $E(X_{jn}|X_{j-1},\ldots,X_1)$ and $\to 0$ by $\to 0$ in probability, and in (iii) replacing $(EX_{jn})^2$ by $E(E(X_{jn}|X_{j-1},\ldots,X_1))^2$, holds in the martingale context. This of course reduces to Loève's condition in the special case of independence. The "only if" part of Loève's proof involves the use of characteristic functions and only this rests on the assumption of independence.

In the case of the *strong law of large numbers*, the most basic tool is probably the Kolmogorov criterion that if the X_i are independent with zero means and $EX_i^2 < \infty$ for each i, then $\sum b_n^{-2}EX_n^2 < \infty$ for some sequence $\{b_n\}$ of positive constants with $b_n \uparrow \infty$ entails $\lim_{n\to\infty} b_n^{-1}S_n = 0$ a.s. [e.g., Loève (1977, p. 250)]. That this result continues to hold in the case when $\{S_n\}$ is a martingale is a simple consequence of the martingale convergence theorem [e.g., Feller (1971, p. 242)].

In the case of the *central limit theorem* there is the following result of Brown (1971). Let $\{S_n,\mathscr{F}_n\}$ denote a zero-mean martingale whose increments have finite variance. Write

$$S_n = \sum_{i=1}^{n} X_i, \qquad V_n^2 = \sum_{i=1}^{n} E(X_i^2|\mathscr{F}_{i-1}), \qquad \text{and} \qquad s_n^2 = EV_n^2 = ES_n^2.$$

If

$$s_n^{-2}V_n^2 \xrightarrow{\text{p}} 1 \tag{1.1}$$

and

$$s_n^{-2} \sum_{i=1}^{n} E(X_i^2 I(|X_i| \geqslant \varepsilon s_n)) \to 0 \tag{1.2}$$

as $n \to \infty$, for all $\varepsilon > 0$ [$I(\cdot)$ denotes the indicator function], then

$$\lim_{n \to \infty} P(s_n^{-1} S_n \leqslant x) = (2\pi)^{-1/2} \int_{-\infty}^{x} e^{-u^2/2}\, du.$$

This result reduces to the sufficiency part of the standard Lindeberg–Feller result [e.g., Loève (1977, p. 292)] in the case of independent random variables, as the condition $s_n^{-2} V_n^2 \xrightarrow{\text{p}} 1$ is then trivially satisfied. The proof is a little more delicate than that of the Lindeberg theorem but uses similar ideas (whose use in the martingale context dates back to Lévy).

The following *law of the iterated logarithm* has been obtained by Stout (1970b). Let $\{S_n, \mathscr{F}_n\}$ denote a zero-mean, square-integrable martingale whose increments form a stationary ergodic sequence and have finite variance σ^2. Then

$$n^{-1} S_n = \sigma \zeta(n)(2 \log \log n/n)^{1/2},$$

where $\limsup_{n \to \infty} \zeta(n) = 1$ a.s., $\liminf_{n \to \infty} \zeta(n) = -1$ a.s. This result contains the classical law of the iterated logarithm for sums of i.i.d. r.v. [due to Hartman and Wintner (1941)], and the proof closely follows that for the independence case. Extensions of all these results are provided in Chapters 2–4.

1.7. Martingale Limit Theorems Viewed as Rate of Convergence Results in the Martingale Convergence Theorem

We introduced the classical CLT and LIL for sums of independent r.v. as rates of convergence in the SLLN. Since the martingale convergence theorem may be regarded as an analog of the SLLN, it may be expected that there will be analogs of the CLT and LIL which can be interpreted as results on rates of convergence in the martingale convergence theorem.

Let $\{Z_n, \mathscr{F}_n, n \geqslant 1\}$ be an L^2-bounded martingale. Then the martingale convergence theorem asserts that there exists a random variable Z such that $\lim_{n \to \infty} Z_n = Z$ a.s. and $E(Z - Z_n)^2 \to 0$ as $n \to \infty$. The r.v. Z_n may be regarded as a sum of martingale differences: $Z_n = \sum_1^n X_i$, where $X_i = Z_i - Z_{i-1}$ $(Z_0 = 0)$. The random series $\sum_1^\infty X_i$ converges, both a.s. and in L^2, to Z, and the difference $Z - Z_n$ can be written as $\sum_{n+1}^\infty X_i$.

Let

$$V_n^2 = \sum_{n+1}^{\infty} E(X_i^2|\mathscr{F}_{i-1}) \qquad \text{and} \qquad s_n^2 = EV_n^2 = E(Z - Z_n)^2.$$

Suppose that the following analogs of conditions (1.1) and (1.2) hold:

$$s_n^{-2}V_n^2 \xrightarrow{\text{p}} 1 \tag{1.3}$$

and

$$s_n^{-2} \sum_{n+1}^{\infty} E(X_i^2 I(|X_i| \geqslant \varepsilon s_n)) \to 0 \tag{1.4}$$

as $n \to \infty$, for all $\varepsilon > 0$. Then a CLT holds:

$$s_n^{-1}(Z - Z_n) \xrightarrow{\text{d}} N(0,1).$$

If the convergence in (1.3) is strengthened to a.s. convergence and if

$$\sum_1^{\infty} s_j^{-1} E[|X_j| I(|X_j| > \varepsilon s_j)] < \infty \qquad \text{for all} \qquad \varepsilon > 0$$

and

$$\sum_1^{\infty} s_j^{-4} E[X_j^4 I(|X_j| \leqslant \delta s_j)] < \infty \qquad \text{for some} \quad \delta > 0,$$

then an LIL holds:

$$Z - Z_n = \zeta(n)(2s_n^2 \log|\log s_n|)^{1/2},$$

where $\zeta(n)$ has its lim sup equal to $+1$ and its lim inf equal to -1 a.s. (Heyde, 1977a). Once again, the CLT and LIL provide, respectively, "weak" and "strong" rates of convergence in the limit theorem $\lim_{n\to\infty} Z_n = Z$ a.s.

The message of this book is that the abovementioned theorems have applicability far beyond that enjoyed by the corresponding results for sums of independent random variables. Chapters 5, 6, and 7 are concerned with a detailed discussion of examples which well illustrate the power of the methodology. Basically, the theory seems relevant in any context in which conditional expectations, given the past, have a simple form. This is a consequence of the simplicity and tractability of the martingale defining property. If $\{Z_n, n \geqslant 0\}$ is any sequence of integrable random variables, then

$$\left\{\sum_{i=1}^{n} [Z_i - E(Z_i|Z_{i-1}, \ldots\ldots, Z_1)]\right\}$$

is a martingale relative to the sequence of σ-fields generated by Z_i, $i \leqslant n$.

2

Inequalities and Laws of Large Numbers

2.1. Introduction

Almost all proofs of martingale limit theorems rely to some extent on inequalities, and the inequalities established here will be used extensively in later work. In this chapter we apply them to obtain laws of large numbers. We present only the basic tools for applications and the prerequisites for the other limit theorems of this book. The reader is referred to Burkholder (1973), Garsia (1973), and Chow and Teicher (1978) for more detailed accounts.

Let $S_n = \sum_1^n X_i$ denote a sum of r.v. In this classical formulation the weak (strong) law of large numbers is said to hold if $n^{-1}(S_n - ES_n) \to 0$ in probability (respectively, almost surely). If $\{S_n,\mathscr{F}_n\}$ is a martingale, this definition has several disadvantages. First of all, it seems natural to replace $E(S_n)$ by the sum of conditional means $\sum_{i=1}^n E(X_i|\mathscr{F}_{i-1})$. And second, the norming by n^{-1} is very restrictive. For a sequence of independent but not identically distributed r.v. X_n it seems appropriate to norm by a different constant, and for a sequence of dependent r.v. we should consider norming by another r.v. For these reasons, and those given in Chapter 1, we have adopted a rather liberal interpretation of the term *law of large numbers*. For example, there is a very strong case for including the martingale convergence theorem as a strong law of large numbers, especially since the classical law for sums of independent and identically distributed r.v. is a corollary.

It is appropriate here to mention some basic martingale properties. If $\{S_n,\mathscr{F}_n, n \geqslant 1\}$ is a submartingale and ϕ is a nondecreasing convex function on $\mathbb{R}$ such that $E|\phi(S_n)| < \infty$ for each n, then $\{\phi(S_n),\mathscr{F}_n, n \geqslant 1\}$ is also a submartingale. [Use Jensen's inequality; see Chung (1974, p. 302).] The requirement that ϕ be nondecreasing may be dropped if $\{S_n,\mathscr{F}_n\}$ is a martingale. For example, if $\{S_n,\mathscr{F}_n\}$ is a submartingale, then so is $\{S_n^+ = \max(S_n,0), \mathscr{F}_n\}$, and if $\{S_n,\mathscr{F}_n\}$ is a martingale with $E|S_n|^p < \infty$ for all $n(p \geqslant 1)$, then $\{|S_n|^p,\mathscr{F}_n\}$

is a submartingale. It follows that the sequence of L^p-norms $\|S_n\|_p = (E|S_n|^p)^{1/p}$, $n \geqslant 1$, is nondecreasing. If $\{S_n, \mathscr{F}_n, n \geqslant 1\}$ is a submartingale (martingale) and $\{\tau_n, n \geqslant 1\}$ is a nondecreasing sequence of stopping times with respect to the σ-fields $\mathscr{F}_n$ (that is, the event $\{\tau_n = i\}$ is in $\mathscr{F}_i$ for each n and i), then $\{S_{\tau_n}, \mathscr{F}_{\tau_n}, n \geqslant 1\}$ is also a submartingale (respectively, martingale). Here $\mathscr{F}_{\tau_n}$ is the σ-field generated by the events $\{\tau_n = i\} \cap E_i$ with $E_i \in \mathscr{F}_i$, $i \geqslant 1$.

2.2. Basic Inequalities

We begin with a generalization and sharpening of Kolmogorov's inequality.

Theorem 2.1. *If $\{S_i, \mathscr{F}_i, 1 \leqslant i \leqslant n\}$ is a submartingale, then for each real λ,*

$$\lambda P\left(\max_{i \leqslant n} S_i > \lambda\right) \leqslant E\left[S_n I\left(\max_{i \leqslant n} S_i > \lambda\right)\right].$$

Proof. Define

$$E = \left\{\max_{i \leqslant n} S_i > \lambda\right\} = \bigcup_{i=1}^{n} \left\{S_i > \lambda;\ \max_{1 \leqslant j < i} S_j \leqslant \lambda\right\} = \bigcup_{i=1}^{n} E_i,$$

say. The events E_i are $\mathscr{F}_i$-measurable and disjoint. Then

$$\begin{aligned}
\lambda P(E) &\leqslant \sum_i E[S_i I(E_i)] \\
&\leqslant \sum_i E[E(S_n|\mathscr{F}_i) I(E_i)] \quad \text{(submartingale property)} \\
&= \sum_i E[E(S_n I(E_i)|\mathscr{F}_i)] \\
&= \sum_i E[S_n I(E_i)] \\
&= E[S_n I(E)]. \quad \blacksquare
\end{aligned}$$

If $\{S_i, 1 \leqslant i \leqslant n\}$ is a martingale, then $\{|S_i|^p, 1 \leqslant i \leqslant n\}$ is a submartingale. By applying Theorem 2.1 to this submartingale we obtain

Corollary 2.1. *If $\{S_i, \mathscr{F}_i, 1 \leqslant i \leqslant n\}$ is a martingale, then for each $p \geqslant 1$ and $\lambda > 0$,*

$$\lambda^p P\left(\max_{i \leqslant n} |S_i| > \lambda\right) \leqslant E|S_n|^p.$$

Theorem 2.1 has an application in another direction, which yields the following result.

Theorem 2.2. (*Doob's inequality*) *If* $\{S_i,\mathscr{F}_i, 1 \leqslant i \leqslant n\}$ *is a martingale, then for* $p > 1$,

$$\|S_n\|_p \leqslant \left\|\max_{i\leqslant n}|S_i|\right\|_p \leqslant q\|S_n\|_p,$$

where $p^{-1} + q^{-1} = 1$.

Proof. The left-hand inequality is trivial. To prove the right-hand one we note that by Theorem 2.1 and Hölder's inequality,

$$\begin{aligned}
E\left(\max_{i\leqslant n}|S_i|^p\right) &= p\int_0^\infty x^{p-1}P\left(\max_{i\leqslant n}|S_i| > x\right)dx \\
&\leqslant p\int_0^\infty x^{p-2}E\left[|S_n|I\left(\max_{i\leqslant n}|S_i| > x\right)\right]dx \\
&= pE\left[|S_n|\int_0^{\max_{i\leqslant n}|S_i|} x^{p-2}\,dx\right] \\
&= qE\left[|S_n|\left(\max_{i\leqslant n}|S_i|^{p-1}\right)\right] \\
&\leqslant q(E|S_n|^p)^{1/p}\left(E\left(\max_{i\leqslant n}|S_i|^p\right)\right)^{1/q},
\end{aligned}$$

which gives the desired result. ▮

Our next result is the upcrossing inequality, which provides an upper bound for the mean number of oscillations of the submartingale $\{S_i,\mathscr{F}_i, 1 \leqslant i \leqslant n\}$ across an interval $[a,b]$. It has found an important application as a tool in the proof of the martingale convergence theorem, although Doob's (1940) original proof was more direct.

If $-\infty < a < b < \infty$, let $v = v(a,b,n)$ denote the number of times that the sequence $\{S_i, 1 \leqslant i \leqslant n\}$ crosses from a value $\leqslant a$ to one $\geqslant b$; v is called the *number of upcrossings* of $[a,b]$ by $\{S_i\}$.

Theorem 2.3. *Let* v *denote the number of upcrossings of the compact interval* $[a,b]$ *by the submartingale* $\{S_i,\mathscr{F}_i, 1 \leqslant i \leqslant n\}$. *Then*

$$(b - a)E(v) \leqslant E(S_n - a)^+ - E(S_1 - a)^+. \tag{2.1}$$

Proof. Since $\{S_i,\mathscr{F}_i, 1 \leqslant i \leqslant n\}$ is a submartingale, then so is $\{(S_i - a)^+ = \max(S_i - a,0), \mathscr{F}_i, 1 \leqslant i \leqslant n\}$, and the number of upcrossings of $[a,b]$ by the sequence $\{S_i\}$ equals that of $[0, b - a]$ by $\{(S_i - a)^+\}$. Hence it suffices to prove that for a nonnegative submartingale $\{S_i,\mathscr{F}_i, 1 \leqslant i \leqslant n\}$ the number of upcrossings v of $[0,b]$ satisfies

$$bE(v) \leqslant E(S_n - S_1). \tag{2.2}$$

Let $\tau_0 = 1$, τ_1 equal the smallest j for which $S_j = 0$, τ_{2i} equal the smallest j in the range $\tau_{2i-1} < j \leqslant n$ for which $S_j \geqslant b$ $(i \geqslant 1)$, and τ_{2i-1} equal the smallest j in the range $\tau_{2i-2} < j \leqslant n$ for which $S_j = 0$ $(i \geqslant 2)$. Let l denote the largest i for which τ_i is well defined $(1 \leqslant l \leqslant n)$, and now define $\tau_i = n$ for $i > l$. Then $\tau_n = n$, and

$$S_n - S_1 = \sum_{i=0}^{n-1} (S_{\tau_{i+1}} - S_{\tau_i}) = \sum_{i \text{ even}} + \sum_{i \text{ odd}}.$$

Suppose that i is odd. If $i < l$, then

$$S_{\tau_{i+1}} \geqslant b > 0 = S_{\tau_i};$$

if $i = l$, then

$$S_{\tau_{i+1}} = S_n \geqslant 0 = S_{\tau_i};$$

and if $i > l$, then

$$S_{\tau_{i+1}} = S_n = S_{\tau_i}.$$

Hence

$$\sum_{i \text{ odd}} (S_{\tau_{i+1}} - S_{\tau_i}) \geqslant \sum_{\substack{i \text{ odd}, \\ i<l}} (S_{\tau_{i+1}} - S_{\tau_i}) \geqslant [l/2]b = vb. \tag{2.3}$$

($[l/2]$ denotes the integer part of $l/2$.) The r.v. τ_i, $1 \leqslant i \leqslant n$, form a nondecreasing sequence of stopping times with respect to the σ-fields $\mathscr{F}_i$, and so $\{S_{\tau_i}, \mathscr{F}_{\tau_i}, 1 \leqslant i \leqslant n\}$ is a submartingale. It follows that each $E(S_{\tau_{i+1}} - S_{\tau_i}) \geqslant 0$, and so

$$E\left[\sum_{i \text{ even}} (S_{\tau_{i+1}} - S_{\tau_i})\right] \geqslant 0.$$

The equality (2.2) follows from this and (2.3). ∎

We apply Theorem 2.3 to obtain a useful alternative to Theorem 2.1 (Brown, (1971)). A multidimensional version of this result is given in Shorack and Smythe (1976).

Theorem 2.4. *If $\{S_i, \mathscr{F}_i, 1 \leqslant i \leqslant n\}$ is a zero-mean martingale, then for each $\lambda > 0$,*

$$\begin{aligned}\lambda P\left(\max_{i \leqslant n} |S_i| > 2\lambda\right) &\leqslant \lambda P(|S_n| > \lambda) + E[(|S_n| - 2\lambda) I(|S_n| \geqslant 2\lambda)] \\ &\leqslant E[|S_n| I(|S_n| > \lambda)]. \end{aligned} \tag{2.4}$$

Proof. Let $E_n = \{\min_{i \leqslant n} S_i \leqslant -2\lambda\}$ and $S_0 = 0$, and let $\mathscr{F}_0$ denote the trivial σ-field. Since $E(S_1) = 0$, the extended sequence $\{S_i, \mathscr{F}_i, 0 \leqslant i \leqslant n\}$ is a martingale. Let v denote the number of upcrossings of $[-2\lambda,-\lambda]$ by $\{S_i, 0 \leqslant i \leqslant n\}$. From Theorem 2.3 we have

$$\begin{aligned}\lambda E(v) &\leqslant E(S_n + 2\lambda)^+ - E(S_0 + 2\lambda)^+ \\ &= E(S_n + 2\lambda)^+ - 2\lambda.\end{aligned}$$

Now,

$$\begin{aligned}P(E_n) &= P(E_n; S_n \geqslant -\lambda) + P(E_n; S_n < -\lambda) \\ &\leqslant P(v > 0) + P(S_n < -\lambda) \\ &\leqslant E(v) + P(S_n < -\lambda).\end{aligned}$$

Hence

$$\lambda P(E_n) \leqslant E(S_n + 2\lambda)^+ - 2\lambda + \lambda P(S_n < -\lambda).$$

By considering the number of upcrossings of $[-2\lambda,-\lambda]$ by $\{-S_i, 0 \leqslant i \leqslant n\}$ we derive

$$\lambda P\left(\max_{i \leqslant n} S_i > 2\lambda\right) \leqslant E(-S_n + 2\lambda)^+ - 2\lambda + \lambda P(S_n > \lambda).$$

Adding the last two inequalities we obtain the first inequality in (2.4), and the second follows after a little manipulation. ∎

2.3 The Martingale Convergence Theorem

Doob's convergence theorem is an easy consequence of Theorem 2.3.

Theorem 2.5. *If $\{S_n, \mathscr{F}_n, n \geqslant 1\}$ is an L^1-bounded submartingale, then S_n converges a.s. to an r.v. X with $E|X| < \infty$.*

Proof. Let v_n denote the v of Theorem 2.3 and let $v_\infty = \lim_{n\to\infty} v_n$ (finite or infinite). The inequality (2.1) and L^1-boundedness imply that

$$(b - a)E(v_\infty) \leqslant \sup_n E|S_n| + |a| < \infty$$

and so $v_\infty < \infty$ a.s. for all values of a and b. It follows that

$$P(\liminf S_n < a < b < \limsup S_n) = 0$$

for all a and b, and summing over rational values we deduce that

$$P(\liminf S_n < \limsup S_n) = 0,$$

so that S_n converges a.s. The limit must be a.s. finite since by Fatou's lemma,

$$E(\lim|S_n|) \leqslant \sup E|S_n| < \infty. \quad ∎$$

Since

$$E|S_n| = 2E(S_n^+) - E(S_n) \leqslant 2E(S_n^+) - E(S_1),$$

the sequence $\{E|S_n|\}$ is bounded if and only if $\{E(S_n^+)\}$ is bounded. If $\{S_n\}$ is uniformly integrable (for example, if $\{S_n\}$ is uniformly bounded), then of course the convergence in Theorem 2.5 is in L^1 as well as with probability 1. The next result is less obvious.

Corollary 2.2. *Let* $1 < p < \infty$. *If* $\{S_n, \mathscr{F}_n, n \geqslant 1\}$ *is a martingale and* $\sup_n E|S_n|^p < \infty$, *then* S_n *converges in* L^p *as well as with probability* 1. (The result is not true if $p = 1$.)

Proof. By Theorem 2.2,

$$E\left(\max_{i \leqslant n}|S_i|^p\right) \leqslant q^p E|S_n|^p$$

and so

$$E\left(\sup_n |S_n|^p\right) < \infty.$$

Furthermore,

$$P(|S_n| > \lambda) \leqslant \lambda^{-p} E|S_n|^p,$$

which converges to zero uniformly in n as $\lambda \to \infty$. Therefore

$$E[|S_n|^p I(|S_n| > \lambda)] \leqslant E\left[\left(\sup_m |S_m|^p\right) I(|S_n| > \lambda)\right]$$

converges to zero uniformly in n as $\lambda \to \infty$, showing that $\{|S_n|^p\}$ is uniformly integrable. Hence if S_n converges in probability, it converges in L^p. Theorem 2.5 ensures that S_n converges a.s. ∎

The reverse martingale versions of Theorem 2.5 and Corollary 2.2 are also true. For example:

Theorem 2.6. *If* $\{S_n, \mathscr{F}_n, n \geqslant 1\}$ *is a reverse martingale, then* S_n *converges a.s. to a finite limit.*

The proof goes through as before, applying the upcrossing inequality to the martingale $\{S_{n-i+1}, \mathscr{F}_{n-i+1}, 1 \leqslant i \leqslant n\}$:

$$(b - a)E(v) \leqslant E(S_1 - a)^+ - E(S_n - a)^+.$$

Note that a reverse martingale must be L^1-bounded since each sequence $\{|S_{n-i+1}|,\mathscr{F}_{n-i+1}, 1 \leqslant i \leqslant n\}$ is a submartingale and so satisfies

$$E|S_n| \leqslant E|S_{n-1}| \leqslant \cdots \leqslant E|S_1|.$$

Much of the modern interest in martingale theory stems from the wide applicability of results like Theorem 2.5. In addition, McLeish (1975b) has given us the concept of *mixingales*—asymptotic martingales—which are sufficiently like martingale differences to satisfy a convergence theorem. Let $\{X_n, n \geqslant 1\}$ be a sequence of square-integrable r.v. on a probability space $(\Omega,\mathscr{F},P)$, and let $\{\mathscr{F}_n, -\infty < n < \infty\}$ be an increasing sequence of sub σ-fields of $\mathscr{F}$. Then $\{X_n,\mathscr{F}_n\}$ is called a *mixingale* (*difference*) *sequence* if, for sequences of nonnegative constants c_n and ψ_m, where $\psi_m \to 0$ as $m \to \infty$, we have

(i) $\|E(X_n|\mathscr{F}_{n-m})\|_2 \leqslant \psi_m c_n$ and
(ii) $\|X_n - E(X_n|\mathscr{F}_{n+m})\|_2 \leqslant \psi_{m+1} c_n$

for all $n \geqslant 1$ and $m \geqslant 0$. The following examples give an idea of the scope of mixingales.

Example 1. (Martingales) Let $\{S_n = \sum_{i=1}^n X_i, \mathscr{F}_n, n \geqslant 1\}$ be a square-integrable martingale. Then $E(X_n|\mathscr{F}_{n-1}) = 0$ a.s. and $E(X_n|\mathscr{F}_{n+m}) = X_n$ a.s., $m \geqslant 0$, so the mixingale definition applies to $\{X_n,\mathscr{F}_n\}$ with $c_n = (EX_n^2)^{1/2}$, $\psi_0 = 1$ and $\psi_m = 0$ for $m \geqslant 1$.

Example 2. (Lacunary functions) Let $(\Omega,\mathscr{F},P)$ be the unit interval $[0,1)$ with Borel measure. Let $f(x)$ be a periodic function on $[0,\infty)$ with period 1 such that $\int f\,dP = 0$ and $\int f^2\,dP < \infty$. Define $X_n(\omega) = f(2^n\omega)$ for $0 \leqslant \omega < 1$, and define $\mathscr{F}_n$ to be the σ-algebra of sets generated by the intervals $[j2^{-n},(j+1)2^{-n})$. Then $E(X_n|\mathscr{F}_{n-1}) = 0$ a.s. and $E[X_n - E(X_n|\mathscr{F}_{n+m})]^2 \to 0$ uniformly in n as $m \to \infty$ since f can be uniformly approximated in L^2 by $\mathscr{F}_m$-measurable functions. Hence $\{X_n,\mathscr{F}_n\}$ is a mixingale.

Example 3. (Linear processes) Let $\{\xi_i, -\infty < i < \infty\}$ be independent r.v. each having mean 0 and finite variance σ^2. Let $\{\alpha_i, -\infty < i < \infty\}$ be a sequence of constants with $\sum_{i=-\infty}^{\infty} \alpha_i^2 < \infty$. Write $X_n = \sum_{i=-\infty}^{\infty} \alpha_{i-n}\xi_i$ and $\mathscr{F}_n$ for the σ-field generated by the sequence $\{\xi_i, i \leqslant n\}$. Then $\{X_n,\mathscr{F}_n\}$ is a mixingale with all $c_n^2 = \sigma^2$ and $\psi_m^2 = \sum_{|i| \geqslant m} \alpha_i^2$.

Example 4. (Uniformly mixing processes) Let $\{X_n, n \geqslant 1\}$ be a process satisfying the uniform mixing condition

$$\sup_{k,\, A \in \mathscr{G}_1^k,\, B \in \mathscr{G}_{k+n}^\infty} P(A)^{-1}|P(A \cap B) - P(A)P(B)| = \phi(n) \downarrow 0$$

as $n \to \infty$, where $\mathscr{G}_1^k$ and $\mathscr{G}_{k+n}^{\infty}$ denote the σ-fields generated by $\{X_i, 1 \leqslant i \leqslant k\}$ and $\{X_i, k+n \leqslant i < \infty\}$, respectively. Write $\mathscr{F}_k$ for $\mathscr{G}_1^k$. Using a Hölder-type inequality for uniformly mixing processes (see Theorem A.6 of Appendix III) we have

$$\begin{aligned} E[(E(X_n|\mathscr{F}_{n-m}))^2] &= E[X_n E(X_n|\mathscr{F}_{n-m})] \\ &\leqslant 2\phi^{1/2}(m)(EX_n^2)^{1/2}[E(E(X_n|\mathscr{F}_{n-m}))^2]^{1/2}, \end{aligned}$$

so that

$$\{E[(E(X_n|\mathscr{F}_{n-m}))^2]\}^{1/2} \leqslant 2\phi^{1/2}(m)(EX_n^2)^{1/2}$$

and $\{X_n, \mathscr{F}_n\}$ is a mixingale with $c_n = 2(EX_n^2)^{1/2}$ and $\psi_m = \phi^{1/2}(m)$. It is known that Markov processes with stationary transition probabilities satisfying a very general condition of Doeblin are uniformly mixing with geometrically decreasing $\phi(m)$ [(e.g., Doob (1953, Chapter 5) and Rosenblatt (1971, Chapter 7)].

The key to a mixingale convergence theorem lies in establishing a mixingale analog of Theorem 2.2 (McLeish, 1975b).

Lemma 2.1. *If $\{X_n, \mathscr{F}_n\}$ is a mixingale such that $\psi_n = O(n^{-1/2}(\log n)^{-2})$ as $n \to \infty$, then there exists a constant K depending only on $\{\psi_n\}$ such that*

$$E\left(\max_{i \leqslant n} S_i^2\right) \leqslant K \sum_{i=1}^{n} c_i^2$$

where $S_i = \sum_{j=1}^{i} X_j$.

Proof. We first show that

$$X_i = \sum_{k=-\infty}^{\infty} [E(X_i|\mathscr{F}_{i+k}) - E(X_i|\mathscr{F}_{i+k-1})] \quad \text{a.s.} \tag{2.5}$$

For any $m, n \geqslant 1$,

$$\sum_{k=-m}^{n} [E(X_i|\mathscr{F}_{i+k}) - E(X_i|\mathscr{F}_{i+k-1})] = E(X_i|\mathscr{F}_{i+n}) - E(X_i|\mathscr{F}_{i-m-1}).$$

For each fixed i the sequence $\{E(X_i|\mathscr{F}_{i+n}), \mathscr{F}_{i+n}, n \geqslant 1\}$ is an L^1-bounded martingale and so by Theorem 2.5 it converges a.s. Condition (ii) above ensures that the limit is X_i.

Similarly, applying Theorem 2.6 to the reverse martingale $\{E(X_i|\mathscr{F}_{i-m-1}), \mathscr{F}_{i-m-1}, m \geqslant 1\}$, we deduce that $E(X_i|\mathscr{F}_{i-m-1}) \xrightarrow{\text{a.s.}} 0$ as $m \to \infty$, which gives (2.5).

Let

$$Y_{nk} = \sum_{i=1}^{n} [E(X_i|\mathscr{F}_{i+k}) - E(X_i|\mathscr{F}_{i+k-1})].$$

Then by (2.5) and after a change of order of summation,

$$S_n = \sum_{k=-\infty}^{\infty} Y_{nk} \quad \text{a.s.}$$

Setting $a_0 = a_1 = a_{-1} = 1$ and $a_n = a_{-n} = (n(\log n)^2)^{-1}$ for $n \geqslant 2$, we have from the Cauchy–Schwarz inequality that

$$\left(\sum_{k=-\infty}^{\infty} Y_{nk}\right)^2 \leqslant \left(\sum_{k=-\infty}^{\infty} a_k\right)\left(\sum_{k=-\infty}^{\infty} a_k^{-1} Y_{nk}^2\right)$$

and hence

$$E\left(\max_{i\leqslant n} S_i^2\right) \leqslant \left(\sum_{k=-\infty}^{\infty} a_k\right)\left(\sum_{k=-\infty}^{\infty} a_k^{-1} E\left(\max_{i\leqslant n} Y_{ik}^2\right)\right).$$

For each k the sequence $\{Y_{ik}, \mathscr{F}_{i+k}, 1 \leqslant i \leqslant n\}$ is a martingale and so by Theorem 2.2,

$$E\left(\max_{i\leqslant n} S_i^2\right) \leqslant 4\left(\sum_{k=-\infty}^{\infty} a_k\right)\left(\sum_{k=-\infty}^{\infty} a_k^{-1} E Y_{nk}^2\right). \tag{2.6}$$

Now put $Z_{ik} = X_i - E(X_i|\mathscr{F}_{i+k})$ for $k \geqslant 0$ and note that

$$\begin{aligned} EY_{nk}^2 &= \sum_{i=1}^{n} [E(E(X_i|\mathscr{F}_{i+k})^2) - E(E(X_i|\mathscr{F}_{i+k-1})^2)] \\ &= \sum_{i=1}^{n} [EZ_{i,k-1}^2 - EZ_{ik}^2]. \end{aligned}$$

Substituting into (2.6) for $k \geqslant 1$ we deduce that

$$\begin{aligned} E\left(\max_{i\leqslant n} S_i^2\right) \leqslant 4\left(\sum_{k=-\infty}^{\infty} a_k\right)\sum_{i=1}^{n}\Big\{&a_0^{-1}E(E(X_i|\mathscr{F}_i)^2) + a_1^{-1}EZ_{i0}^2 \\ &+ \sum_{k=1}^{\infty}(a_{k+1}^{-1} - a_k^{-1})EZ_{ik}^2 + \sum_{k=1}^{\infty}(a_k^{-1} - a_{k-1}^{-1})E(E(X_i|\mathscr{F}_{i-k})^2)\Big\}. \end{aligned}$$

Using the mixingale properties (i) and (ii) we can bound the right side of this inequality by

$$4\left(\sum_{k=-\infty}^{\infty} a_k\right)\left(\sum_{i=1}^{n} c_i^2\right)\left\{a_0^{-1}(\psi_0^2 + \psi_1^2) + 2\sum_{k=1}^{\infty}(a_k^{-1} - a_{k-1}^{-1})\psi_k^2\right\},$$

which establishes the lemma. ∎

The mixingale convergence theorem is a simple corollary.

Theorem 2.7. *If* $\{X_n, \mathscr{F}_n\}$ *is a mixingale such that* $\sum_{i=1}^{\infty} c_i^2 < \infty$ *and* $\psi_n = O(n^{-1/2}(\log n)^{-2})$ *as* $n \to \infty$, *then* $S_n = \sum_{i=1}^{n} X_i$ *converges a.s. to a finite limit as* $n \to \infty$.

Proof. For each $m' > m$ we have from Chebyshev's inequality and Lemma 2.1 that

$$P\left(\max_{m<n\leqslant m'} |S_n - S_m| > \varepsilon\right) \leqslant \varepsilon^{-2} K\left(\sum_{i=m}^{m'} c_i^2\right),$$

so that

$$P\left(\max_{n\leqslant m} |S_n - S_m| > \varepsilon\right) \leqslant \varepsilon^{-2} K\left(\sum_{i=m}^{\infty} c_i^2\right) \to 0$$

as $m \to \infty$. Therefore S_n converges a.s. The limit r.v. has finite variance. ∎

The convergence theorem for an L^2-bounded martingale is an immediate consequence. If the r.v. X_i satisfy less stringent weak dependence conditions than those of a mixingale, it may still be possible to establish strong law results, although these will typically be less powerful. For example, when the X_i have zero means and are orthogonal (i.e., $E(X_i X_j) = 0$ or $E(X_i^2)$ according as $i \neq j$ or $i = j$), Doob's inequality can be replaced by the Rademacher–Mensov inequality:

$$E\left(\max_{i\leqslant n}\left(\sum_{j=1}^{i} X_j\right)^2\right) \leqslant (\log 4n/\log 2)^2 \sum_{i=1}^{n} EX_i^2,$$

which leads, along the lines of Theorem 2.7 above, to the following result.

Theorem 2.8. *Let* $\{X_n, n \geqslant 1\}$ *be a sequence of orthogonal r.v. with zero means. Let* $\{c_n, n \geqslant 1\}$ *be real constants such that*

$$\sum_{n=1}^{\infty} c_n^2 (\log n)^2 EX_n^2 < \infty.$$

Then the series $\sum_{n=1}^{\infty} c_n X_n$ *converges a.s.*

The Rademacher–Mensov inequality and Theorem 2.8 have often been given in books on probability [e.g., Révész (1968, Chapter 3), Loève (1978, Section 36), Doob (1953, Chapter 4), Stout (1974, Chapter 3)] and the details consequently will be omitted.

The result of Theorem 2.8 can be refined even further via an extension of the Rademacher–Mensov inequality to obtain the following result [cf. Stout (1974, pp. 202–203)].

Theorem 2.9. *Let $\{X_n, n \geqslant 1\}$ be a sequence of zero-mean r.v. and suppose that there exist constants $\{p_n, n \geqslant 0\}$ such that $0 \leqslant p_n \leqslant 1$ and*

$$E(X_i X_j) \leqslant p_{j-i}(EX_i^2 EX_j^2)^{1/2}$$

for all $j \geqslant i \geqslant 0$, while $\sum_{n=0}^{\infty} p_n < \infty$. If $\{c_n, n \geqslant 1\}$ is a sequence of real numbers for which

$$\sum_{n=1}^{\infty} c_n^2 (\log n)^2 EX_n^2 < \infty,$$

then $\sum_{n=1}^{\infty} c_n X_n$ converges a.s.

2.4. Square Function Inequalities

The square function inequalities developed by Burkholder and others form a major contemporary addition to our armory of martingale tools. They imply a rather unexpected relationship between the behavior of a martingale and the squares of its differences. This duality was noticed earlier for sums of independent r.v. and orthogonal functions, and many of the inequalities are generalizations of this work. We present here only those results which we shall need later in the book.

Let $X_1 = S_1$ and $X_i = S_i - S_{i-1}$, $2 \leqslant i \leqslant n$, denote the differences of the sequences $\{S_i, 1 \leqslant i \leqslant n\}$.

Theorem 2.10. (*Burkholder's inequality*) *If $\{S_i, \mathscr{F}_i, 1 \leqslant i \leqslant n\}$ is a martingale and $1 < p < \infty$, then there exist constants C_1 and C_2 depending only on p such that*

$$C_1 E\left|\sum_{i=1}^{n} X_i^2\right|^{p/2} \leqslant E|S_n|^p \leqslant C_2 E\left|\sum_{i=1}^{n} X_i^2\right|^{p/2} \tag{2.7}$$

(The proof shows that suitable constants are given by $C_1^{-1} = (18p^{1/2}q)^p$ and $C_2 = (18pq^{1/2})^p$, where $p^{-1} + q^{-1} = 1$.)

Theorem 2.11. *If $\{S_i, \mathscr{F}_i, 1 \leqslant i \leqslant n\}$ is a martingale and $p > 0$, then there exists a constant C depending only on p such that*

$$E\left(\max_{i \leqslant n}|S_i|^p\right) \leqslant C\left\{E\left[\left(\sum_{i=1}^{n} E(X_i^2|\mathscr{F}_{i-1})\right)^{p/2}\right] + E\left(\max_{i \leqslant n}|X_i|^p\right)\right\}.$$

Theorem 2.12. (*Rosenthal's inequality*) *If $\{S_i, \mathscr{F}_i, 1 \leqslant i \leqslant n\}$ is a martingale and $2 \leqslant p < \infty$, then there exist constants C_1 and C_2 depending only* *on p such that*

$$C_1\left\{E\left[\left(\sum_{i=1}^{n} E(X_i^2|\mathscr{F}_{i-1})\right)^{p/2}\right]+\sum_{i=1}^{n} E|X_i|^p\right\}$$
$$\leqslant E|S_n|^p \leqslant C_2\left\{E\left[\left(\sum_{i=1}^{n} E(X_i^2|\mathscr{F}_{i-1})\right)^{p/2}\right]+\sum_{i=1}^{n} E|X_i|^p\right\}.$$

We preface the proofs with three lemmas.

Lemma 2.2. *Suppose that $\{S_i, \mathscr{F}_i, 1 \leqslant i \leqslant n\}$ is an L^1-bounded martingale or nonnegative submartingale. For $\lambda > 0$ define the stopping time τ by*

$$\tau = \begin{cases} \min\{i \leqslant n \mid |S_i| > \lambda\} & \text{if this set is nonempty,} \\ n+1 & \text{otherwise.} \end{cases}$$

Then

$$E\left(\sum_{i=1}^{\tau-1} X_i^2\right) + E(S_{\tau-1}^2) \leqslant 2\lambda E|S_n|.$$

Proof. For any $m \leqslant n+1$,

$$\begin{aligned}\sum_{i=1}^{m-1} X_i^2 + S_{m-1}^2 &= 2S_{m-1}^2 - 2\sum_{1\leqslant i<j\leqslant m-1} X_i X_j \\ &= 2S_{m-1}^2 - 2\sum_{j=2}^{m-1} S_{j-1}X_j \\ &= 2S_m S_{m-1} - 2\sum_{j=2}^{m} S_{j-1}X_j,\end{aligned}$$

where we define $S_{n+1} = S_n$ and $X_{n+1} = 0$. In particular,

$$\sum_{i=1}^{\tau-1} X_i^2 + S_{\tau-1}^2 = 2S_\tau S_{\tau-1} - 2\sum_{i=2}^{\tau} S_{i-1}X_i.$$

Now,

$$E\left(\sum_{i=2}^{\tau} S_{i-1}X_i\right) = \sum_{i=2}^{n} E[S_{i-1}E(X_i|\mathscr{F}_{i-1})I(\tau \geqslant i)] \geqslant 0,$$

with equality holding in the martingale case. Hence

$$E\left(\sum_{i=1}^{\tau-1} X_i^2\right) + E(S_{\tau-1}^2) \leqslant 2E(S_\tau S_{\tau-1}) \leqslant 2\lambda E|S_\tau| \leqslant 2\lambda E|S_n|,$$

the last two inequalities following from the facts that $|S_{\tau-1}| \leqslant \lambda$ and that $\{|S_\tau|, |S_n|\}$ is a submartingale with σ-fields $\{\mathscr{F}_\tau, \mathscr{F}_n\}$.

Lemma 2.3. *Let* $\{S_i,\mathcal{F}_i, 1 \leqslant i \leqslant n\}$ *be a nonnegative submartingale and let*

$$Y = \max\left(\theta\left(\sum_{i=1}^{n} X_i^2\right)^{1/2}, \max_{i\leqslant n} S_i\right),$$

where $\theta > 0$. *Then for each* $\lambda > 0$,

$$\lambda P(Y > \beta\lambda) \leqslant 3E[S_n I(Y > \lambda)], \tag{2.8}$$

where $\beta = (1 + 2\theta^2)^{1/2}$, *and for each* $1 < p < \infty$,

$$\left\|\left(\sum_{i=1}^{n} X_i^2\right)^{1/2}\right\|_p \leqslant 9p^{1/2}q\|S_n\|_p, \tag{2.9}$$

where $p^{-1} + q^{-1} = 1$.

Proof. Since $\beta > 1$, the left side of (2.8) does not exceed

$$\lambda P\left(\max_{i\leqslant n} S_i > \lambda\right) + \lambda P\left(\theta\left(\sum_{i=1}^{n} X_i^2\right)^{1/2} > \beta\lambda, \max_{i\leqslant n} S_i \leqslant \lambda\right). \tag{2.10}$$

Using Theorem 2.1 we can bound the first term by

$$E\left[S_n I\left(\max_{i\leqslant n} S_i > \lambda\right)\right] \leqslant E[S_n I(Y > \lambda)]. \tag{2.11}$$

Let $T_m = S_m I(\theta(\sum_{i=1}^{m} X_i^2)^{1/2} > \lambda)$, $m \leqslant n$. Since

$$\begin{aligned} E(T_m|\mathcal{F}_{m-1}) &\geqslant E\left[S_m I\left(\theta\left(\sum_{i=1}^{m-1} X_i^2\right)^{1/2} > \lambda\right)\middle|\mathcal{F}_{m-1}\right] \\ &= E(S_m|\mathcal{F}_{m-1}) I\left(\theta\left(\sum_{i=1}^{m-1} X_i^2\right)^{1/2} > \lambda\right) \\ &\geqslant T_{m-1}, \end{aligned}$$

$\{T_i,\mathcal{F}_i, 1 \leqslant i \leqslant n\}$ is a nonnegative submartingale. Let $Y_1 = T_1$ and $Y_i = T_i - T_{i-1}$ denote the differences. We shall prove that

$$\left\{\theta\left(\sum_{i=1}^{n} X_i^2\right)^{1/2} > \beta\lambda, \max_{i\leqslant n} S_i \leqslant \lambda\right\} \subseteq \left\{\left(\sum_{i=1}^{n} Y_i^2\right)^{1/2} > \lambda, \max_{i\leqslant n} T_i \leqslant \lambda\right\}. \tag{2.12}$$

From this it follows that the second term in (2.10) is dominated by

$$\begin{aligned} \lambda P\left(\left(\sum_{i=1}^{n} Y_i^2\right)^{1/2} > \lambda, \max_{i\leqslant n} T_i \leqslant \lambda\right) &\leqslant \lambda^{-1} E\left[\left(\sum_{i=1}^{n} Y_i^2\right) I\left(\max_{i\leqslant n} T_i \leqslant \lambda\right)\right] \\ &\leqslant 2E(T_n) \\ &\leqslant 2E[S_n I(Y > \lambda)], \end{aligned}$$

using Lemma 2.2 to obtain the second inequality. Combined with the inequality (2.11) this establishes (2.8). The inequality (2.9) follows from (2.8):

$$\begin{aligned} E(Y^p) &= p \int_0^\infty x^{p-1} P(Y > x)\, dx \\ &= p\beta^p \int_0^\infty y^{p-1} P(Y > \beta y)\, dy \\ &\leqslant 3p\beta^p \int_0^\infty y^{p-2} E[S_n I(Y > y)]\, dy \\ &= 3p\beta^p E\left[S_n \int_0^Y y^{p-2}\, dy \right] \\ &= 3q\beta^p E[S_n Y^{p-1}] \\ &\leqslant 3q\beta^p (ES_n^p)^{1/p} (EY^p)^{1/q}. \end{aligned}$$

Therefore,

$$\theta \left\| \left(\sum_{i=1}^n X_i^2 \right)^{1/2} \right\|_p \leqslant \|Y\|_p \leqslant 3q\beta^p \|S_n\|_p.$$

If $\theta = p^{-1/2}$, then $\beta^p = (1 + 2/p)^{p/2} < e < 3$, and (2.9) follows.

It remains to prove (2.12). Consider the stopping time

$$v = \begin{cases} \min\left\{ i \leqslant n \,\middle|\, \theta \left(\sum_{j=1}^i X_j^2 \right)^{1/2} > \lambda \right\} & \text{if this set is nonempty} \\ n & \text{otherwise.} \end{cases}$$

On the set on the left in (2.12), $\max_{i \leqslant n} T_i \leqslant \lambda$ and

$$\begin{aligned} \beta^2\lambda^2 < \theta^2 \sum_{i=1}^n X_i^2 &= \theta^2 \sum_{i=1}^{v-1} X_i^2 + \theta^2 X_v^2 + \theta^2 \sum_{i=v+1}^n X_i^2 \\ &< \lambda^2 + \theta^2\lambda^2 + \theta^2 \sum_{i=1}^n Y_i^2. \end{aligned}$$

(Note that $|X_v| \leqslant \max(S_{v-1}, S_v) \leqslant \lambda$, and that $X_i = Y_i$ for $i \geqslant v + 1$; if $r < s$, $\sum_s^r$ is taken as null.) Hence on the set in question, $\sum_{i=1}^n Y_i^2 > \lambda^2$, which proves (2.12). ∎

Lemma 2.4. *Let X and Y be nonnegative r.v. and suppose that $\beta > 1$, $\delta > 0$, and $\varepsilon > 0$ are such that for all $\lambda > 0$,*

$$P(X > \beta\lambda;\ Y \leqslant \delta\lambda) \leqslant \varepsilon P(X > \lambda). \tag{2.13}$$

Then if $0 < p < \infty$ and $\varepsilon < \beta^{-p}$,

$$E(X^p) \leqslant \beta^p \delta^{-p} (1 - \varepsilon\beta^p)^{-1} E(Y^p). \tag{2.14}$$

Proof. From (2.13),

$$\begin{aligned} P(X > \beta\lambda) &= P(X > \beta\lambda,\, Y \leqslant \delta\lambda) + P(X > \beta\lambda,\, Y > \delta\lambda) \\ &\leqslant \varepsilon P(X > \lambda) + P(Y > \delta\lambda). \end{aligned}$$

Hence

$$\begin{aligned} E(X^p) &= p\beta^p \int_0^\infty x^{p-1} P(X > \beta x)\, dx \\ &\leqslant \varepsilon p\beta^p \int_0^\infty x^{p-1} P(X > x)\, dx + p\beta^p \int_0^\infty x^{p-1} P(Y > \delta x)\, dx \\ &= \varepsilon\beta^p E(X^p) + \beta^p \delta^{-p} E(Y^p), \end{aligned}$$

which gives (2.14). ∎

Proof of Theorem 2.10. Let $T_i = E(S_n^+ | \mathscr{F}_i)$ and $U_i = E(S_n^- | \mathscr{F}_i)$, $1 \leqslant i \leqslant n$. The sequences $\{T_i, \mathscr{F}_i\}$ and $\{U_i, \mathscr{F}_i\}$ are nonnegative martingales; let $T_0 = U_0 = 0$, $Y_i = T_i - T_{i-1}$, and $Z_i = U_i - U_{i-1}$, $i \geqslant 1$, denote the differences. Since $X_i = Y_i - Z_i$, by Minkowski's inequality,

$$\left(\sum_{i=1}^n X_i^2\right)^{1/2} \leqslant \left(\sum_{i=1}^n Y_i^2\right)^{1/2} + \left(\sum_{i=1}^n Z_i^2\right)^{1/2},$$

and so by Lemma 2.3,

$$\begin{aligned} \left\| \left(\sum_{i=1}^n X_i^2\right)^{1/2} \right\|_p &\leqslant \left\| \left(\sum_{i=1}^n Y_i^2\right)^{1/2} \right\|_p + \left\| \left(\sum_{i=1}^n Z_i^2\right)^{1/2} \right\|_p \\ &\leqslant 9p^{1/2} q(\|T_n\|_p + \|U_n\|_p) \\ &\leqslant 18 p^{1/2} q \|S_n\|_p, \end{aligned}$$

which gives the left-hand side of (2.7). To obtain the right-hand side let

$$R_n = \operatorname{sgn}(S_n)|S_n|^{p-1} / \|S_n\|_p^{p-1}.$$

The sequence $R_i = E(R_n | \mathscr{F}_i)$, $1 \leqslant i \leqslant n$, is a martingale with differences $W_1 = R_1$ and $W_i = R_i - R_{i-1}$, $i \geqslant 1$. Now,

$$\begin{aligned} \|S_n\|_p &= E(R_n S_n) = E\left(\sum_{i=1}^n W_i X_i\right) \\ &\leqslant E\left[\left(\sum_{i=1}^n W_i^2\right)^{1/2} \left(\sum_{i=1}^n X_i^2\right)^{1/2}\right] \\ &\leqslant \left\| \left(\sum_{i=1}^n W_i^2\right)^{1/2} \right\|_q \left\| \left(\sum_{i=1}^n X_i^2\right)^{1/2} \right\|_p, \end{aligned} \tag{2.15}$$

using Hölder's inequality. We then apply the left-hand inequality in (2.7) to the martingale $\{R_i, \mathscr{F}_i\}$ and obtain

$$\left\| \left(\sum_{i=1}^{n} W_i^2 \right)^{1/2} \right\|_q^q \leqslant (18q^{1/2}p)^q E|R_n|^q = (18q^{1/2}p)^q.$$

The right-hand side of (2.7) then follows from the inequalities (2.15). ∎

Proof of Theorem 2.11. The theorem will follow immediately from Lemma 2.3 if we establish the inequality (2.13) with

$$X = \max_{i \leqslant n} |S_i|, \qquad Y = \max\left(\left(\sum_{i=1}^{n} E(X_i^2 | \mathscr{F}_{i-1}) \right)^{1/2}, \max_{i \leqslant n} |X_i| \right)$$

and $\varepsilon = \delta^2/(\beta - \delta - 1)^2$ ($\beta > 1$, $0 < \delta < \beta - 1$). Let I_k be the indicator of the event

$$\left\{ \lambda < \max_{i \leqslant k-1} |S_i| \leqslant \beta\lambda;\ \max_{i \leqslant k-1} |X_i| \leqslant \delta\lambda;\ \sum_{i=1}^{k} E(X_i^2 | \mathscr{F}_{i-1}) \leqslant \delta^2\lambda^2 \right\}$$

and define

$$T_i = \sum_{k=1}^{i} I_k X_k, \qquad 1 \leqslant i \leqslant n.$$

The sequence $\{T_i, \mathscr{F}_i, 1 \leqslant i \leqslant n\}$ is a martingale, and

$$\begin{aligned} P(X > \beta\lambda;\ Y \leqslant \delta\lambda) &\leqslant P\left(\max_{i \leqslant n} |T_i| > (\beta - \delta - 1)\lambda \right) \\ &\leqslant (\beta - \delta - 1)^{-2} \lambda^{-2} E(T_n^2). \end{aligned}$$

But

$$\begin{aligned} E(T_n^2) &= E\left[\sum_{k=1}^{n} I_k E(X_k^2 | \mathscr{F}_{k-1}) \right] \\ &\leqslant E\left[\sum_{k=1}^{n} I\left(\max_{i \leqslant k-1} |S_i| > \lambda;\ \sum_{i=1}^{k} E(X_i^2 | \mathscr{F}_{i-1}) \leqslant \delta^2\lambda^2 \right) E(X_k^2 | \mathscr{F}_{k-1}) \right] \\ &\leqslant \delta^2\lambda^2 P\left(\max_{i \leqslant n} |S_i| > \lambda \right) \end{aligned}$$

and so

$$P(X > \beta\lambda;\ Y \leqslant \delta\lambda) \leqslant (\beta - \delta - 1)^{-2} \delta^2 P(X > \lambda),$$

which is (2.13). ∎

Proof of Theorem 2.12. The right-hand inequality is immediate from Theorem 2.11; we must prove the left. From Theorem A.8 of Appendix III we deduce that there exists a constant K depending only on p such that

$$E\left[\left(\sum_{i=1}^{n} E(X_i^2|\mathscr{F}_{i-1})\right)^{p/2}\right] \leqslant KE\left[\left(\sum_{i=1}^{n} X_i^2\right)^{p/2}\right].$$

It follows that

$$\begin{aligned} E\left[\left(\sum_{i=1}^{n} E(X_i^2|\mathscr{F}_{i-1})\right)^{p/2}\right] &+ E\left(\sum_{i=1}^{n} |X_i|^p\right) \\ &\leqslant KE\left[\left(\sum_{i=1}^{n} X_i^2\right)^{p/2}\right] + E\left[\left(\sum_{i=1}^{n} X_i^2\right)^{p/2}\right] \\ &\leqslant C_1^{-1}(K+1)E|S_n|^p, \end{aligned}$$

using (2.7) and the inequality for real numbers x_i,

$$\sum_1^n |x_i|^p \leqslant \left(\sum_1^n x_i^2\right)^{p/2}, \qquad p \geqslant 2.$$

This establishes the left-hand inequality. ∎

2.5. Weak Law of Large Numbers

We confine ourselves here to a martingale version of the classical degenerate convergence criterion. Further conditions for weak convergence follow from the results of the next two sections.

Theorem 2.13. *Let $\{S_n = \sum_{i=1}^n X_i, \mathscr{F}_n, n \geqslant 1\}$ be a martingale and $\{b_n\}$ a sequence of positive constants with $b_n \uparrow \infty$ as $n \to \infty$. Then, writing $X_{ni} = X_i I(|X_i| \leqslant b_n)$, $1 \leqslant i \leqslant n$, we have that $b_n^{-1} S_n \xrightarrow{\text{p}} 0$ as $n \to \infty$ if*

(i) $\sum_{i=1}^n P(|X_i| > b_n) \to 0$,
(ii) $b_n^{-1} \sum_{i=1}^n E(X_{ni}|\mathscr{F}_{i-1}) \xrightarrow{\text{p}} 0$, *and*
(iii) $b_n^{-2} \sum_{i=1}^n \{EX_{ni}^2 - E[E(X_{ni}|\mathscr{F}_{i-1})]^2\} \to 0$.

Remark. In the particular case when the X_i are independent, the above conditions are necessary as well as sufficient for $b_n^{-1} S_n \xrightarrow{\text{p}} 0$ [see, for example, Loève (1977, pp. 290, 329)]. Characteristic function methods are traditionally used to prove necessity.

In the general case the conditions do not appear to be close to necessity. We give the following simple example. Let Y_i, $i = 1, 2, \ldots$, be independent r.v. with $Y_1 = 1$ and for $i > 1$,

$$P(Y_i = 0) = i^{-1},$$
$$P(Y_i = -2) = \tfrac{1}{2}(1 - i^{-1}),$$
$$P(Y_i = 2 + 2(1 - i^{-1})^{-1}) = \tfrac{1}{2}(1 - i^{-1}),$$

so that $EY_i = 1$ for all i. Let

$$S_n = \left(\prod_{i=1}^{n} Y_i\right) - 1, \qquad n \geqslant 1,$$

and note that $\{S_n, \mathscr{F}_n, n \geqslant 1\}$ is a zero-mean martingale, $\mathscr{F}_n$ denoting the σ-field generated by $\{Y_i, 1 \leqslant i \leqslant n\}$. We have that

$$P\left(\left|\prod_{i=1}^{n} Y_i\right| > 0\right) = \prod_{i=2}^{n} (1 - i^{-1}) = n^{-1} \to 0$$

as $n \to \infty$, so that $S_n \xrightarrow{\text{p}} -1$ and $b_n^{-1} S_n \xrightarrow{\text{p}} 0$ for any sequence $\{b_n\}$ of positive constants with $b_n \uparrow \infty$. Take in particular $b_n = n$, $n \geqslant 1$. If $2^{i-1} \geqslant n$, then

$$\begin{aligned} P(|X_i| > n) &= P(|Y_1 Y_2 \cdots Y_{i-1}||Y_i - 1| > n) \\ &= \prod_{j=2}^{i-1} (1 - j^{-1}) = (i-1)^{-1} \end{aligned}$$

and hence

$$\sum_{i=1}^{n} P(|X_i| > n) \geqslant \sum_{1 + \log n \leqslant i \leqslant n} (i-1)^{-1} \to \infty$$

as $n \to \infty$, so that (i) is not satisfied in this case.

Proof of Theorem 2.13. Let $S_{nn} = \sum_{i=1}^{n} X_{ni}$. On account of (i),

$$P(S_{nn}/b_n \neq S_n/b_n) \leqslant \sum_{i=1}^{n} P(X_{ni} \neq X_i) = \sum_{i=1}^{n} P(|X_i| > b_n) \to 0,$$

and so it suffices to prove that $b_n^{-1} S_{nn} \xrightarrow{\text{p}} 0$. But on account of (ii),

$$b_n^{-1} \sum_{i=1}^{n} E(X_{ni} | \mathscr{F}_{i-1}) \xrightarrow{\text{p}} 0,$$

so that it suffices to prove that

$$b_n^{-1} \sum_{i=1}^{n} [X_{ni} - E(X_{ni} | \mathscr{F}_{i-1})] \xrightarrow{\text{p}} 0.$$

This last result, however, follows from an application of Chebyshev's inequality together with (iii). ▌

2.6. Strong Law of Large Numbers

As a prelude to the results of this section we give the lemmas of Toeplitz and Kronecker which will often be applied in this book.

Toeplitz Lemma. *Let* a_{ni}, $1 \leqslant i \leqslant k_n$, $n \geqslant 1$, *and* x_i, $i \geqslant 1$, *be real numbers such that for every fixed i*, $a_{ni} \to 0$ *and for all n*, $\sum_i |a_{ni}| \leqslant C < \infty$. *If* $x_n \to 0$, *then* $\sum_i a_{ni} x_i \to 0$, *and if* $\sum_i a_{ni} \to 1$, *then* $x_n \to x$ *ensures that* $\sum_i a_{ni} x_i \to x$. *In particular, if* a_i, $i \geqslant 1$, *are positive numbers and* $b_n = \sum_{i=1}^n a_i \uparrow \infty$, *then* $x_n \to x$ *ensures that* $b_n^{-1} \sum_{i=1}^n a_i x_i \to x$.

The proof is simple. If $x_n \to 0$, then for a given $\varepsilon > 0$ and $n \geqslant n_\varepsilon$ sufficiently large, $|x_n| < C^{-1}\varepsilon$, so that

$$\left| \sum_i a_{ni} x_i \right| \leqslant \sum_{i < n_\varepsilon} |a_{ni} x_i| + \varepsilon,$$

and $\sum_i a_{ni} x_i \to 0$ follows upon letting $n \to \infty$ and then $\varepsilon \to 0$. The second assertion follows since

$$\sum_i a_{ni} x_i = x \sum_i a_{ni} + \sum_i a_{ni}(x_i - x) \to x,$$

using the first result. The particular case is obtained by setting $a_{ni} = a_i/b_n$, $1 \leqslant i \leqslant n$.

Kronecker Lemma. *Let* $\{x_n, n \geqslant 1\}$ *be a sequence of real numbers such that* $\sum x_n$ *converges, and let* $\{b_n\}$ *be a monotone sequence of positive constants with* $b_n \uparrow \infty$. *Then* $b_n^{-1} \sum_{i=1}^n b_i x_i \to 0$.

This is easily proved using the particular case cited in the Toeplitz lemma. Set $b_0 = 0$, $a_i = b_i - b_{i-1}$, $i \geqslant 1$, and $s_{n+1} = \sum_{i=1}^n x_i \to s$ (say), as $n \to \infty$. Then as $n \to \infty$,

$$b_n^{-1} \sum_{i=1}^n b_i x_i = b_n^{-1} \sum_{i=1}^n b_i(s_{i+1} - s_i) = s_{n+1} - b_n^{-1} \sum_{i=1}^n a_i s_i \to s - s = 0,$$

as required.

We now embark on a discussion of strong law behavior for martingales. Our emphasis in this book is mostly on martingales whose differences do not become asymptotically negligible, so the results in this chapter are oriented in this direction. No attempt at an exhaustive discussion of the subject is made since it has already been treated comprehensively in book form in the work of Stout (1974), Neveu (1965, 1975) and Chow and Teicher (1978). Many of the results date back to Doob (1953).

Theorem 2.14. *Let* $\{S_n = \sum_{i=1}^n X_i, \mathscr{F}_n, n \geqslant 1\}$ *be a zero-mean martingale with* $E(\sup_n|X_n|) < \infty$*. Then* $\liminf_{n\to\infty} S_n = -\infty$ *a.s. and* $\limsup_{n\to\infty} S_n = +\infty$ *a.s. on the set* $\{\sum_{i=1}^\infty X_i$ *diverges*$\}$.

Proof. Let $\nu_a = \min\{n|S_n > a\}$, with $\nu_a = \infty$ if no such n exists. The variables $\nu_{a\wedge n}$, $n \geqslant 1$, form a nondecreasing sequence of stopping times and so $\{S_{\nu_a\wedge n}, \mathscr{F}_{\nu_a \wedge n}, n \geqslant 1\}$ is a martingale. Moreover,

$$S^+_{\nu_a\wedge n} \leqslant S^+_{\nu_a\wedge(n-1)} + X^+_{\nu_a\wedge n} \leqslant a + \sup_n(X_n^+),$$

and since $E(\sup_n|X_n|) < \infty$,

$$E|S_{\nu_a\wedge n}| = 2E(S^+_{\nu_a\wedge n})$$

is bounded as n $\to \infty$. The martingale convergence theorem (Theorem 2.5) implies that $S_{\nu_a\wedge n}$ converges a.s. to a finite limit as $n \to \infty$, and consequently $\lim_{n\to\infty} S_n$ exists and is a.s. finite on the set $\{\sup S_n \leqslant a\}$. Letting $a \to \infty$ we see that $\lim S_n$ exists and is a.s. finite on the set $\{\sup S_n < \infty\} = \{\limsup S_n < \infty\}$, and therefore that $\limsup S_n = +\infty$ a.s. on the set $\{S_n$ diverges$\}$. The result for $\liminf S_n$ follows on replacing S_n by $-S_n$. ▮

Corollary 2.3. *Let* $\{Z_n, n \geqslant 1\}$ *be a sequence of r.v. such that* $0 \leqslant Z_n \leqslant 1$*, and* $\{\mathscr{F}_n, n \geqslant 1\}$ *be an increasing sequence of* σ*-fields such that each* Z_n *is* $\mathscr{F}_n$*-measurable. Then* $\sum_n Z_n < \infty$ *a.s. if and only if* $\sum_n E(Z_n|\mathscr{F}_{n-1}) < \infty$ *a.s. In particular, the following extension of the Borel–Cantelli lemma holds. If* $\{Y_n, n \geqslant 1\}$ *is any process and* $A_n \in \mathscr{G}_n$ *for all n, where* $\mathscr{G}_n$ *denotes the* σ*-field generated by* $Y_i, i \leqslant n$*, then the sets* $\{\limsup_{n\to\infty} A_n\}$ *and* $\{\sum_n P(A_{n+1}|\mathscr{G}_n) = \infty\}$ *are almost surely equal.*

Proof. Take $\mathscr{F}_0$ as the trivial σ-field and let

$$X_n = Z_n - E(Z_n|\mathscr{F}_{n-1}), \qquad n \geqslant 1.$$

Then $\{S_n = \sum_{i=1}^n X_i, \mathscr{F}_n, n \geqslant 1\}$ is a zero-mean martingale with uniformly bounded differences. Furthermore,

$$\limsup_{n\to\infty} S_n \leqslant \sum_{i=1}^\infty Z_i,$$

so that, via Theorem 2.14, $\lim_{n\to\infty} S_n$ exists and is finite a.s. on the set $\{\sum_{i=1}^\infty Z_i < \infty\}$, while also

$$\sum_{i=1}^\infty E(Z_i|\mathscr{F}_{i-1}) = \sum_{i=1}^\infty Z_i - \lim_{n\to\infty} S_n$$

converges a.s. on $\{\sum_{i=1}^\infty Z_i < \infty\}$. Similar reasoning, using the inequality

$$\liminf_{n\to\infty} S_n \geqslant -\sum_{i=1}^\infty E(Z_i|\mathscr{F}_{i-1}),$$

shows that $\sum_1^\infty Z_i < \infty$ a.s. on the set $\{\sum_{i=1}^\infty E(Z_i|\mathscr{F}_{i-1}) < \infty\}$. The extended Borel–Cantelli lemma follows immediately upon noting that $\{\limsup A_n\} = \{\sum_1^\infty I(A_n) = \infty\}$, while

$$P(A_{n+1}|\mathscr{G}_n) = E(I(A_{n+1})|\mathscr{G}_n). \quad \blacksquare$$

Theorem 2.15. *Let $\{S_n = \sum_1^n X_i, \mathscr{F}_n, n \geqslant 1\}$ be a zero-mean, square-integrable martingale. Then S_n converges a.s. on the set $\{\sum_{i=1}^\infty E(X_i^2|\mathscr{F}_{i-1}) < \infty\}$.*

Proof. Take $\mathscr{F}_0$ as the trivial σ-field. Fix $K > 0$ and let τ be the smallest integer $n \geqslant 1$ such that $\sum_{i=1}^{n+1} E(X_i^2|\mathscr{F}_{i-1}) > K$ if such an n exists; otherwise let $\tau = \infty$. Then $\{S_{\tau\wedge n} = \sum_{i=1}^n I(\tau \geqslant i)X_i, \mathscr{F}_n, n \geqslant 1\}$ is a martingale. Furthermore, using the martingale property and since $I(\tau \geqslant i)$ is $\mathscr{F}_{i-1}$-measurable,

$$\begin{aligned} ES_{\tau\wedge n}^2 &= E\left\{\sum_{i=1}^n I(\tau \geqslant i)X_i^2\right\} \\ &= E\left\{\sum_{i=1}^n E[I(\tau \geqslant i)X_i^2|\mathscr{F}_{i-1}]\right\} \\ &= E\left\{\sum_{i=1}^n I(\tau \geqslant i)E(X_i^2|\mathscr{F}_{i-1})\right\} \\ &= E\left\{\sum_{i=1}^{\tau\wedge n} E(X_i^2|\mathscr{F}_{i-1})\right\} \leqslant K. \end{aligned}$$

It now follows from the martingale convergence theorem that $S_{\tau\wedge n}$ converges a.s. as $n \to \infty$. Thus S_n converges a.s. on $\{\tau = \infty\}$. The required result follows upon letting $K \uparrow \infty$. $\blacksquare$

Theorem 2.15 can be applied to obtain the following conditional version of the three-series theorem.

Theorem 2.16. *Let $\{S_n = \sum_{i=1}^n X_i, n \geqslant 1\}$ be a sequence of r.v. and $\{\mathscr{F}_n, n \geqslant 1\}$ an increasing sequence of σ-fields such that S_n is $\mathscr{F}_n$-measurable. Let c be a positive constant. Then S_n converges a.s. on the set where*

(i) $\sum_{i=1}^\infty P(|X_i| \geqslant c|\mathscr{F}_{i-1}) < \infty$,
(ii) $\sum_{i=1}^\infty E[X_i I(|X_i| \leqslant c)|\mathscr{F}_{i-1}]$ *converges, and*
(iii) $\sum_{i=1}^\infty \{E[X_i^2 I(|X_i| \leqslant c)|\mathscr{F}_{i-1}] - [E(X_i I(|X_i| \leqslant c)|\mathscr{F}_{i-1})]^2\} < \infty$ *hold.*

The three-series theorem is a result of considerable significance. In the case where the X_i are independent it provides a necessary as well as a sufficient condition for a.s. convergence [see, for example, Chung (1974, p. 119)]. The conditional form of the theorem, however, does not in general provide necessary conditions for convergence. Indeed, take the case of uniformly bounded X_i, so that (i) can be automatically satisfied and truncation at c is irrelevant. Then there can be no condition on the conditional means $\{E(X_i|\mathscr{F}_{i-1}), i \geqslant 1\}$ and conditional variances $(E(X_i^2|\mathscr{F}_{i-1}) - (EX_i|\mathscr{F}_{i-1}))^2$, $i \geqslant 1\}$ alone which is necessary and sufficient for $\sum_i X_i$ to converge a.s. To see this, take $\{S_n, n \geqslant 1\}$ to be a sequence of independent r.v. with $P(S_n = n^{-1/2}) = P(S_n = -n^{-1/2}) = \frac{1}{2}$, and set $S_0 = 0$ and $X_i = S_i - S_{i-1}$, $i \geqslant 1$. The X_i are uniformly bounded, $\sum_i X_i$ converges a.s. (via the Borel–Cantelli lemmas), and

$$E(X_n|\mathscr{F}_{n-1}) = -S_{n-1} \qquad \text{and} \qquad E(X_n^2|\mathscr{F}_{n-1}) = 2n^{-1}.$$

On the other hand, if $Y_n = S_n + S_{n-1}$, $n \geqslant 1$, then

$$E(Y_n|\mathscr{F}_{n-1}) = S_{n-1}, \qquad E(Y_n^2|\mathscr{F}_{n-1}) = 2n^{-1},$$

and since $\sum_i E(S_i^2) = \infty$, $\sum_i S_i$ diverges a.s., implying that $\sum_i Y_i$ diverges a.s.

Proof of Theorem 2.16. Let A be the set on which (i), (ii), and (iii) hold. Using (i) together with Corollary 2.3 and then (ii), we have that on A,

$$\left\{\sum_{i=1}^{\infty} X_i \text{ converges}\right\} = \left\{\sum_{i=1}^{\infty} X_i I(|X_i| \leqslant c) \text{ converges}\right\}$$
$$= \left\{\sum_{i=1}^{\infty} Y_i \text{ converges}\right\},$$

where

$$Y_i = X_i I(|X_i| \leqslant c) - E[X_i I(|X_i| \leqslant c)|\mathscr{F}_{i-1}], \qquad i \geqslant 1.$$

Now, $\{\sum_{i=1}^n Y_i, \mathscr{F}_n, n \geqslant 1\}$ is a zero-mean martingale with

$$E(Y_i^2|\mathscr{F}_{i-1}) = E[X_i^2 I(|X_i| \leqslant c)|\mathscr{F}_{i-1}] - [E(X_i I(|X_i| \leqslant c)|\mathscr{F}_{i-1})]^2,$$

and, using Theorem 2.15, $\sum_{i=1}^{\infty} Y_i$ converges a.s. on A. This completes the proof. ∎

It is sometimes useful to turn the three-series theorem into a two-series theorem by truncation "to" the level c rather than "at" the level c. This involves using

$$Y_n = X_n I(|X_n| \leqslant c) + c\,\text{sgn}(X_n) I(|X_n| > c),$$

instead of $X_n I(|X_n| \leqslant c)$. The conditions (i), (ii), and (iii) of Theorem 2.16 can then be replaced by the equivalent forms

(a) $\sum_{i=1}^{\infty} E(Y_i|\mathscr{F}_{i-1})$ converges, and
(b) $\sum_{i=1}^{\infty} E[(Y_i - E(Y_i|\mathscr{F}_{i-1}))^2|\mathscr{F}_{i-1}] < \infty$.

That the condition (i) can be dispensed with follows from

$$\begin{aligned} E[(Y_i - E(Y_i|\mathscr{F}_{i-1}))^2|\mathscr{F}_{i-1}] &\geqslant E[(Y_i - E(Y_i|\mathscr{F}_{i-1}))^2 I(X_i \geqslant c)|\mathscr{F}_{i-1}] \\ &\quad + E[(Y_i - E(Y_i|\mathscr{F}_{i-1}))^2 I(X_i \leqslant -c)|\mathscr{F}_{i-1}] \\ &= [c - E(Y_i|\mathscr{F}_{i-1})]^2 P(X_i \geqslant c|\mathscr{F}_{i-1}) \\ &\quad + [c + E(Y_i|\mathscr{F}_{i-1})]^2 P(X_i \leqslant -c|\mathscr{F}_{i-1}) \\ &\geqslant (c - |E(Y_i|\mathscr{F}_{i-1})|)^2 P(|X_i| \geqslant c|\mathscr{F}_{i-1}). \end{aligned}$$

The use of Theorem 2.16 leads to the following extension of Theorem 2.15, due to Chow (1965).

Theorem 2.17. *Let $\{S_n = \sum_{i=1}^n X_i, \mathscr{F}_n, n \geqslant 1\}$ be a martingale and let $1 \leqslant p \leqslant 2$. Then S_n converges a.s. on the set $\{\sum_{i=1}^\infty E(|X_i|^p|\mathscr{F}_{i-1}) < \infty\}$.*

Proof. We shall verify the conditions of Theorem 2.16 under the condition

$$\sum_{i=1}^{\infty} E(|X_i|^p|\mathscr{F}_{i-1}) < \infty.$$

Condition (i) is checked by noting that

$$\begin{aligned} P(|X_i| \geqslant c|\mathscr{F}_{i-1}) &= E[I(|X_i| \geqslant c)|\mathscr{F}_{i-1}] \\ &\leqslant c^{-p} E[|X_i|^p I(|X_i| \geqslant c)|\mathscr{F}_{i-1}] \\ &\leqslant c^{-p} E(|X_i|^p|\mathscr{F}_{i-1}), \end{aligned}$$

while (ii) holds since, using the martingale property,

$$\begin{aligned} |E[X_i I(|X_i| \leqslant c)|\mathscr{F}_{i-1}]| &= |E[X_i I(|X_i| > c)|\mathscr{F}_{i-1}]| \\ &\leqslant E[|X_i| I(|X_i| > c)|\mathscr{F}_{i-1}] \\ &\leqslant c^{1-p} E(|X_i|^p|\mathscr{F}_{i-1}). \end{aligned}$$

Finally, (iii) holds since

$$\begin{aligned} E[X_i^2 I(|X_i| \leqslant c)|\mathscr{F}_{i-1}] &- [E(X_i I(|X_i| \leqslant c)|\mathscr{F}_{i-1})]^2 \\ &\leqslant E[X_i^2 I(|X_i| \leqslant c)|\mathscr{F}_{i-1}] \\ &\leqslant c^{2-p} E(|X_i|^p|\mathscr{F}_{i-1}). \quad \blacksquare \end{aligned}$$

Since much of the effort in this book, and in applications in general, is concerned with normed sums, we now abstract from the above results some others which are more directly applicable in that context.

Theorem 2.18. *Let $\{S_n = \sum_{i=1}^n X_i, \mathscr{F}_n, n \geqslant 1\}$ be a martingale and $\{U_n, n \geqslant 1\}$ a nondecreasing sequence of positive r.v. such that U_n is $\mathscr{F}_{n-1}$-*

measurable for each n. If $1 \leqslant p \leqslant 2$ *then*

$$\sum_{1}^{\infty} U_i^{-1} X_i \quad \text{converges a.s.} \tag{2.16}$$

on the set $\{\sum_{i=1}^{\infty} U_i^{-p} E(|X_i|^p | \mathscr{F}_{i-1}) < \infty\}$, *and*

$$\lim_{n \to \infty} U_n^{-1} S_n = 0 \quad \text{a.s.} \tag{2.17}$$

on the set $\{\lim_{n\to\infty} U_n = \infty, \sum_{i=1}^{\infty} U_i^{-p} E(|X_i|^p | \mathscr{F}_{i-1}) < \infty\}$. *If* $2 < p < \infty$, *then* (2.16) *and* (2.17) *both hold on the set*

$$\left\{\sum_{i=1}^{\infty} U_i^{-1} < \infty, \sum_{i=1}^{\infty} U_i^{-1-p/2} E(|X_i|^p | \mathscr{F}_{i-1}) < \infty\right\}.$$

Proof. Put $Y_n = U_n^{-1} X_n$, $n \geqslant 1$, and note that $\{\sum_{i=1}^{n} Y_i, \mathscr{F}_n, n \geqslant 1\}$ is a martingale. For the case $1 \leqslant p \leqslant 2$ the result (2.16) follows immediately from an application of Theorem 2.17 to this martingale, while (2.17) follows from (2.16) and an application of Kronecker's lemma.

To deal with the case $p > 2$ we note that if

$$[E(|X_n|^p | \mathscr{F}_{n-1})]^{2/p} > U_n,$$

we have

$$\begin{aligned}[E(|X_n|^p | \mathscr{F}_{n-1})]^{2/p} &= E(|X_n|^p | \mathscr{F}_{n-1})[E(|X_n|^p | \mathscr{F}_{n-1})]^{(2/p)-1} \\ &< U_n^{1-p/2} E(|X_n|^p | \mathscr{F}_{n-1}).\end{aligned}$$

Hence

$$\begin{aligned}E(Y_n^2 | \mathscr{F}_{n-1}) = U_n^{-2} E(X_n^2 | \mathscr{F}_{n-1}) &\leqslant [U_n^{-p} E(|X_n|^p | \mathscr{F}_{n-1})]^{2/p} \\ &\leqslant \max[U_n^{-1}, U_n^{-1-p/2} E(|X_n|^p | \mathscr{F}_{n-1})],\end{aligned}$$

and again (2.16) follows from an application of Theorem 2.17 while (2.17) follows from (2.16) using Kronecker's lemma. ∎

In many cases of interest the increments of the process under study are sufficiently close to stationarity for there to be a natural bounding random variable, and the following convergence result is useful.

Theorem 2.19. *Let* $\{X_n, n \geqslant 1\}$ *be a sequence of r.v. and* $\{\mathscr{F}_n, n \geqslant 1\}$ *an increasing sequence of σ-fields with* X_n *measurable with respect to* $\mathscr{F}_n$ *for each n. Let X be an r.v. and c a constant such that* $E|X| < \infty$ *and* $P(|X_n| > x) \leqslant cP(|X| > x)$ *for each* $x \geqslant 0$ *and* $n \geqslant 1$. *Then*

$$n^{-1} \sum_{i=1}^{n} [X_i - E(X_i | \mathscr{F}_{i-1})] \xrightarrow{\text{p}} 0 \tag{2.18}$$

as $n \to \infty$. If $E(|X|\log^+|X|) < \infty$, or if the X_n are independent, or if $\{X_n, n \geqslant 1\}$ and $\{E(X_n|\mathscr{F}_{n-1}), n \geqslant 2\}$ are stationary sequences, then the convergence in probability in (2.18) *can be strengthened to a.s. convergence.*

Proof. Let $Y_n = X_n I(|X_n| \leqslant n)$, $n \geqslant 1$. Now,

$$\begin{aligned}\sum_{n=1}^{\infty} n^{-2}E[(Y_n - E(Y_n|\mathscr{F}_{n-1}))^2] &\leqslant \sum_{n=1}^{\infty} n^{-2}E(Y_n^2)\\ &\leqslant 2\sum_{n=1}^{\infty} n^{-2}\int_{0<x\leqslant n} xP(|X_n| > x)\,dx\\ &\leqslant 2c\sum_{n=1}^{\infty} n^{-2}\int_{0<x\leqslant n} xP(|X| > x)\,dx\\ &= 2c\sum_{n=1}^{\infty} n^{-2}\sum_{i=1}^{n}\int_{i-1<x\leqslant i} xP(|X| > x)\,dx\\ &\leqslant 2c\sum_{n=1}^{\infty} n^{-2}\sum_{i=1}^{n} iP(|X| > i-1)\\ &= 2c\sum_{i=1}^{\infty} iP(|X| > i-1)\sum_{n=i}^{\infty} n^{-2}\\ &\leqslant 4c\sum_{i=1}^{\infty} P(|X| > i-1) < \infty,\end{aligned}$$

since $E|X| < \infty$. It then follows from Theorem 2.15 that

$$n^{-1}\sum_{i=1}^{n}[Y_i - E(Y_i|\mathscr{F}_{i-1})] \xrightarrow{\text{a.s.}} 0. \tag{2.19}$$

Next we note that

$$\sum_{n=1}^{\infty} P(X_n \neq Y_n) = \sum_{n=1}^{\infty} P(|X_n| > n) \leqslant c\sum_{n=1}^{\infty} P(|X| > n) < \infty,$$

so that the sequences $\{X_n\}$ and $\{Y_n\}$ are tail equivalent, and hence from (2.19),

$$n^{-1}\sum_{i=1}^{n}[X_i - E(Y_i|\mathscr{F}_{i-1})] \xrightarrow{\text{a.s.}} 0. \tag{2.20}$$

Now,

$$n^{-1}E\left|\sum_{i=1}^{n} E[X_i I(|X_i| > i)|\mathscr{F}_{i-1}]\right| \leqslant n^{-1}\sum_{i=1}^{n} E[|X_i| I(|X_i| > i)] \to 0$$

as $n \to \infty$ since

$$E[|X_n|I(|X_n| > n)] = \int_n^\infty P(|X_n| > x)\,dx \leqslant c\int_n^\infty P(|X| > x)\,dx \to 0.$$

Therefore,

$$n^{-1}\sum_{i=1}^{n} E(X_i - Y_i|\mathscr{F}_{i-1}) = n^{-1}\sum_{i=1}^{n} E[X_i I(|X_i| > i)|\mathscr{F}_{i-1}] \xrightarrow{\mathrm{p}} 0,$$

implying (2.18). If the X_n are independent, then each $E(X_n - Y_n|\mathscr{F}_{n-1})$ is constant, and so the a.s. convergence version of (2.18) holds. If $\{X_n\}$ and $\{E(X_n|\mathscr{F}_n)\}$ are stationary, we make two applications of the ergodic theorem (see Appendix IV) and obtain

$$n^{-1}\sum_{i=1}^{n} X_i \xrightarrow{\text{a.s.}} E(X_1|\mathscr{G}) \quad \text{and} \quad n^{-1}\sum_{i=2}^{n} E(X_i|\mathscr{F}_{i-1}) \xrightarrow{\text{a.s.}} E[E(X_2|\mathscr{F}_1)|\mathscr{H}],$$

where $\mathscr{G}$ and $\mathscr{H}$ are the respective σ-fields of invariant events. Thus

$$n^{-1}\sum_{i=1}^{n}[X_i - E(X_i|\mathscr{F}_{i-1})] \xrightarrow{\text{a.s.}} E(X_1|\mathscr{G}) - E[E(X_2|\mathscr{F}_1)|\mathscr{H}] \qquad (2.21)$$

and a.s. convergence to zero results from identifying the limit in (2.18) and (2.21).

To complete the proof we now suppose that $E(|X|\log^+|X|) < \infty$. Then

$$\sum_{n=1}^{\infty} n^{-1}E[|X_n|I(|X_n| > n)|\mathscr{F}_{n-1}] < \infty \quad \text{a.s.}$$

since

$$\begin{aligned}
\sum_{n=1}^{\infty} n^{-1}E[|X_n|I(|X_n| > n)] &= \sum_{n=1}^{\infty} n^{-1}\int_n^\infty P(|X_n| > x)\,dx \\
&\leqslant c\sum_{n=1}^{\infty} n^{-1}\int_n^\infty P(|X| > x)\,dx \\
&= c\sum_{n=1}^{\infty} n^{-1}\sum_{i=n}^{\infty}\int_{i<x\leqslant i+1} P(|X| > x)\,dx \\
&\leqslant c\sum_{i=1}^{\infty} P(|X| > i)\sum_{n=1}^{i} n^{-1} \\
&\leqslant c\sum_{i=1}^{\infty}(1 + \log i)P(|X| > i) < \infty.
\end{aligned}$$

Thus using Kronecker's lemma,

$$n^{-1} \sum_{i=1}^{n} E[|X_i| I(|X_i| > i) | \mathscr{F}_{i-1}] \xrightarrow{\text{a.s.}} 0$$

and hence

$$n^{-1} \sum_{i=1}^{n} E(X_i - Y_i | \mathscr{F}_{i-1}) = n^{-1} \sum_{i=1}^{n} E[X_i I(|X_i| > i) | \mathscr{F}_{i-1}] \xrightarrow{\text{a.s.}} 0,$$

which, together with (2.20), completes the proof. ∎

Remark. Counterexamples related to the problem of Theorem 2.19 are not simple to construct but the following illustration is instructive concerning the possibilities. We exhibit a martingale $\{S_n = \sum_{i=1}^{n} X_i, \mathscr{F}_n, n \geqslant 1\}$ for which $P(S_n = n) = P(S_n = -n) = \frac{1}{2}$ and thus

$$n^{-1} \sum_{i=1}^{n} [X_i - E(X_i | \mathscr{F}_{i-1})] = n^{-1} S_n = \pm 1,$$

each with probability $\frac{1}{2}$. The martingale has increments whose conditional distributions are given by

$$P(X_n = 1 | S_{n-1} = n-1) = P(X_n = -1 | S_{n-1} = -(n-1)) = 1 - 1/2n,$$
$$P(X_n = -(2n-1) | S_{n-1} = n-1) = P(X_n = 2n-1 | S_{n-1} = -(n-1)) = 1/2n,$$

and whose unconditional distributions are given by

$$P(X_n = 1) = P(X_n = -1) = \tfrac{1}{2}(1 - 1/2n),$$
$$P(X_n = 2n-1) = P(X_n = -(2n-1)) = 1/4n.$$

For $1 \leqslant x < 2n - 1$,

$$P(|X_n| > x) = 1/2n < x^{-1},$$

while for $x \geqslant 2n - 1$,

$$P(|X_n| > x) = 0 < x^{-1}.$$

Therefore $P(|X_n| > x) \leqslant P(|X| > x)$ for all $n \geqslant 1$ and $x \geqslant 0$, where the r.v. X has a symmetric distribution with $P(|X| \leqslant 1) = 0$ and $P(|X| > x) = x^{-1}$, $x \geqslant 1$. Note that $E|X| = \infty$. The behavior of S_n contrasts strongly with that of a sum of independent and identically distributed variables, where if $E|X_1| = \infty$, $\limsup n^{-1}|S_n| = \infty$ a.s. [see Chung (1974, Theorem 5.4.3, p. 128)].

Strong laws may be obtained for a wide variety of weak dependence situations by using martingale techniques. The simplest of these occur when so-called mixing conditions are satisfied.

Let $\{X_n, n \geqslant 1\}$ be a sequence of r.v. and write $\mathscr{B}_r^s$ for the σ-field generated by $\{X_r, \ldots, X_s\}$, $1 \leqslant r \leqslant s \leqslant \infty$. The sequence $\{X_n\}$ will be called **-mixing* if there exists a positive integer N and a function f defined on the integers $n \geqslant N$ such that $f \downarrow 0$ and for all $n \geqslant N$, $m \geqslant 1$, $A \in \mathscr{B}_1^m$, and $B \in \mathscr{B}_{m+n}^\infty$,

$$|P(A \cap B) - P(A)P(B)| \leqslant f(n)P(A)P(B).$$

Clearly this is equivalent to the condition

$$\text{for all} \quad B \in \mathscr{B}_{m+n}^\infty, \qquad |P(B|\mathscr{B}_1^m) - P(B)| \leqslant f(n)P(B) \quad \text{a.s.,}$$

which implies, when the X_n are integrable, that

$$|E(X_{m+n}|\mathscr{B}_1^m) - E(X_{m+n})| \leqslant f(n)E|X_{m+n}| \quad \text{a.s.} \tag{2.22}$$

The *-mixing condition is the most stringent of a hierarchy of mixing conditions. The following strong law holds for *-mixing sequences [see Blum *et al.* (1963)].

Theorem 2.20. *Let $\{X_n, n \geqslant 1\}$ be a *-mixing sequence such that $E(X_n) = 0$ and $E(X_n^2) < \infty$, $n \geqslant 1$. Suppose that*

$$\sum_{n=1}^{\infty} b_n^{-2} E(X_n^2) < \infty \qquad \textit{and} \qquad \sup_n b_n^{-1} \sum_{i=1}^{n} E|X_i| < \infty,$$

where $\{b_n\}$ is a sequence of positive constants increasing to ∞. Then

$$b_n^{-1} \sum_{i=1}^{n} X_i \xrightarrow{\text{a.s.}} 0.$$

Proof. We may suppose that (2.22) holds for all $n \geqslant N$ and $m \geqslant 1$. Given $\varepsilon > 0$, choose $n_0 \geqslant N$ so large that $f(n_0) < \varepsilon$. From (2.22) we deduce that for all positive integers i and j,

$$\begin{aligned}
&|E(X_{in_0+j}|X_{n_0+j}, X_{2n_0+j}, \ldots, X_{(i-1)n_0+j})| \\
&\quad = |E[E(X_{in_0+j}|X_1, X_2, \ldots, X_{(i-1)n_0+j})|X_{n_0+j}, X_{2n_0+j}, \ldots, X_{(i-1)n_0+j}]| \\
&\quad \leqslant f(n_0)E|X_{in_0+j}| \quad \text{a.s.}
\end{aligned}$$

If $n \geqslant n_0$, choose nonnegative integers q and r such that $0 \leqslant r \leqslant n_0 - 1$ and $n = qn_0 + r$. Then

$$b_n^{-1} \sum_{i=1}^{n} X_i = b_n^{-1} \sum_{i=1}^{n_0} X_i + b_n^{-1} \sum_{i=1}^{q-1} \sum_{j=1}^{n_0-1} X_{in_0+j} + b_n^{-1} \sum_{j=1}^{r} X_{qn_0+j}. \tag{2.23}$$

The first term on the right-hand side converges a.s. to zero as $n \to \infty$. The sum of the last two terms is dominated by

$$\sum_{j=1}^{n_0-1} b_n^{-1} \sum_{i=1}^{q-1} |[X_{in_0+j} - E(X_{in_0+j}|X_{n_0+j}, \ldots, X_{(i-1)n_0+j})]|$$

$$+ \sum_{j=1}^{r} b_n^{-1}|X_{qn_0+j} - E(X_{qn_0+j}|X_{n_0+j}, \ldots, X_{(q-1)n_0+j})|$$

$$+ f(n_0)b_n^{-1} \sum_{i=n_0+1}^{n} E|X_i|.$$

The fact that $\sum_{n=1}^{\infty} b_n^{-2}E(X_n^2) < \infty$ and an application of Theorem 2.18 imply that the first two terms here converge a.s. to zero. The second term converges to zero since r is fixed, and the last term is dominated by $(\sup_n b_n^{-1} \sum_{i=1}^{n} E|X_i|)$. Substituting into (2.23) we deduce that for all $\varepsilon > 0$,

$$\limsup_n \left| b_n^{-1} \sum_{i=1}^{n} X_i \right| < \varepsilon \left(\sup_n b_n^{-1} \sum_{i=1}^{n} E|X_i| \right) \quad \text{a.s.,}$$

which proves the theorem. ∎

The concept of *-mixing fitted conveniently into the martingale framework because of the concrete bound that it was possible to obtain for $E(X_{m+n}|\mathscr{B}_1^m)$. The definition of a mixingale (see Section 2.3) also rests heavily on conditional moments and has considerable applicability. Our last result in this section is a simple corollary to the mixingale convergence theorem, Theorem 2.7.

Theorem 2.21. *If $\{X_n, \mathscr{F}_n\}$ is a mixingale and $\{b_n\}$ is a sequence of positive constants increasing to ∞ such that*

$$\sum_{n=1}^{\infty} b_n^{-2}c_n^2 < \infty \qquad \text{and} \qquad \psi_n = O(n^{-1/2}(\log n)^{-2}) \qquad \text{as} \quad n \to \infty,$$

then $b_n^{-1} \sum_{i=1}^{n} X_i \xrightarrow{\text{a.s.}} 0.$

2.7 Convergence in L^p

Let $\{S_n = \sum_{i=1}^{n} X_i, \mathscr{F}_n, n \geqslant 1\}$ be a martingale and let $1 \leqslant p < 2$. Suppose that the sequence $\{|X_n|^p, n \geqslant 1\}$ is uniformly integrable. We shall verify the conditions of Theorem 2.13, with $b_n = n^{1/p}$, to prove that $n^{-1/p}S_n \xrightarrow{\text{p}} 0$ as $n \to \infty$. First of all note that

$$\sum_{i=1}^{n} P(|X_i| > n^{1/p}) \leqslant n^{-1} \sum_{i=1}^{n} E[|X_i|^p I(|X_i| > n^{1/p})] \to 0$$

under the condition of uniform integrability. Second,

$$n^{-1/p}E\left\{\sum_{i=1}^{n}|E[X_iI(|X_i|\leqslant n^{1/p})|\mathscr{F}_{i-1}]|\right\}$$

$$= n^{-1/p}E\left\{\sum_{i=1}^{n}|E[X_iI(|X_i|> n^{1/p})|\mathscr{F}_{i-1}]|\right\}$$

$$\leqslant n^{-1}\sum_{i=1}^{n}E[|X_i|^pI(|X_i|>n^{1/p})]$$

$$\to 0.$$

This verifies condition (ii). Condition (iii) will hold if we show that

$$n^{-2/p}\sum_{i=1}^{n}E[X_i^2I(|X_i|\leqslant n^{1/p})]\to 0.$$

For any $0<\varepsilon<1$ the left-hand side above equals

$$n^{-2/p}\sum_{i=1}^{n}E[X_i^2I(|X_i|\leqslant \varepsilon n^{1/p})+X_i^2I(\varepsilon n^{1/p}\leqslant|X_i|\leqslant n^{1/p})]$$

$$\leqslant n^{-2/p}(\varepsilon n^{1/p})^{2-p}\sum_{i=1}^{n}E|X_i|^p+n^{-2/p}(n^{1/p})^{2-p}\sum_{i=1}^{n}E[|X_i|^pI(|X_i|>\varepsilon n^{1/p})]$$

$$\leqslant \varepsilon^{2-p}\left(\max_{i\leqslant n}E|X_i|^p\right)+n^{-1}\sum_{i=1}^{n}E[|X_i|^pI(|X_i|>\varepsilon n^{1/p})],$$

which can be made arbitrarily small by choosing ε sufficiently small and then n sufficiently large. Therefore (i), (ii), and (iii) hold, and so $n^{-1/p}S_n\xrightarrow{\text{p}}0$.

Our next result (Chow, 1971) allows us to strengthen this convergence in probability to convergence in L^p.

Theorem 2.22. *Let $\{S_n=\sum_{i=1}^n X_i,\mathscr{F}_n,n\geqslant 1\}$ be a martingale and let $1\leqslant p<2$. If $\{|X_n|^p,n\geqslant 1\}$ is uniformly integrable, then $n^{-1}E|S_n|^p\to 0$.*

Proof. Let $\varepsilon>0$ and choose $\lambda>0$ so large that for all n,

$$E[|X_n|^pI(|X_n|>\lambda)]<\varepsilon.$$

Let $Y_n=X_nI(|X_n|\leqslant\lambda)$ and $Z_n=X_n-Y_n$. Suppose first that $1<p<2$. By Theorem 2.10 there exists a constant C depending only on p such that for all n,

$$E|S_n|^p\leqslant CE\left|\sum_{i=1}^{n}X_i^2\right|^{p/2}=CE\left|\sum_{i=1}^{n}(Y_i^2+Z_i^2)\right|^{p/2}$$

$$\leqslant C\left[E\left|\sum_{i=1}^{n}Y_i^2\right|^{p/2}+E\left|\sum_{i=1}^{n}Z_i^2\right|^{p/2}\right].$$

Since each $|Y_i| \leqslant \lambda$ and

$$\left|\sum_{i=1}^{n} Z_i^2\right|^{p/2} \leqslant \sum_{i=1}^{n} |Z_i|^p = \sum_{i=1}^{n} |X_i|^p I(|X_i| > \lambda),$$

it follows that

$$E|S_n|^p \leqslant C[(n\lambda^2)^{p/2} + n\varepsilon].$$

Hence,

$$n^{-1}E|S_n|^p \leqslant C[\lambda^p n^{(p/2)-1} + \varepsilon],$$

which can be made arbitrarily small by choosing ε sufficiently small and then n sufficiently large.

Now suppose that $p = 1$. Then

$$E\left|\sum_{i=1}^{n} [Y_i - E(Y_i|\mathscr{F}_{i-1})]\right|^2 \leqslant \sum_{i=1}^{n} E(Y_i^2) \leqslant n\lambda^2$$

and so

$$n^{-1}\sum_{i=1}^{n} [Y_i - E(Y_i|\mathscr{F}_{i-1})] \xrightarrow{L^2} 0.$$

Since

$$E\left|\sum_{i=1}^{n} [Z_i - E(Z_i|\mathscr{F}_{i-1})]\right| \leqslant 2\sum_{i=1}^{n} E|Z_i| \leqslant 2n\varepsilon,$$

it follows that

$$\begin{aligned} n^{-1}E|S_n| &\leqslant n^{-1}E\left|\sum_{i=1}^{n} [Y_i - E(Y_i|\mathscr{F}_{i-1})]\right| + n^{-1}E\left|\sum_{i=1}^{n} [Z_i - E(Z_i|\mathscr{F}_{i-1})]\right| \\ &\leqslant o(1) + 2\varepsilon \end{aligned}$$

as $n \to \infty$. The right side can be made arbitrarily small by choosing ε sufficiently small and then n sufficiently large, and so $n^{-1}E|S_n| \to 0$. ∎

We define the *conditional variance* of a zero-mean martingale $\{S_n = \sum_{i=1}^{n} X_i, \mathscr{F}_n, n \geqslant 1\}$ by

$$V_n^2 = \sum_{i=1}^{n} E(X_i^2|\mathscr{F}_{i-1}),$$

and its *squared variation* by

$$U_n^2 = \sum_{i=1}^{n} X_i^2.$$

These two quantities play a major role in martingale limit theory. They are both estimates of the variance $E(S_n^2)$. It is important to have conditions under which V_n^2 and U_n^2 are asymptotically equivalent, either in probability or in L^p. These are given in Theorem 2.23, which is stated for a general martingale triangular array. It has applications in Chapter 3. Parts of the theorem were established by Dvoretzky (1972), Scott (1973), and McLeish (1974).

Theorem 2.23. *Let* $\{S_{ni} = \sum_{j=1}^{i} X_{nj}, \mathscr{F}_{ni}, 1 \leqslant i \leqslant k_n\}$ *be a zero-mean martingale for each* $n \geqslant 1$, *and define*

$$V_{ni}^2 = \sum_{j=1}^{i} E(X_{nj}^2 | \mathscr{F}_{n,j-1}) \qquad \textit{and} \qquad U_{ni}^2 = \sum_{j=1}^{i} X_{nj}^2, \qquad 1 \leqslant i \leqslant k_n.$$

Suppose that the conditional variances $V_{nk_n}^2$ *are tight, that is,*

$$\sup_n P(V_{nk_n}^2 > \lambda) \to 0 \qquad \text{as} \quad \lambda \to \infty, \tag{2.24}$$

and that the conditional Lindeberg condition holds:

$$\textit{for all} \quad \varepsilon > 0, \qquad \sum_i E[X_{ni}^2 I(|X_{ni}| > \varepsilon) | \mathscr{F}_{n,i-1}] \xrightarrow{\text{p}} 0. \tag{2.25}$$

Then

$$\max_i |U_{ni}^2 - V_{ni}^2| \xrightarrow{\text{p}} 0. \tag{2.26}$$

If (2.25) *holds and, instead of* (2.24),

$$\{V_{nk_n}^2, n \geqslant 1\} \qquad \textit{is uniformly integrable,} \tag{2.27}$$

then

$$E|U_{nk_n}^2 - V_{nk_n}^2| \to 0. \tag{2.28}$$

Let $p > 1$. *If* (2.25) *holds and*

$$\{V_{nk_n}^{2p}, n \geqslant 1\} \qquad \textit{is uniformly integrable} \tag{2.29}$$

then the following two conditions are equivalent:

$$E\left(\max_i |U_{ni}^2 - V_{ni}^2|^p\right) \to 0 \tag{2.30}$$

and

$$\sum_i E|X_{ni}|^{2p} \to 0. \tag{2.31}$$

Remarks. Suppose that $p > 1$. By Doob's inequality (Theorem 2.2) applied to the martingale $\{U_{ni}^2 - V_{ni}^2, \mathscr{F}_{ni}, 1 \leqslant i \leqslant k_n\}$, condition (2.30) is equivalent to

$$E|U_{nk_n}^2 - V_{nk_n}^2|^p \to 0.$$

If $\sup_n E(S_{nk_n}^2) < \infty$ (which is implied by (2.27)), then condition (2.31) is equivalent to the apparently weaker condition:

$$\text{for all} \quad \varepsilon > 0, \qquad \sum_i E[|X_{ni}|^{2p} I(|X_{ni}| > \varepsilon)] \to 0.$$

To see this note that for any $\varepsilon > 0$,

$$\begin{aligned}\sum_i E|X_{ni}|^{2p} &= \sum_i E[|X_{ni}|^{2p} I(|X_{ni}| \leqslant \varepsilon)] + \sum_i E[|X_{ni}|^{2p} I(|X_{ni}| > \varepsilon)] \\ &\leqslant \varepsilon^{2p-2} E(S_{nk_n}^2) + \sum_i E[|X_{ni}|^{2p} I(|X_{ni}| > \varepsilon)],\end{aligned}$$

which can be made arbitrarily small by choosing ε small and then n large. Finally, if (2.27) holds, then (2.25) is equivalent to the unconditional Lindeberg condition:

$$\text{for all} \quad \varepsilon > 0, \qquad \sum_i E[X_{ni}^2 I(|X_{ni}| > \varepsilon)] \to 0.$$

Proof of Theorem 2.23. We first need a curious result due to Dvoretzky (1972).

Lemma 2.5. *If $\{\mathscr{F}_i, 0 \leqslant i \leqslant n\}$ is an increasing sequence of σ-fields ($\mathscr{F}_0$ is not necessarily the trivial σ-field) and A_i is an event in $\mathscr{F}_i$ for each $1 \leqslant i \leqslant n$, then for any nonnegative $\mathscr{F}_0$-measurable function η,*

$$P\left(\bigcup_1^n A_i \middle| \mathscr{F}_0\right) \leqslant \eta + P\left[\sum_1^n P(A_i|\mathscr{F}_{i-1}) > \eta \middle| \mathscr{F}_0\right].$$

Proof. Clearly

$$P(A_1|\mathscr{F}_0) \leqslant \eta + P[P(A_1|\mathscr{F}_0) > \eta|\mathscr{F}_0],$$

since the last term equals 1 if $P(A_1|\mathscr{F}_0) > \eta$. This establishes the assertion when $n = 1$; we now procede by induction. Once again, the result is obvious if $P(A_1|\mathscr{F}_0) > \eta$. Suppose that $P(A_1|\mathscr{F}_0) \leqslant \eta$. If the lemma is true for $n-1$ sets, then for any nonnegative $\mathscr{F}_1$-measurable function η',

$$P\left(\bigcup_2^n A_i \middle| \mathscr{F}_1\right) \leqslant \eta' + P\left(\sum_2^n P(A_i|\mathscr{F}_{i-1}) > \eta' \middle| \mathscr{F}_1\right).$$

Let $\eta' = \eta - P(A_1|\mathscr{F}_0)$. Then we have

$$P(A_1|\mathscr{F}_0) + P\left(\bigcup_2^n A_i \middle| \mathscr{F}_1\right) \leqslant \eta + P\left(\sum_1^n P(A_i|\mathscr{F}_{i-1}) > \eta \middle| \mathscr{F}_1\right)$$

(conditional, of course, on the set $\{P(A_1|\mathscr{F}_0) \leqslant \eta\}$). But

$$\begin{aligned} P\left(\bigcup_1^n A_i|\mathscr{F}_0\right) &\leqslant P(A_1|\mathscr{F}_0) + E\left[P\left(\bigcup_2^n A_i|\mathscr{F}_1\right)\Big|\mathscr{F}_0\right] \\ &= E\left[P(A_1|\mathscr{F}_0) + P\left(\bigcup_2^n A_i|\mathscr{F}_1\right)\Big|\mathscr{F}_0\right] \\ &\leqslant \eta + P\left(\sum_1^n P(A_i|\mathscr{F}_{i-1}) > \eta|\mathscr{F}_0\right), \end{aligned}$$

using the last proved inequality. This completes the proof. ∎

By taking $\mathscr{F}_0$ equal to the trivial σ-field and $A_i = \{|X_{ni}| > \varepsilon\}$, we deduce that for any constants ε and $\eta > 0$,

$$P\left(\max_i |X_{ni}| > \varepsilon\right) \leqslant \eta + P\left(\sum_{i=1}^n P(|X_{ni}| > \varepsilon|\mathscr{F}_{n,i-1}) > \eta\right).$$

From the conditional version of Chebyshev's inequality,

$$P(|X_{ni}| > \varepsilon|\mathscr{F}_{n,i-1}) \leqslant \varepsilon^{-2} E[X_{ni}^2 I(|X_{ni}| > \varepsilon)|\mathscr{F}_{n,i-1}],$$

and so

$$P\left(\max_i |X_{ni}| > \varepsilon\right) \leqslant \eta + P\left(\sum_{i=1}^n E[X_{ni}^2 I(|X_{ni}| > \varepsilon)|\mathscr{F}_{n,i-1}] > \varepsilon^2\eta\right). \quad (2.32)$$

Suppose first that (2.24) and (2.25) hold. Let ε, δ, and $\lambda > 0$ and define

$$Y_{ni} = X_{ni} I(|X_{ni}| \leqslant \varepsilon \text{ and } V_{ni}^2 \leqslant \lambda),$$

$$C_{ni}^2 = \sum_{j=1}^i Y_{nj}^2,$$

$$D_{ni}^2 = \sum_{j=1}^i E(Y_{nj}^2|\mathscr{F}_{n,j-1}) = \sum_{j=1}^i E[X_{nj}^2 I(|X_{nj}| \leqslant \varepsilon)|\mathscr{F}_{n,j-1}] I(V_{nj}^2 \leqslant \lambda),$$

and

$$\begin{aligned} p_n &= P(U_{ni}^2 \neq C_{ni}^2 \text{ or } |V_{ni}^2 - D_{ni}^2| > \delta \text{ for some } 1 \leqslant i \leqslant n) \\ &\leqslant P\left(\max_i |X_{ni}| > \varepsilon\right) + P(V_{nk_n}^2 > \lambda) \\ &\quad + P\left(\sum_i E[X_{ni}^2 I(|X_{ni}| > \varepsilon)|\mathscr{F}_{n,i-1}] > \delta\right). \end{aligned}$$

Condition (2.25) and the inequality (2.32) imply that the first and last terms in this last expression above converge to zero as $n \to \infty$. Now,

$$P\left(\max_i |U_{ni}^2 - V_{ni}^2| > 2\delta\right) \leqslant p_n + P\left(\max_i |C_{ni}^2 - D_{ni}^2| > \delta\right)$$
$$\leqslant p_n + \delta^{-2} \sum_i E[Y_{ni}^2 - E(Y_{ni}^2|\mathscr{F}_{n,i-1})]^2,$$

using Corollary 2.1. The last term does not exceed

$$\delta^{-2} \sum_i E(Y_{ni}^4) \leqslant \delta^{-2}\varepsilon^2 \sum_i E(Y_{ni}^2) = \delta^{-2}\varepsilon^2 E\left[\sum_i E(Y_{ni}^2|\mathscr{F}_{n,i-1})\right] \leqslant \delta^{-2}\varepsilon^2\lambda.$$

Hence for all ε, δ and $\lambda > 0$,

$$\limsup_{n\to\infty} P\left(\max_i |U_{ni}^2 - V_{ni}^2| > 2\delta\right) \leqslant \limsup_{n\to\infty} P(V_{nk_n}^2 > \lambda) + \delta^{-2}\varepsilon^2\lambda.$$

Condition (2.26) follows on letting $\varepsilon \to 0$ and then $\lambda \to \infty$.

Fix $\lambda > 0$ and define

$$Y_{ni} = X_{ni}I(|X_{ni}| \leqslant 1) - E[X_{ni}I(|X_{ni}| \leqslant 1)|\mathscr{F}_{n,i-1}],$$
$$Z_{ni} = X_{ni} - Y_{ni},$$
$$A_{ni} = Y_{ni}I(V_{ni}^2 \leqslant \lambda),$$
$$B_{ni} = Y_{ni} - A_{ni},$$

noting that Y_{ni}, Z_{ni}, A_{ni}, and B_{ni} are all martingale differences. For any $r > \frac{1}{2}$ we deduce from Theorems 2.10 and 2.11 that there exist constants K_1 and K_2 such that

$$E\left(\sum_i A_{ni}^2\right)^r \leqslant K_1 E\left|\sum_i A_{ni}\right|^{2r}$$
$$\leqslant K_2\left\{E\left[\left(\sum_i E(A_{ni}^2|\mathscr{F}_{n,i-1})\right)^r\right] + E\left(\max_i |A_{ni}|^{2r}\right)\right\}$$
$$\leqslant K_2(\lambda^r + 2^{2r}) < \infty.$$

Hence for any $p > 0$,

$$\left\{\left(\sum_i A_{ni}^2\right)^p, n \geqslant 1\right\} \quad \text{and} \quad \left\{\left|\sum_i A_{ni}\right|^{2p}, n \geqslant 1\right\} \quad \text{are uniformly integrable.} \tag{2.33}$$

Moreover,

$$\begin{aligned} E\left(\sum_i B_{ni}^2\right)^p &\leqslant K_1 E\left|\sum_i B_{ni}\right|^{2p} \\ &\leqslant K_2\left\{E\left[\left(\sum_i E(B_{ni}^2|\mathscr{F}_{n,i-1})\right)^p + E\left(\max_i |B_{ni}|^{2p}\right)\right\}\right. \\ &\leqslant K_2\{E[V_{nk_n}^{2p} I(V_{nk_n}^2 > \lambda)] + 2^{2p} P(V_{nk_n}^2 > \lambda)\}. \end{aligned}$$

We deduce from (2.27) if $p = 1$ and (2.29) if $p > 1$ that

$$\sup_n E\left(\sum_i B_{ni}^2\right)^p \to 0 \quad \text{and} \quad \sup_n E\left|\sum_i B_{ni}\right|^{2p} \to 0 \quad \text{as} \quad \lambda \to \infty. \tag{2.34}$$

Conditions (2.25) and (2.27) together imply the usual Lindeberg condition:

$$\text{for all} \quad \varepsilon > 0, \quad \sum_i E[X_{ni}^2 I(|X_{ni}| > \varepsilon)] \to 0$$

(see Theorem A.7 of Appendix V), which in turn implies

$$E\left|\sum_i Z_{ni}\right|^2 = E\left(\sum_i Z_{ni}^2\right) \leqslant \sum_i E[X_{ni}^2 I(|X_{ni}| > 1)] \to 0.$$

Therefore

$$E\left(\sum_i Z_{ni}^2\right) = E\left|\sum_i Z_{ni}\right|^2 \to 0. \tag{2.35}$$

We are now in a position to prove the next part of the theorem. Suppose first that (2.25) and (2.27) hold. In view of (2.26) it suffices to prove that $\{U_{nk_n}^2, n \geqslant 1\}$ is uniformly integrable. Let $\delta > 0$ and let E be any event with $P(E) < \delta$. Since

$$U_{nk_n}^2 \leqslant 4\left\{\sum_i A_{ni}^2 + \sum_i B_{ni}^2 + \sum_i Z_{ni}^2\right\},$$

it follows that

$$E[U_{nk_n}^2 I(E)] \leqslant 4\left\{E\left[\left(\sum_i A_{ni}^2\right) I(E)\right] + E\left(\sum_i B_{ni}^2\right) + E\left[\left(\sum_i Z_{ni}^2\right) I(E)\right]\right\},$$

and in view of (2.33), (2.34), and (2.35), the right-hand side can be made arbitrarily small uniformly in n by choosing λ sufficiently large and then δ sufficiently small. Hence $\{U_{nk_n}^2\}$ is uniformly integrable, and (2.28) follows from (2.26) and (2.27).

Suppose next that (2.25), (2.29), and (2.30) hold. Then

$$E|U_{nk_n}^2 - V_{nk_n}^2|^p \to 0$$

and so $\{U_{nk_n}^{2p}\}$ is uniformly integrable. Since

$$\left(\sum_i Z_{ni}^2\right)^p \leqslant 2^{6p-2}\left\{U_{nk_n}^{2p} + \left(\sum_i A_{ni}^2\right)^p + \left(\sum_i B_{ni}^2\right)^p\right\},$$

the uniform integrability of $(\sum_i Z_{ni}^2)^p$ follows from (2.33) and (2.34). (Note that Z_{ni} does not depend on λ.) Condition (2.35) implies that $\sum_i Z_{ni}^2 \xrightarrow{p} 0$, and so

$$E\left(\sum_i Z_{ni}^2\right)^p \to 0,$$

and from Theorems 2.10 and 2.12 we deduce that

$$\sum_i E|Z_{ni}|^{2p} \to 0. \tag{2.36}$$

Now,

$$\begin{aligned}\max_i Y_{ni}^2 &\leqslant 2\left(\max_i X_{ni}^2 + \max_i \{E[X_{ni}I(|X_{ni}| > 1)|\mathscr{F}_{n,i-1}]\}^2\right)\\ &\leqslant 2\left(\max_i X_{ni}^2 + \sum_i E[X_{ni}^2 I(|X_{ni}| > 1)|\mathscr{F}_{n,i-1}]\right)\\ &\xrightarrow{p} 0,\end{aligned}$$

using (2.25) and the inequality (2.32). Therefore $\max_i|Y_{ni}|$ is bounded by 2 and converges to zero in probability. Furthermore,

$$\sum_i Y_{ni}^2 \leqslant 2\left\{\sum_i X_{ni}^2 + \sum_i E(X_{ni}^2|\mathscr{F}_{n,i-1})\right\} = 2(U_{nk_n}^2 + V_{nk_n}^2),$$

and so $\{\sum_i Y_{ni}^2, n \geqslant 1\}$ is uniformly integrable. It follows that

$$\sum_i E|Y_{ni}|^{2p} \leqslant E\left[\left(\max_i |Y_{ni}|^{2p-2}\right)\left(\sum_i Y_{ni}^2\right)\right] \to 0.$$

Using this and (2.36) we deduce that

$$\sum_i E|X_{ni}|^{2p} \leqslant 2^{2p-1}\left(\sum_i E|Y_{ni}|^{2p} + \sum_i E|Z_{ni}|^{2p}\right) \to 0,$$

which gives (2.31).

Finally, suppose that (2.25), (2.29), and (2.31) hold. To establish (2.30) it suffices, using the remarks after the statement of the theorem, to prove that

$$E|U_{nk_n}^2 - V_{nk_n}^2|^p \to 0,$$

which will follow if we prove that $\{U_{nk_n}^{2p}\}$ is uniformly integrable. The uniform integrability of $\{U_{nk_n}^2\}$ comes from (2.28), and it follows as above that

$$\sum_i E|Y_{ni}|^{2p} \to 0.$$

Combined with (2.31) this implies (2.36). Applying Theorems 2.10 and 2.12 to the martingale with differences Z_{ni} we deduce that

$$\begin{aligned}
E\Big(\sum_i Z_{ni}^2\Big)^p &\leqslant K_1 E\Big|\sum_i Z_{ni}\Big|^{2p} \\
&\leqslant K_2\Big\{E\Big[\Big(\sum_i E(Z_{ni}^2|\mathscr{F}_{n,i-1})\Big)^p\Big] + \sum_i E|Z_{ni}|^{2p}\Big\} \\
&\leqslant K_2\Big\{E\Big[\Big(\sum_i E[X_{ni}^2 I(|X_{ni}| > 1)|\mathscr{F}_{n,i-1}]\Big)^p\Big] + \sum_i E|Z_{ni}|^{2p}\Big\} \\
&\to 0
\end{aligned}$$

under (2.25), (2.29), and (2.36). Since

$$U_{nk_n}^{2p} \leqslant 2^{4p-2}\Big\{\Big(\sum_i A_{ni}^2\Big)^p + \Big(\sum_i B_{ni}^2\Big)^p + \Big(\sum_i Z_{ni}^2\Big)^p\Big\},$$

the last result combined with (2.33) and (2.34) implies that $\{U_{nk_n}^{2p}\}$ is uniformly integrable. ∎

3
The Central Limit Theorem

3.1. Introduction

Perhaps the greatest achievement of modern probability is the elegant and unified theory of limits for sums of independent r.v. In fact, the development of mathematical statistics is often considered to have begun with the early limit laws of Bernoulli and de Moivre. The mathematical theory of martingales may be regarded as an extension of the independence theory, and it too has its origins in limit results, beginning with Bernstein's (1927) and Lévy's (1935a,b, 1937) early central limit theorems.

Let $\{S_n, \mathscr{F}_n, n \geqslant 1\}$ be a zero-mean, square-integrable martingale and let $X_n = S_n - S_{n-1}$, $n \geqslant 2$, and $X_1 = S_1$ denote the martingale differences. Lévy introduced the *conditional variance* for martingales

$$V_n^2 = \sum_1^n E(X_i^2 | \mathscr{F}_{i-1}),$$

which plays an important role in modern martingale limit theory. His early results required the strong assumption that for each n, V_n^2 is a.s. constant, and this assumption recurs even in contemporary work—see, for example, Csörgö (1968). Lévy's proofs involved a direct estimation of the difference between a martingale's distribution and the standard normal distribution.

Doob (1953) sketched a characteristic function proof of Lévy's (1937) Theorem 67.1. Billingsley (1961b), and independently Ibragimov (1963), established the CLT for martingales with stationary and ergodic differences. For such martingales the conditional variance is asymptotically constant:

$$s_n^{-2} V_n^2 \xrightarrow{\text{p}} 1, \tag{3.1}$$

where $s_n^2 = E(V_n^2) = E(S_n^2)$.

Further extensions were provided by Rosén (1967a,b,c), Dvoretzky (1969, 1971, 1972), Loynes (1969, 1970) and Bergström (1970). Brown (1971) showed that it is condition (3.1) and not stationarity or ergodicity which is crucial.

He established a martingale analog of the Lindeberg–Feller theorem. His techniques have been further developed by Gänssler *et al.* (1978). Scott (1973) gave an alternative proof together with applications and extensions, and in 1971 provided an analog for reverse martingales.

McLeish (1974) introduced an elegant method of proof which provided new CLTs and invariance principles. The conditions for McLeish's invariance principles have been slightly weakened by Aldous (1978). Chatterji (1974b), Eagleson (1975a), Hall (1977), and Rootzén (1977b) explained the behavior of the martingale when condition (3.1) is relaxed. The question of martingale convergence to more general distributions was examined by Brown and Eagleson (1971), Eagleson (1976), Kłopotowski (1977a,b), Adler (1978), and Beśka (1978). Martingale invariance principles were studied in greater detail by Drogin (1972), Rootzén (1977a, 1980), and Gänssler and Hänsler (1980) who obtained certain sets of necessary and sufficient conditions. Durrett and Resnick (1978) have presented a unified approach for obtaining invariance principles in the convergence to several classes of distributions. Adler and Scott (1975, 1978) and Scott (1977) have proved martingale CLTs without asymptotic negligibility assumptions, extending some of Zolotarev's (1967) results for sums of independent r.v. Recently Johnstone (1978) has obtained limit theorems for martingales which are exactly equivalent to Zolotarev's in the case of independence.

An impressive armory of techniques has been developed to handle the technical difficulties in proofs of martingale limit theorems. Most have their origins in independence theory and mimic classical methods, although others are more novel. One powerful technique is Strassen's (1967) martingale extension of Skorokhod's representation for sums of independent r.v. (see Appendix I).

Our aim in this chapter is to present those results which are most relevant from the point of view of applications. These are not always the most general results and we use examples to give an idea of the sort of generalizations which are possible. Our proofs are often our own, although we have drawn several useful techniques from McLeish (1974) and Aldous and Eagleson (1978). In Sections 3.2 and 3.3 we present sufficient conditions for the martingale CLT. The problem of convergence of moments is dealt with in Section 3.4, and Section 3.5 treats the CLT for reverse martingales and martingale tail sums. In Section 3.6 we investigate the rate of convergence in the CLT.

3.2. The Central Limit Theorem

Let $\{S_{ni}, \mathscr{F}_{ni}, 1 \leqslant i \leqslant k_n\}$ be a zero-mean, square-integrable martingale for each $n \geqslant 1$, and let $X_{ni} = S_{ni} - S_{n,i-1}$, $1 \leqslant i \leqslant k_n$ ($S_{n0} = 0$) denote the martingale differences. (It is assumed that $k_n \uparrow \infty$ as $n \to \infty$.) We shall call the double sequence $\{S_{ni}, \mathscr{F}_{ni}, 1 \leqslant i \leqslant k_n, n \geqslant 1\}$ a *martingale array*. Let $V_{ni}^2 = \sum_{j=1}^{i} E(X_{nj}^2 | \mathscr{F}_{n,j-1})$ be the conditional variance of S_{ni} and let $U_{ni}^2 = \sum_{j=1}^{i} X_{nj}^2$ be the squared variation.

Martingale arrays are frequently derived from ordinary martingales $\{S_n, \mathscr{F}_n, 1 \leqslant n < \infty\}$ in the following way: define $k_n = n$, $\mathscr{F}_{ni} = \mathscr{F}_i$, and $S_{ni} = s_n^{-1} S_i$, $1 \leqslant i \leqslant n$, where s_n is the standard deviation of S_n. In this case $E(S_{nk_n}^2) = 1$, and it is not uncommon to make this assumption for an arbitrary martingale array.

Our main results are Theorem 3.3 and its corollary, which appear at the end of this section. Subsections 3.2(i)–3.2(vi) serve to introduce the various sufficient conditions for these central limit theorems. Theorem 3.2 and its corollary give sufficient conditions for a martingale array to converge to a mixture of normal distributions.

(i) *The assumption of asymptotic negligibility.* A large variety of negligibility assumptions have been made about the differences X_{ni} during the formulation of martingale central limit theorems. The classic condition of negligibility in the theory of sums of independent r.v. asks that the X_{ni} be uniformly asymptotically negligible:

$$\text{for all} \quad \varepsilon > 0, \qquad \max_i P(|X_{ni}| > \varepsilon) \to 0 \qquad \text{as} \quad n \to \infty. \tag{3.2}$$

This is generally a little weaker than the summation condition,

$$\text{for all} \quad \varepsilon > 0, \qquad \sum_i P(|X_{ni}| > \varepsilon) \to 0, \tag{3.3}$$

although when the sums S_{nk_n} converge in distribution, (3.2) and (3.3) are frequently equivalent. When the X_{ni} are independent, (3.3) is equivalent to

$$\max_i |X_{ni}| \xrightarrow{\text{p}} 0. \tag{3.4}$$

Since $P(\max_i |X_{ni}| > \varepsilon) = P(\sum_i X_{ni}^2 I(|X_{ni}| > \varepsilon) > \varepsilon^2)$, (3.4) is equivalent to the weak Lindeberg condition:

$$\text{for all} \quad \varepsilon > 0, \qquad \sum_i X_{ni}^2 I(|X_{ni}| > \varepsilon) \xrightarrow{\text{p}} 0. \tag{3.5}$$

McLeish (1974) proved CLTs under an L^2-boundedness condition and (3.4). Condition (3.5) is generally a little weaker than the Lindeberg condition:

$$\text{for all} \quad \varepsilon > 0, \qquad \sum_i E[X_{ni}^2 I(|X_{ni}| > \varepsilon)] \to 0. \tag{3.6}$$

However, if $E(S_{nk_n}^2) = 1$ and $\sum_i X_{ni}^2 \xrightarrow{\text{p}} 1$ —or more generally, if $\{U_{nk_n}^2\}$ is uniformly integrable—then (3.4) and (3.6) are equivalent. To see this, note that for any $\lambda > 0$,

$$\sum_i E[X_{ni}^2 I(|X_{ni}| > \varepsilon)] \leqslant E[U_{nk_n}^2 I(U_{nk_n}^2 > \lambda)] + \lambda P\left(\max_i |X_{ni}| > \varepsilon\right).$$

Given any $\delta > 0$, choose λ so large that for all n, the first term on the right-hand side is bounded by $\delta/2$. Now choose N so large that for all $n \geqslant N$, the second term is bounded by $\delta/2$. Then for all $n \geqslant N$,

$$\sum_i E[X_{ni}^2 I(|X_{ni}| > \varepsilon)] \leqslant \delta,$$

proving (3.6).

Some authors have imposed a conditional Lindeberg condition instead of (3.4) or (3.6):

$$\text{for all} \quad \varepsilon > 0, \qquad \sum_i E[X_{ni}^2 I(|X_{ni}| > \varepsilon) | \mathscr{F}_{n,i-1}] \xrightarrow{p} 0. \tag{3.7}$$

In the majority of cases (3.6) and (3.7) are equivalent.

(ii) *The conditional variance.* Here we shall discard the triangular array format for clarity. Let $\{S_n = \sum_{i=1}^n X_i, \mathscr{F}_n, n \geqslant 1\}$ be a zero-mean square-integrable martingale and put $V_n^2 = \sum_{i=1}^n E(X_i^2|\mathscr{F}_{i-1})$. The conditional variance V_n^2 is one of several "estimates" of the variance ES_n^2. It is an intrinsic measure of time for the martingale. For many purposes the time taken for a martingale to cross a level is best represented through its conditional variance rather than the number of increments up to the crossing [see Blackwell and Freedman (1973) and Freedman (1975)]. Also, in some circumstances the conditional variance can represent an "amount of information" contained in the past history of the process. This interpretation of V_n^2 is particularly relevant in the theory of inference for stochastic processes. Let $Y_1, Y_2, \ldots$ be a sequence of observations of a stochastic process whose distribution depends on a single parameter θ, and let $L_n(\theta)$ be the likelihood function associated with $Y_1, Y_2, \ldots, Y_n$. Under very mild conditions, the variables $S_n = \partial \log L_n/\partial\theta$ form a martingale whose conditional variance $V_n^2 = I_n(\theta)$ is a generalized form of the standard Fisher information (see Chapter 6).

Bearing the above ideas in mind, it is perhaps not surprising that the limiting behavior of S_n is closely tied up with that of V_n^2. If the behavior of V_n^2 is very erratic, then so is that of S_n, and it may not be possible to obtain a CLT.

Another convenient way of introducing V_n^2 is via Doob's decomposition of the submartingale $\{S_n^2, \mathscr{F}_n\}$ (Doob, (1953, p. 297). We can write

$$S_n^2 = M_n + A_n,$$

where $\{M_n, \mathscr{F}_n\}$ is a martingale and $\{A_n\}$ is an increasing sequence of non-negative r.v. If we stipulate that A_n be $\mathscr{F}_{n-1}$-measurable, this decomposition is uniquely determined a.s. by the relationship

$$\begin{aligned} A_n - A_{n-1} &= E(S_n^2|\mathscr{F}_{n-1}) - S_{n-1}^2 \\ &= E[(S_n - S_{n-1})^2|\mathscr{F}_{n-1}] \\ &= E(X_n^2|\mathscr{F}_{n-1}). \end{aligned}$$

Hence $A_n = V_n^2$.

(iii) *The relationship between V_{ni}^2 and U_{ni}^2.* The conditional variance V_{ni}^2 may often be approximated by the sum of squares U_{ni}^2. For example, suppose that (3.7) holds and that the sequence $\{V_{nk_n}^2, n \geqslant 1\}$ is tight:

$$\sup_{n \geqslant 1} P(V_{nk_n}^2 > \lambda) \to 0 \qquad \text{as} \quad \lambda \to \infty. \tag{3.8}$$

Theorem 2.23 then implies that

$$\max_i |U_{ni}^2 - V_{ni}^2| \xrightarrow{\text{p}} 0, \tag{3.9}$$

and that if (3.8) is strengthened to the uniform integrability of $\{V_{nk_n}^2, n \geqslant 1\}$,

$$E|U_{nk_n}^2 - V_{nk_n}^2| \to 0.$$

In view of (3.9), the normalizing condition (3.1) may frequently be replaced by

$$s_n^{-2} \sum_1^n X_i^2 \xrightarrow{\text{p}} 1.$$

Sometimes the sum of squares $\sum_1^n X_i^2$ is easier to deal with than the conditional variance.

The relationship (3.9) has considerable theoretical importance. Suppose for the time being that the r.v. X_{ni} are independent and that $\mathscr{F}_{ni}$ is the σ-field generated by $S_{n1}, S_{n2}, \ldots, S_{ni}$. Then the variables V_{ni}^2 are a.s. constant, and if $E(S_{nk_n}^2) = 1$, then $V_{nk_n}^2 = 1$ a.s. In this case, (3.9) implies that

$$\sum_i X_{ni}^2 \xrightarrow{\text{p}} 1. \tag{3.10}$$

Under very mild conditions, (3.10) is both necessary and sufficient for the CLT

$$S_{nk_n} = \sum_i X_{ni} \xrightarrow{\text{d}} N(0,1). \tag{3.11}$$

Assuming only the negligibility condition (3.2), and that the X_{ni}, $1 \leqslant i \leqslant k_n$, are independent with zero means and variances adding up to 1, (3.10) and (3.11) are equivalent [Raikov (1938); see Gnedenko and Kolmogorov (1954, Theorem 4, p. 143) and Loève (1977, Exercise 5, p. 349)].

Extrapolating to the martingale case, we may hope that a condition like (3.9) would be necessary and sufficient for a limit law like (3.11). This is not the case, but Raikov's result nevertheless has a generalization to martingales; see Section 3.4.

(iv) *Stability, weak L^1-convergence, and mixing.* If $\{Y_n\}$ is a sequence of r.v. on a probability space $(\Omega,\mathscr{F},P)$ converging in distribution to an r.v. Y, we say that the convergence is *stable* if for all continuity points y of Y and all events $E \in \mathscr{F}$, the limit

$$\lim_{n\to\infty} P(\{Y_n \leqslant y\} \cap E) = Q_y(E)$$

exists, and if $Q_y(E) \to P(E)$ as $y \to \infty$. (Clearly Q_y, if it exists, is a probability measure on $(\Omega,\mathscr{F})$.) We designate the convergence by writing

$$Y_n \xrightarrow{\text{d}} Y \quad \text{(stably)}.$$

The concept of stability was introduced by Rényi (1963), and many martingale limit theorems are stable (Aldous and Eagleson, 1978). This fact enables us to interchange normings in the central limit theorem with relative ease. Indeed, the theory of stable convergence provides a particularly elegant approach to martingale central limit theory.

A sequence of integrable r.v. $\{Z_n\}$ on $(\Omega,\mathscr{F},P)$ is said to converge *weakly in* L^1 to an integrable r.v. Z on $(\Omega,\mathscr{F},P)$ if for all $E \in \mathscr{F}$,

$$E[Z_n I(E)] \to E[ZI(E)];$$

the convergence is designated by writing

$$Z_n \to Z \quad (\text{weakly in } L^1).$$

If $\exp(itY_n) \to \exp(itY)$ (weakly in L^1) for each real t, then clearly $Y_n \xrightarrow{\text{d}} Y$.

The condition of weak convergence in L^1 is stronger than uniform integrability but weaker than L^1-convergence. The r.v. Z_n converge to the r.v. Z in L^1 if and only if $E[Z_n I(E)] \to E[ZI(E)]$ *uniformly* in $E \in \mathscr{F}$; see Neveu (1965, Proposition II.5.3). And if $Z_n \to Z$ (weakly in L^1), then the sequence $\{Z_n\}$ is uniformly integrable; see Neveu (1965, Proposition IV.2.2).

Our next result connects the concepts of weak L^1-convergence and stability.

Theorem 3.1. *Suppose that $Y_n \xrightarrow{\text{d}} Y$, where all the Y_n are on the same space $(\Omega,\mathscr{F},P)$. Then $Y_n \to Y$ (stably) if and only if there exists a variable Y' on an extension of $(\Omega,\mathscr{F},P)$, with the same distribution as Y, such that for all real t,*

$$\exp(itY_n) \to Z(t) = \exp(itY') \quad (\textit{weakly in } L^1) \qquad \text{as} \quad n \to \infty,$$

and $E[Z(t)I(E)]$ is a continuous function of t for all $E \in \mathscr{F}$.

Theorem 3.1 is a trivial consequence of the convergence theorem for characteristic functions. It allows us to identify stably convergent sequences in terms of characteristic functions, which is usually much simpler than examining the distribution functions.

If the r.v. Y' in Theorem 3.1 can be taken to be independent of each $E \in \mathscr{F}$, then the limit theorem is said to be *mixing* (in the sense of Rényi). The r.v. Y_n are asymptotically independent of each event E. This condition can be restated by asking that for all $E \in \mathscr{F}$ and all continuity points y of Y,

$$P(\{Y_n \leqslant y\} \cap E) \to P(Y \leqslant y)P(E).$$

We write

$$Y_n \xrightarrow{\text{d}} Y \quad \text{(mixing)}.$$

The mixing condition appears at face value to be a distributional limit property, but it actually implies strong law results; see Rootzén (1976).

(v) *The stability of martingale limit theorems.* Let $\{X_{ni}, 1 \leqslant i \leqslant k_n, n \geqslant 1\}$ be any array for r.v. defined on a common probability space $(\Omega, \mathscr{F}, P)$. For real t let

$$T_n(t) = \prod_{j=1}^{k_n} (1 + itX_{nj}).$$

Lemma 3.1. *Let η^2 be an a.s. finite r.v. and suppose that*

$$\max_i |X_{ni}| \xrightarrow{\text{p}} 0, \tag{3.12}$$

$$\sum_i X_{ni}^2 \xrightarrow{\text{p}} \eta^2, \tag{3.13}$$

and

$$\textit{for all real} \quad t, \qquad T_n(t) \to 1 \quad (\textit{weakly in } L^1) \qquad \text{as} \quad n \to \infty. \tag{3.14}$$

Then $S_{nk_n} \xrightarrow{\text{d}} Z$ (stably) where the r.v. Z has characteristic function $E\exp(-\frac{1}{2}\eta^2 t^2)$.

Proof. [After McLeish (1974)] Define $r(x)$ by

$$e^{ix} = (1 + ix)\exp(-\tfrac{1}{2}x^2 + r(x))$$

and note that $|r(x)| \leqslant |x|^3$ for $|x| \leqslant 1$. Let $I_n = \exp(itS_{nk_n})$ and

$$W_n = \exp\left(-\tfrac{1}{2}t^2 \sum_i X_{ni}^2 + \sum_i r(tX_{ni})\right).$$

Then

$$I_n = T_n \exp(-\eta^2 t^2/2) + T_n(W_n - \exp(-\eta^2 t^2/2)).$$

In view of Theorem 3.1 it suffices to prove that for all $E \in \mathscr{F}$,

$$E(I_n I(E)) \to E(\exp(-\eta^2 t^2/2) I(E)). \tag{3.15}$$

Since $\exp(-\eta^2t^2/2)I(E)$ is bounded, (3.14) ensures that

$$E(T_n \exp(-\eta^2t^2/2)I(E)) \to E(\exp(-\eta^2t^2/2)I(E)). \tag{3.16}$$

Moreover, any sequence of r.v. which converges weakly in L^1 is uniformly integrable, and so the sequence

$$T_n(W_n - \exp(-\eta^2t^2/2)) = I_n - T_n \exp(-\eta^2t^2/2)$$

is uniformly integrable. (The uniform integrability of I_n follows from the fact that $|I_n| = 1$.) Conditions (3.12) and (3.13) imply that when $\max_i|X_{ni}| \leqslant 1$,

$$\left|\sum_i r(X_{ni}t)\right| \leqslant |t|^3 \sum_i |X_{ni}|^3$$

$$\leqslant |t|^3\left(\max_i|X_{ni}|\right)\left(\sum_i X_{ni}^2\right) \xrightarrow{p} 0 \qquad \text{as} \quad n \to \infty.$$

It follows that $W_n - \exp(-\eta^2t^2/2) \xrightarrow{p} 0$, and in view of the uniform integrability,

$$E[T_n(W_n - \exp(-\eta^2t^2/2))I(E)] \to 0. \tag{3.17}$$

Conditions (3.16) and (3.17) imply (3.15). ∎

Now we specialize the array $\{X_{ni}\}$ to a martingale difference array.

Theorem 3.2. *Let $\{S_{ni}, \mathscr{F}_{ni}, 1 \leqslant i \leqslant k_n, n \geqslant 1\}$ be a zero-mean, square-integrable martingale array with differences X_{ni}, and let η^2 be an a.s. finite r.v. Suppose that*

$$\max_i |X_{ni}| \xrightarrow{p} 0, \tag{3.18}$$

$$\sum_i X_{ni}^2 \xrightarrow{p} \eta^2, \tag{3.19}$$

$$E\left(\max_i X_{ni}^2\right) \quad \text{is bounded in } n, \tag{3.20}$$

and

$$\text{the } \sigma\text{-fields are nested:} \quad \mathscr{F}_{n,i} \subseteq \mathscr{F}_{n+1,i} \quad \text{for} \quad 1 \leqslant i \leqslant k_n, \quad n \geqslant 1. \tag{3.21}$$

Then $S_{nk_n} = \sum_i X_{ni} \xrightarrow{d} Z$ (stably), where the r.v. Z has characteristic function $E\exp(-\frac{1}{2}\eta^2t^2)$.

Corollary 3.1. *If* (3.18) *and* (3.20) *are replaced by the conditional Lindeberg condition* (3.7):

$$\text{for all} \quad \varepsilon > 0, \qquad \sum_i E[X_{ni}^2 I(|X_{ni}| > \varepsilon)|\mathscr{F}_{n,i-1}] \xrightarrow{p} 0,$$

if (3.19) *is replaced by an analogous condition on the conditional variance*:

$$V_{nk_n}^2 = \sum E(X_{ni}^2|\mathscr{F}_{n,i-1}) \xrightarrow{\text{p}} \eta^2,$$

and if (3.21) *holds, then the conclusion of Theorem* 3.2 *remains true*.

[Various versions of these results were discovered by Chatterji (1974b), Hall (1977), and Rootzén (1977b). The stability part is due to Aldous and Eagleson (1978).]

Remarks. It is possible to replace (3.21) by a measurability condition on η^2. If (3.18)–(3.20) (or the similar conditions of the corollary) hold and if η^2 is measurable in the completions of all the σ-fields $\mathscr{F}_{ni}$, then the conclusion of Theorem 3.2 remains true, provided the stability is dropped (Eagleson (1975a)). For example, if η^2 is constant, then it is trivially measurable in all of the $\mathscr{F}_{ni}$. However, conditions (3.18)–(3.20) alone are not sufficient to imply that $S_{nk_n} \xrightarrow{\text{d}} Z$; see Example 4 in this chapter. Condition (3.21) is satisfied in most applications. For example, if the martingale array is obtained from an ordinary martingale, then $k_n = n$ and $\mathscr{F}_{ni} = \mathscr{F}_i$ for all $i \leqslant n$ and all n, so that (3.21) holds.

Let N be a standard normal variable independent of η. Then ηN has characteristic function $E\exp(-\frac{1}{2}\eta^2 t^2)$, and the distribution of ηN is said to be a *mixture of normal distributions*. In the special case $\eta^2 = 1$ a.s. the limit variable Z has the $N(0,1)$ distribution. This special case of Theorem 3.2 was discovered by Brown (1971) and McLeish (1974), except for the aspect of stability.

It is illuminating to give an example of a martingale satisfying the conditions of Theorem 3.2, but for which η^2 is not a.s. constant. Let $\{Y_n, n \geqslant 1\}$ be independent r.v. each with the symmetric distribution on the points ± 1. Define $X_1 = Y_1$,

$$X_n = Y_n\left(\sum_{i=1}^{n-1} Y_i/i\right), \qquad n \geqslant 2,$$

let $k_n = n$, $X_{ni} = n^{-1/2}X_i$, and let $\mathscr{F}_{ni}$ be the σ-field generated by $Y_1, Y_2, \ldots, Y_i$. Then $\{S_{ni}, \mathscr{F}_{ni}\}$ is a martingale array. Since

$$\sum_{i=1}^{n} Y_i/i \xrightarrow{\text{a.s.}} Y = \sum_{i=1}^{\infty} Y_i/i \qquad \text{(a.s. finite by Theorem 2.5)},$$

it follows that

$$U_{nn}^2 = \sum_{i=1}^{n} X_{ni}^2 = n^{-1}\sum_{i=1}^{n}\left(\sum_{j=1}^{i-1} Y_j/j\right)^2 \xrightarrow{\text{a.s.}} Y^2,$$

and so (3.19) holds with $\eta^2 = Y^2$. Furthermore,

$$\max_i |X_{ni}| = n^{-1/2} \max_i \left| \sum_{j=1}^{i-1} Y_j/j \right| \xrightarrow{\text{a.s.}} 0,$$

$$\max_i X_{ni}^2 \leqslant U_{nn}^2 \xrightarrow{\text{a.s.}} Y^2,$$

and

$$E(U_{nn}^2) = n^{-1} \sum_{i=1}^{n} E\left(\sum_{j=1}^{i-1} Y_j/j \right)^2 = n^{-1} \sum_{i=1}^{n} \sum_{j=1}^{i-1} j^{-2}$$

$$\to \sum_{j=1}^{\infty} j^{-2} = E(Y^2).$$

The dominated convergence theorem (see Theorem A.7 in Appendix V) now implies that

$$E\left(\max_i X_{ni}^2 \right) \to 0,$$

from which follow (3.18) and (3.20). Condition (3.21) holds and so $n^{-1/2}(\sum_{i=1}^{n} X_i) \xrightarrow{\text{d}} Z$ (stably), where Z has characteristic function

$$E \exp(-\tfrac{1}{2} Y^2 t^2).$$

Proof of Theorem 3.2. Suppose first that η^2 is a.s. bounded, so that for some $C\ (>1)$,

$$P(\eta^2 < C) = 1. \tag{3.22}$$

Let $X'_{ni} = X_{ni} I(\sum_{j=1}^{i-1} X_{nj}^2 \leqslant 2C)$ and $S'_{ni} = \sum_{j=1}^{i} X'_{nj}$. Then $\{S'_{ni}, \mathscr{F}_{ni}\}$ is a martingale array. Since

$$P(X'_{ni} \neq X_{ni} \text{ for some } i \leqslant k_n) \leqslant P(U_{nk_n}^2 > 2C) \to 0 \tag{3.23}$$

we have $P(S'_{nk_n} \neq S_{nk_n}) \to 0$, and so

$$E\left|\exp(itS'_{nk_n}) - \exp(itS_{nk_n})\right| \to 0.$$

Hence $S_{nk_n} \xrightarrow{\text{d}} Z$ (stably) if and only if $S'_{nk_n} \xrightarrow{\text{d}} Z$ (stably). In view of (3.23), the martingale differences $\{X'_{ni}\}$ satisfy conditions (3.12) and (3.13) of Lemma 3.1; we must check (3.14).

Let

$$T'_n = \prod_j (1 + itX'_{nj})$$

and

$$J_n = \begin{cases} \min\{i \leqslant k_n | U_{ni}^2 > 2C\} & \text{if } U_{nk_n}^2 > 2C \\ k_n & \text{otherwise.} \end{cases}$$

Then

$$E|T'_n|^2 = E\prod_j(1 + t^2X'^2_{nj}) \leqslant E\left[\left\{\exp\left(t^2 \sum_{j=1}^{J_n-1} X'^2_{nj}\right)\right\}(1 + t^2X^2_{nJ_n})\right]$$
$$\leqslant \{\exp(2Ct^2)\}(1 + t^2EX^2_{nJ_n}),$$

which is bounded uniformly in n, by (3.20). Consequently $\{T'_n\}$ is uniformly integrable.

Let $m \geqslant 1$ be fixed and let $E \in \mathscr{F}_{mk_m}$; then by (3.21), $E \in \mathscr{F}_{nk_n}$ for all $n \geqslant m$. For such an n,

$$E[T'_nI(E)] = E\left[I(E)\prod_{j=1}^{k_n}(1 + itX'_{nj})\right]$$
$$= E\left[I(E)\prod_{j=1}^{k_m}(1 + itX'_{nj})\prod_{j=k_m+1}^{k_n}E(1 + itX'_{nj}|\mathscr{F}_{n,j-1})\right]$$
$$= E\left[I(E)\prod_{j=1}^{k_m}(1 + itX'_{nj})\right] = P(E) + R_n,$$

where the remainder term R_n consists of at most $2^{k_m} - 1$ terms of the form

$$E[I(E)(it)^rX'_{nj_1}X'_{nj_2}\cdots X'_{nj_r}],$$

where $1 \leqslant r \leqslant k_m$ and $1 \leqslant j_1 \leqslant j_2 \leqslant \cdots \leqslant j_r \leqslant k_m$. Since

$$|X'_{nj_1}\cdots X'_{nj_r}|^2 \leqslant \left(\sum_{j=1}^{J_n-1} X'^2_{nj}\right)^{r-1}\left(\max_i X'^2_{ni}\right)$$
$$\leqslant (2C)^{r-1}\left(\max_i X'^2_{ni}\right),$$

it follows that

$$|R_n| \leqslant (2^{k_m} - 1)(2C)^{k_m/2}E\left(\max_i |X'_{ni}|\right).$$

But, for any $\varepsilon > 0$,

$$E\left(\max_i|X_{ni}|\right) \leqslant \varepsilon + E\left[\max_i|X_{ni}|I(|X_{ni}| > \varepsilon)\right]$$
$$= \varepsilon + E\left[\left(\max_i|X_{ni}|\right)I\left(\max_i|X_{ni}| > \varepsilon\right)\right]$$
$$\leqslant \varepsilon + \left[E\left(\max_i X^2_{ni}\right)P\left(\max_i|X_{ni}| > \varepsilon\right)\right]^{1/2}$$
$$\to \varepsilon \quad \text{as} \quad n \to \infty.$$

It follows that $E(\max_i|X_{ni}|) \to 0$ and so $R_n \to 0$. Therefore

$$E[T'_n I(E)] \to P(E). \tag{3.24}$$

Let $\mathscr{F}_\infty = \bigvee_1^\infty \mathscr{F}_{nk_n}$ be the σ-field generated by $\bigcup_1^\infty \mathscr{F}_n$. For any $E' \in \mathscr{F}_\infty$ and any $\varepsilon > 0$ there exists an m and an $E \in \mathscr{F}_{mk_m}$ such that $P(E \,\Delta\, E') < \varepsilon$ (Δ denotes symmetric difference). Since $\{T'_n\}$ is uniformly integrable and

$$|E[T'_n I(E')] - E[T'_n I(E)]| \leqslant E[|T'_n| I(E \,\Delta\, E')],$$

$\sup_n|E[T'_n I(E')] - E[T'_n I(E)]|$ can be made arbitrarily small by choosing ε sufficiently small. It now follows from (3.24) that for any $E' \in \mathscr{F}_\infty$, $E[T'_n I(E')] \to P(E')$. This in turn implies that for any bounded $\mathscr{F}_\infty$-measurable r.v. X, $E[T'_n X] \to E(X)$. Finally, if $E \in \mathscr{F}$, then

$$E[T'_n I(E)] = E[T'_n E(I(E)|\mathscr{F}_\infty)] \to E[E(I(E)|\mathscr{F}_\infty)] = P(E).$$

This establishes (3.14) and completes the proof in the special case where (3.22) holds.

It remains only to remove the boundedness condition (3.22). If η^2 is not a.s. bounded, then given $\varepsilon > 0$, choose a continuity point C of η^2 such that $P(\eta^2 > C) > \varepsilon$. Let

$$\eta_C^2 = \eta^2 I(\eta^2 \leqslant C) + C I(\eta^2 > C),$$

$$X''_{ni} = X_{ni} I\left(\sum_{j=1}^{i-1} X_{nj}^2 \leqslant C\right) \quad \text{and} \quad S''_{ni} = \sum_{j=1}^{i} X''_{nj}.$$

Then $\{S''_{ni}, \mathscr{F}_{ni}\}$ is a martingale array and conditions (3.18), (3.20), and (3.21) are satisfied. Now,

$$\left(\sum_i X_{ni}^2\right) I\left(\sum_i X_{ni}^2 \leqslant C\right) + C I\left(\sum_i X_{ni}^2 > C\right) \leqslant \sum_i X_{ni}''^2$$
$$\leqslant \left(\sum_i X_{ni}^2\right) I\left(\sum_i X_{ni}^2 \leqslant C\right) + \left(C + \max_i X_{ni}^2\right) I\left(\sum_i X_{ni}^2 > C\right).$$

Since C is a continuity point of the distribution of η^2,

$$I\left(\sum_i X_{ni}^2 \leqslant C\right) \xrightarrow{\text{p}} I(\eta^2 \leqslant C),$$

and so

$$\sum_i X_{ni}''^2 \xrightarrow{\text{p}} \eta_C^2.$$

As η_C^2 is a.s. bounded, the first part of the proof tells us that $S''_n \xrightarrow{\text{d}} Z_C$ (stably), where the r.v. Z_C has characteristic function $E\exp(-\frac{1}{2}\eta_C^2 t^2)$. If $E \in \mathscr{F}$, then

$$\begin{aligned}&|E[I(E)\exp(itS_{nk_n})] - E[I(E)\exp(-\tfrac{1}{2}\eta^2t^2)]|\\&\quad\leqslant E|\exp(itS_{nk_n}) - \exp(itS''_{nk_n})| + |E[I(E)\exp(itS''_{nk_n})]\\&\qquad - E[I(E)\exp(-\tfrac{1}{2}\eta_C^2t^2)]| + E|\exp(-\tfrac{1}{2}\eta_C^2t^2) - \exp(-\tfrac{1}{2}\eta^2t^2)|.\end{aligned}$$

Since $S''_{nk_n} \to Z_C$ (stably), the second term on the right-hand side converges to zero as $n \to \infty$. The first and third terms are each less than 2ε in the limit since

$$\begin{aligned}P(S_{nk_n} \neq S''_{nk_n}) &\leqslant P(X''_{ni} \neq X_{ni} \text{ for some } i)\\&\leqslant P(U^2_{nk_n} > C) \to P(\eta^2 > C) < \varepsilon.\end{aligned}$$

Hence for all $\varepsilon > 0$,

$$\limsup_{n\to\infty} |E[I(E)\exp(itS_{nk_n})] - E[I(E)\exp(-\tfrac{1}{2}\eta^2t^2)]| \leqslant 4\varepsilon.$$

It follows that the limit equals zero, and in view of Theorem 3.1 this completes the proof. ∎

The corollary to Theorem 3.2 follows immediately from the remarks in Subsection 3.2(iii), where it is shown that conditions (3.7) and (3.8) imply that

$$|U^2_{nk_n} - V^2_{nk_n}| \xrightarrow{\text{p}} 0.$$

(vi) *The martingale CLT with random norming.* Let $\{S_{ni}, \mathscr{F}_{ni}\}$ be a martingale array and suppose that the r.v. η^2 is measurable in all of the σ-fields $\mathscr{F}_{ni}$ and that $\eta^2 > 0$ a.s. If $2 \leqslant i \leqslant k_n$, then

$$E(\eta^{-1}S_{ni}|\mathscr{F}_{n,i-1}) = \eta^{-1}E(S_{ni}|\mathscr{F}_{n,i-1}) = \eta^{-1}S_{n,i-1}.$$

Hence $\{\eta^{-1}S_{ni}, \mathscr{F}_{ni}, 2 \leqslant i \leqslant k_n, n \geqslant 2\}$ is also a martingale array. If conditions (3.18)–(3.20) hold for the original martingale, then their analogs hold for the new one, provided that we condition on η. For example,

$$\sum_{i=2}^{k_n} \eta^{-2}X^2_{ni} = \eta^{-2}\sum_{i=1}^{k_n} X^2_{ni} - \eta^{-2}X^2_{n1} \xrightarrow{\text{p}} 1 \tag{3.25}$$

and

$$E\left(\max_{2\leqslant i\leqslant k_n} \eta^{-2}X^2_{ni}|\eta\right) = \eta^{-2}E\left(\max_{2\leqslant i\leqslant k_n} X^2_{ni}|\eta\right),$$

which is a.s. bounded. Since the limit in (3.25) is a constant, then it is a straightforward matter to condition on η and apply Theorem 3.2 to our new martingale, proving that

$$\sum_{i=2}^{k_n} \eta^{-1}X_{ni} \xrightarrow{\text{d}} N(0,1).$$

But $\eta^{-1}X_{n1} \xrightarrow{\text{p}} 0$ and so, conditional on η, $\eta^{-1}S_n \xrightarrow{\text{d}} N(0,1)$. The limit does not depend on η and so the result holds unconditionally. Moreover, since $U_{nk_n}^{-1}\eta \xrightarrow{\text{p}} 1$,

$$U_{nk_n}^{-1}S_{nk_n} = (U_{nk_n}^{-1}\eta)(\eta^{-1}S_{nk_n}) \xrightarrow{\text{d}} N(0,1). \tag{3.26}$$

Norming by the r.v. U_{nk_n} allows us to obtain a CLT.

This argument is basically the one used by Eagleson (1975a), who treated a rather more general problem. The condition that η be measurable in all of the $\mathscr{F}_{ni}$ is quite severe, and usually difficult to check. Condition (3.21) is much more natural, and together with (3.18)–(3.20) and the assumption that $P(\eta^2 > 0) = 1$, it is sufficient for (3.26). If we divide S_{nk_n} by U_{nk_n} we effectively cancel the fluctuations which cause S_{nk_n} to deviate from the $N(0,1)$ distribution.

We can prove even more than this. The limit theorem (3.26) is *mixing* (see Subsection 3.2(iv)). The concepts of mixing and random normalization are closely connected; see Smith (1945) and Takahashi (1951a,b).

Mixing convergence is a special case of stable convergence. Recall from Subsection 3.2(iv) that a sequence $\{Y_n\}$ of r.v. on a probability space $(\Omega,\mathscr{F},P)$ converges stably to an r.v. Y if for all continuity points y of Y and all $E \in \mathscr{F}$,

$$\lim_{n\to\infty} P(\{Y_n \leqslant y\} \cap E) = Q_y(E)$$

exists, and $Q_y(E) \to P(E)$ as $y \to \infty$. The convergence of Y_n to Y is mixing if (and only if) the subprobability measure Q_y takes the form

$$Q_y(E) = P(Y \leqslant y)P(E)$$

for all continuity points y and all events E.

Theorem 3.3. *Suppose that the conditions of Theorem* 3.2 *hold and that* $P(\eta^2 > 0) = 1$. *Then*

$$S_{nk_n}/U_{nk_n} \xrightarrow{\text{d}} N(0,1) \quad (\textit{mixing}).$$

Corollary 3.2. *Suppose that the conditions of Corollary* 3.1 *hold and that* $P(\eta^2 > 0) = 1$. *Then*

$$S_{nk_n}/V_{nk_n} \xrightarrow{\text{d}} N(0,1) \quad (\textit{mixing}).$$

Remarks. As we pointed out at the start of this subsection and in the remarks following Corollary 3.1, the "nested σ-field" condition (3.21) may be replaced by the condition that η be measurable in all of the $\mathscr{F}_{ni}$, although the convergence may no longer be mixing. For example, if η is a.s. constant then (3.21) may be dropped.

Proof of Theorem 3.3. Since $S_{nk_n} \xrightarrow{\text{d}} Z$ (stably), for any real t

$$\exp(itS_{nk_n}) \to \exp(-\tfrac{1}{2}\eta^2t^2) \quad (\text{weakly in } L^1).$$

Hence, for any bounded r.v. X measurable in our σ-field $\mathscr{F}$,

$$E[\exp(itS_{nk_n})X] \to E[\exp(-\tfrac{1}{2}\eta^2t^2)X].$$

Let $X = \exp(iu\eta + ivI(E))$, where u and v are fixed real numbers and $E \in \mathscr{F}$. It follows that the joint characteristic function of $(S_{nk_n},\eta,I(E))$ converges to that of $(\eta N,\eta,I(E))$, where N is a standard normal variable independent of $(\eta,I(E))$. This in turn implies that

$$(\eta^{-1}S_{nk_n},I(E)) \xrightarrow{\text{d}} (N,I(E)),$$

and so

$$(U_{nk_n}^{-1}S_{nk_n},I(E)) \xrightarrow{\text{d}} (N,I(E)),$$

completing the proof. ▮

The corollary is proved in the same way, making use of Corollary 3.1.

3.3. Toward a General Central Limit Theorem

We shall give some examples to motivate the search for a more general CLT. Let $\{Y_n, n \geq 1\}$ be independent r.v. each with the symmetric distribution on the points ± 1. Define $k_n = 2n$, let $\mathscr{F}_{ni}$ be the σ-field generated by $Y_1, Y_2, \ldots, Y_i$, $1 \leq i \leq 2n$, let $m(n)$ be a (nonrandom) integer between 1 and n, and set

$$X_{ni} = \begin{cases} n^{-1/2}Y_i & \text{if } 1 \leq i \leq n \\ n^{-1/2}Y_iI\left(\sum_{j=1}^{m(n)} Y_j > 0\right) & \text{if } n+1 \leq i \leq 2n. \end{cases}$$

Then $\{S_{ni},\mathscr{F}_{ni}\}$ is a martingale array. The sum $S_{n,2n}$ is basically a sum of $2n$ independent r.v. except that each of the last n terms is skewed by a common factor based on some of the first n terms.

Example 1. Suppose that $m(n) \equiv n$. Then

$$S_{n,2n} = n^{-1/2}\sum_{i=1}^{n} Y_i + \left(n^{-1/2}\sum_{i=n+1}^{2n} Y_i\right)I\left(n^{-1/2}\sum_{i=1}^{n} Y_i > 0\right)$$
$$\xrightarrow{\text{d}} Z = N_1 + N_2I(N_1 > 0),$$

where N_1 and N_2 are independent $N(0,1)$ variables. The distribution of Z is not a mixture of normal distributions. Further,

$$U_{n,2n}^2 = 1 + I\left(\sum_{i=1}^{n} Y_i > 0\right) \xrightarrow{\text{d}} \eta^2,$$

where η^2 has a two-point distribution: $P(\eta^2 = 1) = P(\eta^2 = 2) = \frac{1}{2}$. The convergence in distribution here cannot be strengthened to convergence in probability. Finally,

$$S_{n,2n}/U_{n,2n} \xrightarrow{d} Z' = N_1 I(N_1 \leqslant 0) + 2^{-1/2}(N_1 + N_2)I(N_1 > 0);$$

Z' does not have the $N(0,1)$ distribution.

Example 2. Let $m(n) = [\sqrt{n}]$ (the integer part of $\sqrt{n}$). Then

$$\begin{aligned} S_{n,2n} &= n^{-1/2} \sum_{i=1}^{m(n)} Y_i + n^{-1/2} \sum_{i=m(n)+1}^{n} Y_i \\ &\quad + \left(n^{-1/2} \sum_{i=n+1}^{2n} Y_i\right) I\left(m(n)^{-1/2} \sum_{i=1}^{m(n)} Y_i > 0\right) \\ &\xrightarrow{d} Z'' = N_1 + N_2 I(N_3 > 0), \end{aligned}$$

where N_1, N_2, and N_3 are independent $N(0,1)$ variables. It is easy to see that $U_{n,2n}^2 \xrightarrow{d} \eta^2$, where η^2 is defined above, and that Z'' has characteristic function $E\exp(-\frac{1}{2}\eta^2 t^2)$. The convergence in distribution to η^2 cannot be strengthened to convergence in probability. Finally,

$$S_{n,2n}/U_{n,2n} \xrightarrow{d} Z''' = N_1 I(N_3 \leqslant 0) + 2^{-1/2}(N_1 + N_2)I(N_3 > 0),$$

and Z''' is distributed as $N(0,1)$.

In both these examples the martingale arrays satisfy conditions (3.18), (3.20), and (3.21) of Theorems 3.2 and 3.3, but not (3.19). It seems that (3.19) is not the right type of condition to impose. If $\{m(n)\}$ is any sequence diverging to ∞, then $U_{n,2n}^2 \xrightarrow{d} \eta^2$, although

$$S_{n,2n}/U_{n,2n} \xrightarrow{d} N(0,1) \tag{3.27}$$

if and only if $m(n)/n \to 0$.

The norming variable

$$U_{n,2n} = \left[1 + I\left(\sum_{i=1}^{m(n)} Y_i > 0\right)\right]^{1/2}$$

depends on $m(n)$ r.v. from the nth row—that is, on a proportion $m(n)/2n$ of the nth row. Since (3.27) is true if and only if $m(n)/n \to 0$, we might deduce that in general a CLT will hold if and only if, in the limit, U_{nk_n} depends on a "negligible amount" of information about $X_{n1}, X_{n2}, \ldots, X_{nk_n}$. While it is possible to formulate and prove a CLT along these lines, as is done in Theorem 3.4 below, the conditions it involves are usually difficult to check. For this reason Theorem 3.3 has many practical advantages, despite its restricted generality.

Theorem 3.4 *Suppose that the probability space* $(\Omega_n, \mathscr{F}_n, P_n)$ *supports square-integrable r.v.* $S_{n1}, S_{n2}, \ldots, S_{nk_n}$, *and that the* S_{ni} *are adapted to the σ-fields* $\mathscr{F}_{ni}$, *where* $\mathscr{F}_{n1} \subseteq \mathscr{F}_{n2} \subseteq \cdots \subseteq \mathscr{F}_{nk_n} \subseteq \mathscr{F}_n$. *Let* $X_{ni} = S_{ni} - S_{n,i-1}$ $(S_{n0} = 0)$ *and* $U_{ni}^2 = \sum_{j=1}^{i} X_{nj}^2$. *If* $\mathscr{G}_n$ *is a sub-σ-field of* $\mathscr{F}_n$, *let* $\mathscr{G}_{ni} = \mathscr{F}_{ni} \vee \mathscr{G}_n$ *(the σ-field generated by* $\mathscr{F}_{ni} \cup \mathscr{G}_n$*) and let* $\mathscr{G}_{n0} = \{\Omega_n, \phi\}$ *denote the trivial σ-field.*

Suppose further that

$$\max_i |X_{ni}| \xrightarrow{\text{p}} 0, \qquad E\left(\max_i X_{ni}^2\right) \quad \text{is bounded in } n,$$

and that there exist σ-fields $\mathscr{G}_n \subseteq \mathscr{F}_n$ *and* $\mathscr{G}_n$*-measurable r.v.* u_n^2 *such that*

$$U_{nk_n}^2 - u_n^2 \xrightarrow{\text{p}} 0, \tag{3.28}$$

$$\sum_i E(X_{ni} | \mathscr{G}_{n,i-1}) \xrightarrow{\text{p}} 0, \qquad \text{and} \qquad \sum_i |E(X_{ni} | \mathscr{G}_{n,i-1})|^2 \xrightarrow{\text{p}} 0. \tag{3.29}$$

If

$$\lim_{\delta \to 0} \liminf_{n \to \infty} P(U_{nk_n} > \delta) = 1, \tag{3.30}$$

then $S_{nk_n}/U_{nk_n} \xrightarrow{\text{d}} N(0,1)$. *If instead of* (3.30), $U_{nk_n}^2 \xrightarrow{\text{d}} \eta^2$, *then* $S_{nk_n} \xrightarrow{\text{d}} Z$, *where the r.v. Z has characteristic function* $\exp(-\frac{1}{2}\eta^2 t^2)$.

(The r.v. $Z_1, \ldots, Z_n$ are said to be *adapted* to the σ-fields $\mathscr{A}_1, \ldots, \mathscr{A}_n$ if each Z_i is $\mathscr{A}_i$-measurable. The proof of Theorem 3.4 is very similar to that of Hall's (1977) theorem, and will be omitted.)

The first two conditions of Theorem 3.4 are just the negligibility conditions of Theorems 3.2 and 3.3. Condition (3.28) says that $U_{nk_n}^2$ really only depends on information contained in the σ-field $\mathscr{G}_n$. If $\{S_{ni}, \mathscr{F}_{ni}\}$ is a zero-mean martingale array, then of course each $E(X_{ni} | \mathscr{F}_{n,i-1}) = 0$. In this case (3.29) tells us that increasing the σ-fields $\mathscr{F}_{ni}$ to $\mathscr{G}_{ni}$ hardly upsets the martingale property. That is, $\mathscr{G}_n$ does not contain very much important information about $X_{n1}, X_{n2}, \ldots, X_{nk_n}$.

Let us apply Theorem 3.4 to Examples 1 and 2. Define

$$u_n^2 = U_{nk_n}^2 = 1 + I\left(\sum_{i=1}^{m(n)} Y_i > 0\right)$$

and let $\mathscr{G}_n$ be the σ-field generated by $Y_1, \ldots, Y_{m(n)}$. Then

$$\mathscr{G}_{ni} = \begin{cases} \mathscr{F}_{n,m(n)} & \text{if} \quad 1 \leqslant i \leqslant m(n) \\ \mathscr{F}_{ni} & \text{if} \quad m(n) + 1 \leqslant i \leqslant 2n, \end{cases}$$

and so

$$E(X_{ni} | \mathscr{G}_{n,i-1}) = \begin{cases} X_{ni} & \text{if} \quad 1 \leqslant i \leqslant m(n) \\ 0 & \text{if} \quad m(n) + 1 \leqslant i \leqslant 2n. \end{cases}$$

Condition (3.28) holds trivially, and (3.29) holds if and only if $m(n)/n \to 0$. Since $U_{n,2n}^2 \xrightarrow{d} \eta^2 > 0$ a.s., (3.30) is true. Theorem 3.4 now implies that $S_{n,2n} \xrightarrow{d} Z$ and $S_{n,2n}/U_{n,2n} \xrightarrow{d} N(0,1)$ if $m(n)/n \to 0$.

Theorems 3.2 and 3.3 (without stable or mixing convergence) can be obtained from Theorem 3.4 in the following way. If (3.18) holds we can choose a sequence of integers $l_n \leqslant k_n$ such that $l_n \uparrow \infty$ but $l_n \max_i |X_{ni}| \xrightarrow{p} 0$. Let $\mathscr{G}_n = \mathscr{F}_{nl_n}$. If $U_{nk_n}^2 \xrightarrow{p} \eta^2$, then η^2 is measurable in the σ-field generated by $\bigcup \mathscr{F}_{nk_n}$,

$$\mathscr{F}_\infty = \bigvee_n \mathscr{F}_{nk_n} = \bigvee_n \mathscr{F}_{nl_n}$$

(or at least, in the completion of $\mathscr{F}_\infty$). The unions above are increasing, and so given $\varepsilon > 0$ there exists an n and a $\mathscr{G}_n$-measurable variable u^2 such that $P(|\eta^2 - u^2| > \varepsilon) < \varepsilon$. Hence we can choose $\mathscr{G}_n$-measurable r.v. u_n^2 such that $u_n^2 \xrightarrow{p} \eta^2$. Condition (3.28) follows. Moreover, since $\mathscr{G}_{ni} = \mathscr{F}_{nl_n}$ for $1 \leqslant i \leqslant l_n$ and $\mathscr{G}_{ni} = \mathscr{F}_{ni}$ for $i > l_n$, then

$$\sum_i |E(X_{ni}|\mathscr{G}_{n,i-1})| = \sum_{i=1}^{l_n} |X_{ni}| \leqslant l_n \max_i |X_{ni}| \xrightarrow{p} 0,$$

implying (3.29). Finally, if $U_{nk_n}^2 \xrightarrow{p} \eta^2$, then (3.30) holds if and only if $\eta^2 > 0$ a.s.

The next two examples will further explain the behavior of randomly normed martingales.

Example 3. Let the r.v. Y_n be as in Examples 1 and 2, let $\{m_j, j \geqslant 1\}$ be the sequence of integers defined by $m_1 = 1$ and $m_{j+1} = m_j + 2m_j^2$, and set $X_1 = Y_1$ and

$$X_i = \begin{cases} Y_i & \text{if} \quad m_j < i \leqslant m_j + m_j^2 \\ Y_i I(Y_{m_j} > 0) & \text{if} \quad m_j + m_j^2 < i \leqslant m_{j+1}. \end{cases}$$

Then $S_n = \sum_{i=1}^n X_i$ defines a martingale with respect to the σ-fields $\mathscr{F}_n$ generated by $Y_1, \ldots, Y_n$. By applying Theorem 3.4 to the triangular array with $k_n = n$ and $S_{ni} = n^{-1/2} S_i$, $1 \leqslant i \leqslant n$, or by using the argument of Example 2, it is easy to see that

$$S_n/U_n = \left(\sum_{i=1}^n X_i\right) \Big/ \left(\sum_{i=1}^n X_i^2\right)^{1/2} \xrightarrow{d} N(0,1).$$

However, note that for no series of constants c_n does U_n^2/c_n converge weakly to a nondegenerate distribution. This follows from the fact that

$$\begin{aligned} m_k^{-1} U_{m_k}^2 &= m_k^{-1}(U_{m_{k-1}}^2 + m_{k-1}^2 + m_{k-1}^2 I(Y_{m_{k-1}} > 0)) \\ &\xrightarrow{d} \tfrac{1}{2}[1 + I(Y_1 > 0)] \qquad \text{as} \quad k \to \infty, \end{aligned}$$

while

$$(m_k + m_k^2)^{-1} U^2_{m_k + m_k^2} = (m_k + m_k^2)^{-1}(U^2_{m_k} + m_k^2) \xrightarrow{\text{p}} 1.$$

In the same way we can show that for no constants c_n does S_n/c_n have a nondegenerate limiting distribution.

Example 4. Let $W(t)$, $t \geqslant 0$, be standard Brownian motion,

$$t_{ni} = \begin{cases} 1/n & \text{if} \quad 1 \leqslant i \leqslant n \\ I(W(1) > 0)/n & \text{if} \quad n+1 \leqslant i \leqslant 2n, \end{cases}$$

and set $S_{ni} = W(\sum_{j=1}^{i} t_{nj})$, $1 \leqslant i \leqslant 2n$. If $\mathscr{F}_{ni}$ is the σ-field generated by $W(t)$ for $0 \leqslant t \leqslant i/n$, then $\{S_{ni}, \mathscr{F}_{ni}, 1 \leqslant i \leqslant 2n, n \geqslant 1\}$ is a martingale array. Conditions (3.18)–(3.20) are satisfied, but not (3.21). To check (3.19), note that

$$U^2_{n,2n} = \sum_{i=1}^{n} Z^2_{ni} + \left(\sum_{i=n+1}^{2n} Z^2_{ni} \right) I(W(1) > 0),$$

where the Z_{ni} are independent $N(0,1/n)$ variables; clearly

$$U^2_{n,2n} \xrightarrow{\text{p}} 1 + I(W(1) > 0).$$

Furthermore,

$$S_{n,2n} = W(1)I(W(1) \leqslant 0) + W(2)I(W(1) > 0),$$

which is not a mixture of normal distributions, and

$$S_{n,2n}/U_{n,2n} \xrightarrow{\text{p}} W(1)I(W(1) \leqslant 0) + 2^{-1/2}W(2)I(W(1) > 0),$$

which does not have the $N(0,1)$ distribution. Hence (3.18)–(3.20) are not sufficient for the conclusions of either of Theorems 3.2 or 3.3.

3.4. Raikov-Type Results in the Martingale CLT

Burkholder's square function inequalities are an important and comparatively recent addition to the probabilist's armory of martingale tools. They imply a close and rather unexpected relationship between the behavior of a martingale and the sum of the squares of its differences. However, a certain duality between the behavior of sums and sums of squares of independent r.v. was noticed considerably earlier by Raikov (1938). For each $n \geqslant 1$ let X_{ni}, $1 \leqslant i \leqslant k_n$, be independent r.v. with zero means, variances summing to 1, and satisfying the asymptotic negligibility condition

$$\text{for all} \quad \varepsilon > 0, \qquad \max_i P(|X_{ni}| > \varepsilon) \to 0. \tag{3.31}$$

Raikov showed that

$$\sum_i X_{ni} \xrightarrow{\text{d}} N(0,1)$$

if and only if

$$\sum_i X_{ni}^2 \xrightarrow{\text{p}} 1.$$

Given the discoveries of Raikov and Burkholder it is hardly surprising that Raikov's results have an analog for martingales. However, the analog is perhaps closer than expected. Raikov's results can be extended to the convergence of moments as well as to convergence in distribution. If μ_{2p} $(p \geqslant 1)$ denotes the $(2p)$th absolute moment of an $N(0,1)$ variable, then in the notation of the previous paragraph,

$$\sum_i X_{ni} \xrightarrow{\text{d}} N(0,1) \qquad \text{and} \qquad E\left|\sum_i X_{ni}\right|^{2p} \to \mu_{2p}$$

if and only if

$$E\left|\sum_i X_{ni}^2 - 1\right|^p \to 0$$

(which is easily seen to be equivalent to the condition $E|\sum_i X_{ni}^2|^p \to 1$) (Hall, 1978a). All of these results carry over to the martingale case (Hall, 1978b), although in a slightly unexpected way.

The convergence of moments in the CLT for sums of independent r.v. was investigated by Bernstein (1939), who provided necessary and sufficient conditions. His results were rediscovered by Brown (1969, 1970), who gave an alternative proof. Bernstein and Brown showed that in the case of independent summands satisfying the negligibility condition (3.31) and such that each $E(X_{ni}) = 0$ and $\sum_i E(X_{ni}^2) = 1$,

$$\sum_i X_{ni} \xrightarrow{\text{d}} N(0,1) \qquad \text{and} \qquad E\left|\sum_i X_{ni}\right|^{2p} \to \mu_{2p}, \qquad p \geqslant 1,$$

if and only if the Lindeberg condition of order $2p$ holds:

$$\text{for all } \varepsilon > 0, \qquad \sum_i E[|X_{ni}|^{2p} I(|X_{ni}| > \varepsilon)] \to 0. \tag{3.32}$$

Let $\{S_{ni}, \mathscr{F}_{ni}, 1 \leqslant i \leqslant k_n, n \geqslant 1\}$ be a zero-mean, square-integrable martingale array, and let $\{X_{ni}\}$ denote the martingale difference sequence. Define

$$V_{ni}^2 = \sum_{j=1}^{i} E(X_{nj}^2|\mathscr{F}_{n,j-1}) \qquad \text{and} \qquad U_{ni}^2 = \sum_{j=1}^{i} X_{nj}^2, \qquad 1 \leqslant j \leqslant k_n.$$

Suppose that $E(S_{nk_n}^2) = 1$ for each n. Brown (1971) defined the *conditional characteristic function* $f_n(t)$ by

$$f_n(t) = \prod_j E[\exp(itX_{nj})|\mathscr{F}_{n,j-1}], \qquad n \geqslant 1,$$

where $i = \sqrt{-1}$. In the independence case f_n is nonrandom and is the characteristic function of S_{nk_n}, while $V_{nk_n}^2 = 1$ a.s. Moreover, when the X_{ni} are independent and $f_n(t) \to e^{-t^2/2}$ for each real t, then obviously the CLT holds.

Let η denote a finite valued r.v. and let ν_{2p} denote the pth absolute moment of a variable with characteristic function $E\exp(-\frac{1}{2}\eta^2 t^2)$.

Theorem 3.5. *Suppose that*

$$E|V_{nk_n}^2 - \eta^2| \to 0 \tag{3.33}$$

and that

$$\max_i E(X_{ni}^2 | \mathscr{F}_{n,i-1}) \xrightarrow{p} 0. \tag{3.34}$$

Then the following three conditions are equivalent:

$$E|U_{nk_n}^2 - \eta^2| \to 0, \tag{3.35}$$

$$\textit{for all} \quad \varepsilon > 0, \qquad \sum_i E[X_{ni}^2 I(|X_{ni}| > \varepsilon)] \to 0, \tag{3.36}$$

and

$$f_n(t) \xrightarrow{p} e^{-\eta^2 t^2/2}. \tag{3.37}$$

Let $p > 1$ and suppose that (3.34) *holds and that*

$$E|V_{nk_n}^2 - \eta^2|^p \to 0. \tag{3.38}$$

If η^2 is not a.s. constant, assume also that $\mathscr{F}_{n,i} \subseteq \mathscr{F}_{n+1,i}$ for $1 \leqslant i \leqslant k_n$ and $n \geqslant 1$ [*condition* (3.21)]. *Then the following three conditions are equivalent:*

$$E|U_{nk_n}^2 - \eta^2|^p \to 0, \tag{3.39}$$

$$\sum_i E|X_{ni}|^{2p} \to 0, \tag{3.40}$$

and

$$f_n(t) \xrightarrow{p} e^{-\eta^2 t^2/2} \qquad \textit{and} \qquad E|S_{nk_n}|^{2p} \to \nu_{2p}. \tag{3.41}$$

Remarks. Convergence in L^1 in conditions (3.33) and (3.35) is equivalent to convergence in probability, since $E(U_{nk_n}^2) = E(V_{nk_n}^2) = 1$; see Billingsley (1968, Theorem 5.4) or Chung (1974, Theorem 4.5.4). If $p > 1$, (3.40) is equivalent to the apparently weaker condition (3.32); see the remark after Theorem 2.23.

Conditions (3.33) and (3.36) (together with (3.21) if η^2 is not a.s. constant) imply that $S_{nk_n} \xrightarrow{d} Z$, where Z has characteristic function $E\exp(-\frac{1}{2}\eta^2 t^2)$; see Corollary 3.1 and the ensuing remarks. Hence Theorem 3.5 gives us simple sufficient conditions for the convergence of moments in the CLT.

Corollary 3.3. *Suppose that* (3.38) *and* (3.40) *hold, and assume also* (3.21) *if* η^2 *is not a.s. constant. Then*

$$S_{nk_n} \xrightarrow{d} Z \qquad \textit{and} \qquad E|S_{nk_n}|^{2p} \to \nu_{2p}. \tag{3.42}$$

In the independence case $\eta^2 = 1$ a.s. and (3.38) is trivially satisfied, while (3.34) and (3.42) imply (3.40).

Proof of Theorem 3.5. Suppose first that (3.33) and (3.34) hold. Define the functions A and B of the real variable x by

$$e^{ix} = 1 + ix - \tfrac{1}{2}x^2 + \tfrac{1}{2}x^2 A(x)$$

and

$$B(x) = \min(x/3,2).$$

Then [see, e.g., Feller (1971, Lemma 1, p. 512)]

$$|A(x)| \leqslant B(|x|) \tag{3.43}$$

and

$$E[\exp(itX_{nj})|\mathscr{F}_{n,j-1}] = 1 - \tfrac{1}{2}t^2 E(X_{nj}^2|\mathscr{F}_{n,j-1}) + \tfrac{1}{2}t^2 E[X_{nj}^2 A(tX_{nj})|\mathscr{F}_{n,j-1}].$$

The inequality (3.43) implies that

$$|-\tfrac{1}{2}t^2 E(X_{nj}^2|\mathscr{F}_{n,j-1}) + \tfrac{1}{2}t^2 E[X_{nj}^2 A(tX_{nj})|\mathscr{F}_{n,j-1}]| \leqslant (3/2)t^2 E(X_{nj}^2|\mathscr{F}_{n,j-1})$$

and so, since $|\log(1+z) - z| \leqslant \tfrac{1}{2}|z|^2/(1-|z|)$ for $|z| < 1$,

$$\begin{aligned}\log f_n(t) &= \sum_j \log\{E[\exp(itX_{nj})|\mathscr{F}_{n,j-1}]\} \\ &= -\tfrac{1}{2}t^2 V_{nk_n}^2 + \tfrac{1}{2}t^2 \sum_j E[X_{nj}^2 A(tX_{nj})|\mathscr{F}_{n,j-1}] + R_n,\end{aligned}$$

where

$$\begin{aligned}|R_n| &\leqslant \tfrac{1}{2}\sum_j [\tfrac{3}{2}t^2 E(X_{nj}^2|\mathscr{F}_{n,j-1})]^2 \Big/ \Big[1 - \tfrac{3}{2}t^2 \max_k E(X_{nk}^2|\mathscr{F}_{n,k-1})\Big] \\ &\leqslant (9t^4/8)V_{nk_n}^2 \Big[\max_j E(X_{nj}^2|\mathscr{F}_{n,j-1})\Big] \Big/ \Big[1 - \tfrac{3}{2}t^2 \max_k E(X_{nk}^2|\mathscr{F}_{n,k-1})\Big] \\ &\xrightarrow{p} 0\end{aligned}$$

under (3.33) and (3.34). Therefore

$$\log f_n(t) + \tfrac{1}{2}t^2 V_{nk_n}^2 - \tfrac{1}{2}t^2 \sum_j E[X_{nj}^2 A(tX_{nj})|\mathscr{F}_{n,j-1}] \xrightarrow{p} 0. \tag{3.44}$$

Using (3.43) we have for any $\varepsilon > 0$,

$$\begin{aligned}
E\left|\sum_j E[X_{nj}^2 A(tX_{nj})|\mathscr{F}_{n,j-1}]\right| &\leqslant \sum_j E[X_{nj}^2 B(|tX_{nj}|)] \\
&= \sum_j E[X_{nj}^2 B(|tX_{nj}|)\{I(|tX_{nj}| \leqslant \varepsilon) + I(|tX_{nj}| > \varepsilon)\}] \\
&\leqslant (\varepsilon/3)\sum_j E(X_{nj}^2) + 2\sum_j E[X_{nj}^2 I(|tX_{nj}| > \varepsilon)] \\
&= \varepsilon/3 + 2\sum_j E[X_{nj}^2 I(|X_{nj}| > \varepsilon|t|^{-1})] \to \varepsilon/3
\end{aligned}$$

if (3.36) holds. Using this together with (3.44) we see that (3.36) implies (3.37).

Now suppose that (3.33), (3.34), and (3.37) hold. By taking real parts in (3.44) we deduce that for $t \neq 0$,

$$\operatorname{Re}\sum_j E[X_{nj}^2 A(tX_{nj})|\mathscr{F}_{n,j-1}] \xrightarrow{p} 0. \tag{3.45}$$

The real part of $A(x)$ satisfies

$$0 \leqslant \operatorname{Re} A(x) = 1 - 2x^{-2}(1 - \cos x) < 1,$$

and so the left-hand side of (3.45) is dominated by $V_{nk_n}^2$, which converges to η^2 in L^1. The dominated convergence theorem (see Appendix V) now allows us to strengthen the convergence in probability in (3.45) to convergence in L^1. $\operatorname{Re} A(x)$ is greater than 3/4 for $|x| > 4$, and so for all $t \neq 0$,

$$(3/4)\sum_j E[X_{nj}^2 I(|tX_{nj}| > 4)] \leqslant \operatorname{Re}\sum_j E[X_{nj}^2 A(tX_{nj})] \to 0,$$

which gives (3.36).

Suppose now that (3.33), (3.34), and (3.35) hold. Applying Corollary 2.1 to the martingale $\{U_{nj}^2 - V_{nj}^2, 1 \leqslant j \leqslant k_n\}$, we deduce that for all $\varepsilon > 0$,

$$P\left(\max_j |U_{nj}^2 - V_{nj}^2| > \varepsilon\right) \leqslant \varepsilon^{-1} E|U_{nk_n}^2 - V_{nk_n}^2| \to 0,$$

which implies that

$$\max_j |X_{nj}^2 - E(X_{nj}^2|\mathscr{F}_{n,j-1})| \leqslant 2\max_j |U_{nj}^2 - V_{nj}^2| \xrightarrow{p} 0.$$

Combined with (3.34) this shows that

$$\max_j X_{nj}^2 \xrightarrow{p} 0,$$

which is equivalent to the condition

$$\text{for all} \quad \varepsilon > 0, \qquad \sum_j X_{nj}^2 I(|X_{nj}| > \varepsilon) \xrightarrow{p} 0. \tag{3.46}$$

The left-hand side of the expression in (3.46) is dominated by $U_{nk_n}^2$, which converges to η^2 in L^1. The dominated convergence theorem now allows us to strengthen the convergence in (3.46) to convergence in L^1, which immediately gives us (3.36).

If (3.33) and (3.36) hold, then (3.35) follows via Theorem 2.23. This completes the proof of the first part of the theorem.

Now let $p > 1$ and assume (3.34) and (3.38). Then $\{V_{nk_n}^{2p}, n \geqslant 1\}$ is uniformly integrable. If (3.39) holds, then so does (3.35), and consequently (3.36). Theorem 2.23 now implies (3.40). Conversely if (3.40) holds, then so does (3.36), and then Theorem 2.23 implies (3.39). It remains only to prove that (3.39) and (3.41) are equivalent.

Define the r.v. Y_{ni}, Z_{ni}, A_{ni}, and B_{ni} as in the proof of Theorem 2.23, namely,

$$\begin{aligned} Y_{ni} &= X_{ni}I(|X_{ni}| \leqslant 1) - E[X_{ni}I(|X_{ni}| \leqslant 1)|\mathscr{F}_{n,i-1}], \\ Z_{ni} &= X_{ni} - Y_{ni}, \\ A_{ni} &= Y_{ni}I(V_{ni}^2 \leqslant \lambda), \end{aligned}$$

and

$$B_{ni} = Y_{ni} - A_{ni}.$$

These quantities are all martingale differences. For any $r > \frac{1}{2}$ we have from Theorems 2.10 and 2.11 that there exist constants K_1 and K_2 such that

$$\begin{aligned} E\Big(\sum_i A_{ni}^2\Big)^r &\leqslant K_1 E\Big|\sum_i A_{ni}\Big|^{2r} \\ &\leqslant K_2\Big\{E\Big[\Big(\sum_i E(A_{ni}^2|\mathscr{F}_{n,i-1})\Big)^r\Big] + E\Big(\max_i |A_{ni}|^{2r}\Big)\Big\} \\ &\leqslant K_2(\lambda^{2r} + 2^{2r}) < \infty. \end{aligned}$$

Consequently for any p,

$$\Big\{\Big(\sum_i A_{ni}^2\Big)^p, n \geqslant 1\Big\} \quad \text{and} \quad \Big\{\Big|\sum_i A_{ni}\Big|^{2p}, n \geqslant 1\Big\} \quad \text{are uniformly integrable.} \tag{3.47}$$

Furthermore,

$$\begin{aligned} E\Big(\sum_i B_{ni}^2\Big)^p &\leqslant K_1 E\Big|\sum_i B_{ni}\Big|^{2p} \\ &\leqslant K_2\Big\{E\Big[\Big(\sum_i E(B_{ni}^2|\mathscr{F}_{n,i-1})\Big)^p\Big] + E\Big(\max_i |B_{ni}|^{2p}\Big)\Big\} \\ &\leqslant K_2\Big\{E\Big[\Big(\sum_i E(X_{ni}^2|\mathscr{F}_{n,i-1})I(V_{ni}^2 > \lambda)\Big)^p\Big] + E\Big[\max_i 2^{2p}I(V_{ni}^2 > \lambda)\Big]\Big\} \\ &\leqslant K_2\{E[V_{nk_n}^{2p}I(V_{nk_n}^2 > \lambda)] + 2^{2p}P(V_{nk_n}^2 > \lambda)\}. \end{aligned}$$

We deduce from (3.38) that

$$\sup_n E\Big(\sum_i B_{ni}^2\Big)^p \to 0 \quad \text{and} \quad \sup_n E\Big|\sum_i B_{ni}\Big|^{2p} \to 0 \quad \text{as} \quad \lambda \to \infty. \tag{3.48}$$

Also,

$$E\Big|\sum_i Z_{ni}\Big|^2 = E\Big(\sum_i Z_{ni}^2\Big) \leqslant \sum_i E[X_{ni}^2 I(|X_{ni}| > 1)] \to 0 \tag{3.49}$$

under either (3.39) or (3.41), since each condition implies (3.36).

If (3.39) holds, then (3.37) follows right away, and this is the first part of (3.41). Condition (3.39) implies that $\{U_{nk_n}^{2p}\}$ is uniformly integrable, and since

$$\Big(\sum_i Z_{ni}^2\Big)^p \leqslant 2^{4p-2}\Big(U_n^{2p} + \Big(\sum_i A_{ni}^2\Big)^p + \Big(\sum_i B_{ni}^2\Big)^p\Big),$$

the uniform integrability of $\{(\sum_i Z_{ni}^2)^p\}$ follows from (3.47) and (3.48). Condition (3.49) now implies that

$$E\Big(\sum_i Z_{ni}^2\Big)^p \to 0, \tag{3.50}$$

and in view of Theorem 2.10,

$$E\Big|\sum_i Z_{ni}\Big|^{2p} \to 0. \tag{3.51}$$

Choose λ to be a continuity point of η^2. Now,

$$\begin{aligned} E&\Big|\sum_i E(A_{ni}^2|\mathscr{F}_{n,i-1}) - \sum_i E(X_{ni}^2|\mathscr{F}_{n,i-1})I(V_{ni}^2 \leqslant \lambda)\Big| \\ &= E\Big|\sum_i E[Y_{ni}^2 - X_{ni}^2|\mathscr{F}_{n,i-1}]I(V_{ni}^2 \leqslant \lambda)\Big| \\ &\leqslant 2\sum_i E[X_{ni}^2 I(|X_{ni}| > 1)] \to 0, \end{aligned}$$

and since

$$\sum_i E(X_{ni}^2|\mathscr{F}_{n,i-1})I(V_{ni}^2 \leqslant \lambda) \xrightarrow{\text{p}} \eta_\lambda^2,$$

where $\eta_\lambda^2 = \eta^2 I(\eta^2 \leqslant \lambda) + \lambda I(\eta^2 > \lambda)$, it follows that

$$\sum_i E(A_{ni}^2|\mathscr{F}_{n,i-1}) \xrightarrow{\text{p}} \eta_\lambda^2.$$

Corollary 3.1 and the ensuing remarks imply that $\sum_i A_{ni} \xrightarrow{\text{d}} Z_\lambda$, where Z_λ has characteristic function $E\exp(-\frac{1}{2}\eta_\lambda^2 t^2)$. Since $\{|\sum_i A_{ni}|^{2p}\}$ is uniformly integrable,

$$E\Big|\sum_i A_{ni}\Big|^{2p} \to E|Z_\lambda|^{2p}. \tag{3.52}$$

But

$$\left|(E|S_{nk_n}|^{2p})^{1/2p} - \left(E\left|\sum_i A_{ni}\right|^{2p}\right)^{1/2p}\right| \leqslant \left\{2^{2p-1}\left(E\left|\sum_i B_{ni}\right|^{2p} + E\left|\sum_i Z_{ni}\right|^{2p}\right)\right\}^{1/2p},$$

and so in view of (3.48), (3.51), and (3.52),

$$\lim_{\lambda\to\infty}\limsup_{n\to\infty}|(E|S_{nk_n}|^{2p})^{1/2p} - (E|Z_\lambda|^{2p})^{1/2p}| = 0,$$

which establishes (3.41).

Finally, assume (3.41). Condition (3.35) follows at once from the first part of the theorem. The sequence $\{|S_{nk_n}|^{2p}\}$ is uniformly integrable, and since

$$\left|\sum_i Z_{ni}\right|^{2p} \leqslant 2^{4p-2}\left\{|S_{nk_n}|^{2p} + \left|\sum_i A_{ni}\right|^{2p} + \left|\sum_i B_{ni}\right|^{2p}\right\},$$

the uniform integrability of $\{|\sum_i Z_{ni}|^{2p}\}$ follows from (3.47) and (3.48). Condition (3.49) now implies (3.51), from which follows (3.50). But

$$U_{nk_n}^{2p} \leqslant 2^{2p-1}\left\{\left(\sum_i A_{ni}^2\right)^p + \left(\sum_i B_{ni}^2\right)^p + \left(\sum_i Z_{ni}^2\right)^p\right\},$$

and conditions (3.47), (3.48), and (3.50) imply the uniform integrability of $\{U_{nk_n}^{2p}\}$. Together with (3.35) this implies (3.39). ∎

Corollary 3.3 follows immediately from Theorem 3.5 and the remarks after the statement of that theorem.

3.5. Reverse Martingales and Martingale Tail Sums

The duality between the definitions of forward and reverse martingales suggests that forward martingale limit theorems should have reverse martingale duals. In the case of the central limit theorem the analog is perhaps best presented by considering infinite martingale arrays.

Let $\{S_{ni}, \mathscr{F}_{ni}, -\infty < i < \infty\}$ be a zero-mean, square-integrable martingale for each $n \geqslant 1$, and suppose that

$$\sup_{n,i} E(S_{ni}^2) < \infty. \tag{3.53}$$

Let $X_{ni} = S_{ni} - S_{n,i-1}$ denote the differences. The forward and reverse martingale convergence theorems (Theorems 2.5 and 2.6) imply that the respective limits

$$S_{n\infty} = \lim_{i\to\infty} S_{ni} \qquad \text{and} \qquad S_{n,-\infty} = \lim_{i\to-\infty} S_{ni}$$

exist (finite) a.s. Consequently, the doubly infinite series $\sum_{i=-\infty}^{\infty} X_{ni}$ converges a.s. to $S_{n\infty} - S_{n,-\infty}$. By subtracting $S_{n,-\infty}$ from each of the S_{ni} we may assume without loss of generality that $S_{n,-\infty} = 0$ a.s., so that

$$S_{n\infty} = \sum_{i=-\infty}^{\infty} X_{ni} \quad \text{a.s.}$$

Condition (3.53) implies that the convergence here is in L^2 as well as with probability 1, and that

$$U_{n\infty}^2 = \sum_{i=-\infty}^{\infty} X_{ni}^2 \qquad \text{and} \qquad V_{n\infty}^2 = \sum_{i=-\infty}^{\infty} E(X_{ni}^2 | \mathscr{F}_{n,i-1})$$

exist (finite) a.s. Our next result is a corollary to Theorems 3.2 and 3.3.

Theorem 3.6. *Let $\{S_{ni}, \mathscr{F}_{ni}, -\infty < i < \infty\}$ be a zero-mean, square-integrable martingale with differences X_{ni}, and let η^2 be an a.s. finite r.v. Suppose that* (3.53) *holds and that $S_{n,-\infty} = 0$ a.s. If*

$$\sup_i |X_{ni}| \xrightarrow{\text{p}} 0,$$

$$\sum_i X_{ni}^2 \xrightarrow{\text{p}} \eta^2,$$

$$E\left(\sup_i X_{ni}^2\right) \qquad \text{is bounded in } n,$$

and

$$\text{for all } n \text{ and } i, \qquad \mathscr{F}_{ni} \subseteq \mathscr{F}_{n+1,i} \tag{3.54}$$

then $S_{n\infty} \xrightarrow{\text{d}} Z$ as $n \to \infty$, where the r.v. Z has characteristic function $E\exp(-\frac{1}{2}\eta^2 t^2)$. If $\eta^2 > 0$ a.s., then

$$S_{n\infty}/U_{n\infty} \xrightarrow{\text{d}} N(0,1).$$

As in the case of Theorems 3.2 and 3.3, condition (3.54) may be dropped if η^2 is constant a.s.

Proof. In view of (3.53) we can choose sequences of positive integers $\{a_n\}$ and $\{b_n\}$ increasing to ∞ such that

$$E\left(\sum_{i \leqslant -a_n} X_{ni}^2 + \sum_{i > b_n} X_{ni}^2\right) \to 0 \tag{3.55}$$

as $n \to \infty$. Let $k_n = a_n + b_n$ and consider the martingale array $\{T_{ni}, \mathscr{G}_{ni}, 1 \leqslant i \leqslant k_n, n \geqslant 1\}$ defined by

$$T_{ni} = \sum_{j=1-a_n}^{i-a_n} X_{nj} \qquad \text{and} \qquad \mathscr{G}_{ni} = \mathscr{F}_{n,i-a_n}, \qquad 1 \leqslant i \leqslant k_n.$$

The array satisfies the conditions of Theorem 3.2 and so

$$T_{nk_n} = S_{n,b_n} - S_{n,-a_n} \xrightarrow{\text{d}} Z.$$

Condition (3.55) implies that

$$S_{n\infty} - T_{nk_n} = \sum_{i \leqslant -a_n} X_{ni} + \sum_{i > b_n} X_{ni} \xrightarrow{\text{p}} 0,$$

and so $S_{n\infty} \xrightarrow{\text{d}} Z$. If $P(\eta^2 > 0) = 1$, then the array satisfies the conditions of Theorem 3.3 and so

$$T_{nk_n} \Bigg/ \left(\sum_{i=1-a_n}^{b_n} X_{ni}^2 \right)^{1/2} \xrightarrow{\text{d}} N(0,1).$$

In view of (3.55) this is sufficient to imply that

$$S_{n\infty}/U_{n\infty} \xrightarrow{\text{d}} N(0,1). \quad \blacksquare$$

Corollary 3.4. *Let $\{S_n, \mathscr{F}_n, n \geqslant 1\}$ be a zero-mean, square-integrable reverse martingale and let $s_n^2 = E(S_n^2)$ and $X_n = S_n - S_{n+1}$, $n \geqslant 1$. As before we may suppose without loss of generality that*

$$S_\infty = \lim_{n\to\infty} S_n = 0 \quad a.s.$$

If

$$E\left(s_n^{-2} \sup_{i \geqslant n} X_i^2 \right) \to 0 \qquad \text{and} \qquad s_n^{-2} \sum_{i=n}^{\infty} X_i^2 \xrightarrow{\text{p}} \eta^2,$$

then $s_n^{-1} S_n \xrightarrow{\text{d}} Z$, where the r.v. Z has characteristic function $E\exp(-\frac{1}{2}\eta^2 t^2)$. If $P(\eta^2 > 0) = 1$ then

$$S_n \Bigg/ \left(\sum_{i=n}^{\infty} X_i^2 \right)^{1/2} \xrightarrow{\text{d}} N(0,1).$$

[Cf. Loynes (1969) and Eagleson and Weber (1978, Theorem 1).]

Proof. Consider the doubly infinite martingale array $\{T_{ni}, \mathscr{G}_{ni}, -\infty < i < \infty, n \geqslant 1\}$ defined by

$$T_{ni} = \begin{cases} s_n^{-1} S_{-i} & \text{and} \quad \mathscr{G}_{ni} = \mathscr{F}_{-i} \quad \text{if } i \leqslant -n \\ s_n^{-1} S_{-n} & \text{and} \quad \mathscr{G}_{ni} = \mathscr{F}_{-n} \quad \text{if } i > -n. \end{cases}$$

Corollary 3.4 follows immediately from an application of Theorem 3.6. $\blacksquare$

Other reverse martingale limit theorems (such as invariance principles in the CLT) may be obtained from forward martingale results using techniques similar to the above. Scott (1979) has shown that if $\{S_n, \mathscr{F}_n, n \geqslant 1\}$ is a square-integrable reverse martingale with $S_n \xrightarrow{\text{a.s.}} 0$, then $\{S_1 - S_n, \mathscr{F}_n^*, n \geqslant 1\}$ is a forward martingale, where $\mathscr{F}_n^*$ is the σ-field generated by $S_1 - S_i$, $1 \leqslant i \leqslant n$. This makes it almost a trivial matter to obtain a law of the iterated logarithm and a form of the Skorokhod representation for reverse martingales.

Another application of Theorem 3.6 concerns the situation of martingale tail sums. Suppose that $\{S_n = \sum_{i=1}^n X_i, \mathscr{F}_n, n \geqslant 1\}$ is a zero-mean, square-integrable martingale with

$$\lim_{n\to\infty} E\left(\sum_{i=1}^{n} X_i\right)^2 = \sum_{i=1}^{\infty} EX_i^2 < \infty.$$

Then the martingale convergence theorem gives $\sum_{i=1}^n X_i \xrightarrow{\text{a.s.}} \sum_{i=1}^\infty X_i$, and in seeking rate results for this convergence we are led to ask for a CLT for $\sum_{i=n+1}^\infty X_i$, when appropriately normalized. Such a result is given in the following corollary.

Corollary 3.5. *Let $\{S_n = \sum_{i=1}^n X_i, \mathscr{F}_n, n \geqslant 1\}$ be a zero-mean, square-integrable martingale and suppose that $\sum_{n=1}^\infty EX_n^2 < \infty$. Write $s_n^2 = \sum_{i=n}^\infty EX_i^2$. If*

$$s_n^{-1} \sup_{i\geqslant n}|X_i| \xrightarrow{\text{p}} 0, \qquad E\left(s_n^{-2} \sup_{i\geqslant n} X_i^2\right) \quad \textit{is bounded in } n,$$

and

$$s_n^{-2} \sum_{i=n}^{\infty} X_i^2 \xrightarrow{\text{p}} \eta^2,$$

then

$$s_n^{-1} \sum_{i=n}^{\infty} X_i \xrightarrow{\text{d}} Z,$$

where the r.v. Z has characteristic function $E\exp(-\frac{1}{2}\eta^2 t^2)$. If $P(\eta^2 > 0) = 1$, then

$$\left(\sum_{i=n}^{\infty} X_i^2\right)^{-1/2} \sum_{i=n}^{\infty} X_i \xrightarrow{\text{d}} N(0,1).$$

Proof. The result follows immediately from Theorem 3.6 upon taking

$$X_{ni} = \begin{cases} s_n^{-1} X_{n+i}, & \mathscr{F}_{ni} = \mathscr{F}_{n+i}, \quad i \geqslant 1, \\ 0, & \mathscr{F}_{ni} = \{\phi,\Omega\}, \quad i \leqslant 0. \end{cases} \qquad \blacksquare$$

Remark. The duality between the results for tail sums in Corollary 3.5 and corresponding results for ordinary sums (e.g., from Theorem 3.2) should be noted. A corresponding duality also holds for the law of the iterated logarithm, as does an invariance principle form for Corollary 3.5 (Heyde, 1977a).

In order to give an idea of the significance of tail sum limit results and the scope of Corollary 3.5 we give an example dealing with the well-known Polya urn scheme.

Suppose we have an urn initially containing b black and r red balls. At each draw we remove a ball at random and then replace it, together with c balls of the colour drawn. Let b_n and r_n denote, respectively, the numbers of black and red balls in the urn after the nth drawing and write $Y_n = b_n/(b_n + r_n)$, $n \geqslant 0$, the proportion of black balls. Note also that $b_n + r_n = b + r + nc$.

Now it is well known, and easily checked, that $\{Y_n, n \geqslant 0\}$ is a martingale with respect to the past history σ-fields, and the martingale convergence theorem is applicable since $0 < Y_n < 1$. Indeed, it is also well known that the limit random variable Y, say, has a beta distribution with parameters b/c and r/c (see Section 7.1). The CLT for martingale tail sums gives us a result on the rate of convergence of Y_n to Y.

Write $K_n = c/[b + r + (n + 1)c]$ and let $\mathscr{F}_n$ denote the σ-field generated by Y_j, $0 \leqslant j \leqslant n$. It is easy to see that, conditional on Y_n, the distribution of the difference $X_{n+1} = Y_{n+1} - Y_n$ is given by

$$X_{n+1} = \begin{cases} (1 - Y_n)K_n & \text{with probability } \; Y_n \\ -Y_nK_n & \text{with probability } \; 1 - Y_n. \end{cases}$$

Then

$$E(X_{n+1}^2|\mathscr{F}_n) = Y_n(1 - Y_n)K_n^2 \sim Y(1 - Y)K_n^2 \quad \text{as} \quad n \to \infty, \tag{3.56}$$

while

$$\begin{aligned} EX_{n+1}^2 &= K_n^2 EY_n(1 - Y_n) \sim K_n^2 EY(1 - Y) \\ &= K_n^2 br(b + r)^{-1}(b + r + c)^{-1} \end{aligned} \tag{3.57}$$

as $n \to \infty$, using dominated convergence, since $Y_n(1 - Y_n) \leqslant 1/4$. Furthermore,

$$E(X_{n+1}^4|\mathscr{F}_n) = Y_n(1 - Y_n)K_n^4\{(1 - Y_n)^3 + Y_n^3\} \leqslant \tfrac{1}{2}K_n^4. \tag{3.58}$$

Next, using (3.57),

$$s_n^2 = \sum_{j=n}^{\infty} EX_j^2 = \sum_{j=n-1}^{\infty} K_j^2 EY_j(1 - Y_j) \sim n^{-1}EY(1 - Y) \tag{3.59}$$

as $n \to \infty$, and

$$\sum_{1}^{\infty} s_n^{-2}\{X_n^2 - E(X_n^2|\mathscr{F}_{n-1})\} \qquad \text{converges a.s.}$$

using Theorem 2.15, since

$$\sum_1^\infty s_n^{-4} E X_n^4 < \infty$$

by virtue of (3.58) and (3.59). Therefore

$$s_n^{-2} \sum_{k=n}^{\infty} \{X_k^2 - E(X_k^2|\mathcal{F}_{k-1})\} \xrightarrow{\text{a.s.}} 0$$

using Abel's lemma, and hence, from (3.56) and (3.59),

$$s_n^{-2} \sum_{k=n}^{\infty} X_k^2 \xrightarrow{\text{a.s.}} Y(1-Y)/EY(1-Y).$$

Corollary 3.5 is clearly applicable and gives, in particular,

$$n^{1/2}(Y_n - Y) \xrightarrow{\text{d}} [Y'(1-Y')/EY'(1-Y')]^{1/2} N(0,1)$$

where $Y'(1-Y')$ is distributed as $Y(1-Y)$ and is independent of the $N(0,1)$ random variable.

It worth remarking that a law of the iterated logarithm for tail sums of martingale differences can also be applied to $Y_n - Y$, to yield (Heyde, 1977a)

$$\limsup_{n\to\infty} n^{1/2}(2\log\log n)^{-1/2}(Y_n - Y) = [Y(1-Y)]^{1/2}[EY(1-Y)]^{-1/2} \quad \text{a.s.},$$

$$\liminf_{n\to\infty} n^{1/2}(2\log\log n)^{-1/2}(Y_n - Y) = -[Y(1-Y)]^{1/2}[EY(1-Y)]^{-1/2} \quad \text{a.s.}$$

3.6. Rates of Convergence in the CLT

Let $X_1, X_2, \ldots$ be independent r.v. with zero means, unit variances, and bounded third moments: $\sup_{1 \leqslant n < \infty} E|X_n|^3 \leqslant \rho$. Then for all n,

$$\sup_{-\infty < x < \infty} \left| P\left(n^{-1/2}\sum_1^n X_i \leqslant x\right) - \Phi(x)\right| \leqslant Cn^{-1/2}\rho,$$

where Φ denotes the standard normal distribution function and C is an absolute constant. This is one version of the celebrated Berry–Esseen theorem, discovered by Berry (1941) and Esseen (1942). It provides a rate of convergence in the CLT which is generally the best possible, in the sense that for certain distributions satisfying the above conditions and certain constants $C' > 0$, we have

$$\sup_{-\infty < x < \infty} \left| P\left(n^{-1/2}\sum_1^n X_i \leqslant x\right) - \Phi(x)\right| \geqslant C'n^{-1/2}\rho.$$

For example, if the X_i are i.i.d. with $P(X_i = -1) = P(X_i = +1) = \frac{1}{2}$, then

$$\left|P\left(\sum_1^n X_i < 0\right) - \frac{1}{2}\right| = \frac{1}{2} P\left(\sum_1^n X_i = 0\right) = \binom{2n}{n} 2^{-(2n+1)} \sim \frac{1}{2}(\pi n)^{-1/2}$$

as $n \to \infty$.

The Berry–Esseen theorem and its extensions are of great significance in probability and statistics. There is a very extensive literature relating to rates of convergence in the CLT for sums of independent r.v. Comprehensive accounts are given in, for example, Gnedenko and Kolmogorov (1954), Ibragimov and Linnik (1971), and Petrov (1975). The results provide a neat and accurate estimate of the error term in the statistician's normal approximations, and the rate of convergence of order $n^{-1/2}$ is often fast enough to justify his testing procedures.

Let $\{S_{ni}, \mathscr{F}_{ni}\}$ be a zero-mean, square-integrable martingale array. According to Corollary 3.1 and the remarks immediately following it, if

$$V_{nk_n}^2 = \sum_i E(X_{ni}^2 | \mathscr{F}_{n,i-1}) \xrightarrow{\text{p}} 1$$

and if the Lindeberg condition (3.7) holds, then $S_{nk_n} = \sum_i X_{ni} \xrightarrow{\text{d}} N(0,1)$. It turns out that the rate of convergence in this CLT is partly determined by the rate of convergence of $V_{nk_n}^2$ to 1, and is generally not better than $(E|V_{nk_n}^2 - 1|)^{1/2}$. This fact is borne out by the following example, which shows that even when a martingale array is only slightly different from a sequence of sums of i.i.d. normal variables, the rate of convergence in the CLT may be arbitrarily slow.

Example 5. Let $1 \geqslant \delta_n \downarrow 0$ be a sequence of constants and set $m_n = [n(1-\delta_n)]$ (the integer part of $n(1-\delta_n)$). Let Z_n, $n \geqslant 1$, be independent $N(0,1)$ variables and define

$$I_n = I\left(\sum_1^{m_n} Z_i > 0\right)$$

and

$$X_{ni} = \begin{cases} n^{-1/2} Z_i & \text{if } \ 1 \leqslant i \leqslant m_n \\ 2^{1/2} n^{-1/2} I_n Z_i & \text{if } \ m_n < i \leqslant n. \end{cases}$$

Let $\mathscr{F}_{ni}$ be the σ-field generated by $X_{n1}, \ldots, X_{ni}$ and set

$$S_{ni} = \sum_1^i X_{nj}, \qquad 1 \leqslant i \leqslant n.$$

Then $\{S_{ni}, \mathscr{F}_{ni}\}$ is a zero-mean, square-integrable martingale array. Each $E(X_{ni}^2) = n^{-1}$, $E|X_{ni}|^3 \leqslant 8\pi^{-1/2}n^{-3/2}$, and

$$V_{nn}^2 = \sum_1^n E(X_{ni}^2|\mathscr{F}_{n,i-1}) = n^{-1}m_n + 2n^{-1}(n - m_n)I_n \xrightarrow{\text{p}} 1,$$

since $n^{-1}(n - m_n) \to 0$. The Lindeberg condition (3.7) is easily verified, and so by Corollary 3.1, $S_{nn} \xrightarrow{\text{d}} N(0,1)$. We shall show that the rate of convergence is no better than the order of

$$(E|V_{nn}^2 - 1|)^{1/2} = \delta_n^{1/2} + O(\delta_n).$$

We can write

$$S_{nn} = n^{-1/2}m_n^{1/2}N_1 + 2^{1/2}n^{-1/2}(n - m_n)^{1/2}I(N_1 > 0)N_2,$$

where N_1 and N_2 are independent $N(0,1)$ variables, depending on n. Now, for any absolutely continuous r.v. X with bounded density,

$$P(X \leqslant x + \varepsilon) = P(X \leqslant x) + O(\varepsilon)$$

uniformly in x as $\varepsilon \to 0$, and since $n^{-1/2}m_n^{1/2} = 1 + O(\delta_n)$ and $n^{-1/2}(n - m_n)^{1/2} = \delta_n^{1/2} + O(\delta_n)$,

$$P(S_{nn} \leqslant x) = P(N_1 + 2^{1/2}\delta_n^{1/2}I(N_1 > 0)N_2 \leqslant x) + O(\delta_n).$$

Set $x = 0$. Since

$$P(N_1 + 2^{1/2}\delta_n^{1/2}I(N_1 > 0)N_2 \leqslant 0) = P(N_1 \leqslant 0) + P(0 < N_1 \leqslant -2^{1/2}\delta_n^{1/2}N_2),$$

it follows that

$$\begin{aligned}\sup_{-\infty<x<\infty} |P(S_{nn} \leqslant x) - \Phi(x)| &\geqslant P(0 < N_1 \leqslant -2^{1/2}\delta_n^{1/2}N_2) + O(\delta_n) \\ &= \tfrac{1}{2}P(0 < N_1 \leqslant 2^{1/2}\delta_n^{1/2}|N_2|) + O(\delta_n) \\ &= \tfrac{1}{2}\pi^{-1/2}\delta_n^{1/2}E|N_2| + O(\delta_n) \\ &= 2^{-1/2}\pi^{-1}\delta_n^{1/2} + O(\delta_n).\end{aligned}$$

(Here we have used the fact that

$$P(0 < N_1 < \varepsilon) = (2\pi)^{-1/2}\varepsilon + O(\varepsilon^2)$$

as $\varepsilon \downarrow 0$.) That is,

$$\sup_{-\infty<x<\infty} |P(S_{nn} \leqslant x) - \Phi(x)| \geqslant 2^{-1/2}\pi^{-1}(E|V_{nn}^2 - 1|)^{1/2} + O(E|V_{nn}^2 - 1|). \quad \blacksquare$$

Let $\{X_n, n \geqslant 1\}$ be a zero-mean martingale difference sequence with each $|X_n| \leqslant M < \infty$, and define

$$S_n = \sum_1^n X_i \qquad \text{and} \qquad v(n) = \inf\left\{j \,\middle|\, \sum_1^j E(X_i^2|X_1, \ldots, X_{i-1}) \geqslant n\right\}.$$

If $\sum_1^\infty E(X_i^2|X_1, \ldots, X_{i-1}) = \infty$ a.s., then

$$\sup_{-\infty<x<\infty} |P(n^{-1/2}S_{\nu(n)} \leqslant x) - \Phi(x)| = O(n^{-1/4})$$

as $n \to \infty$ [Ibragimov (1963); see also Nakata (1976) and Kato (1978)]. If we ask that each $E(X_i^2|X_1, \ldots, X_{i-1}) = 1$ a.s. (this condition holds trivially if the X_i are independent with unit variances), then $\nu(n) = n$ a.s. and so

$$\sup_{-\infty<x<\infty} |P(n^{-1/2}S_n \leqslant x) - \Phi(x)| = O(n^{-1/4}).$$

The restriction that the conditional variances be a.s. constant is far too stringent for most applications. However, as we saw in Example 5, a fast rate of convergence can only be achieved by imposing some type of restriction on the conditional variances. Our next theorem establishes a rate of convergence of almost $n^{-1/4}$ under more realistic conditions.

Theorem 3.7. *Let $\{S_i = \sum_1^i X_j, \mathscr{F}_i, 1 \leqslant i \leqslant n\}$ be a zero-mean martingale with $\mathscr{F}_i$ equal to the σ-field generated by $X_1, \ldots, X_i$. Let*

$$V_i^2 = \sum_1^i E(X_j^2|\mathscr{F}_{j-1}), \qquad 1 \leqslant i \leqslant n,$$

and suppose that

$$\max_{i\leqslant n}|X_i| \leqslant n^{-1/2}M \quad \text{a.s.} \tag{3.60}$$

and

$$P(|V_n^2 - 1| > 9M^2Dn^{-1/2}(\log n)^2) \leqslant Cn^{-1/4}\log n \tag{3.61}$$

for constants M, C, and D ($\geqslant e$). Then for $n \geqslant 2$,

$$\sup_{-\infty<x<\infty} |P(S_n \leqslant x) - \Phi(x)| \leqslant (2 + C + 7MD^{1/2})n^{-1/4}\log n. \tag{3.62}$$

Remarks. (i) Let Z_n, $n \geqslant 1$, be independent r.v. each with the symmetric distribution on ± 1. Let $\delta_n = n^{-1/2}(\log n)^2$ and define the martingale array $\{S_{ni}, \mathscr{F}_{ni}\}$ as in Example 5. Arguing as in that example we can show that there exists a constant $C > 0$ such that

$$\sup_{-\infty<x<\infty} |P(S_n \leqslant x) - \Phi(x)| \geqslant Cn^{-1/4}\log n + O(n^{-1/2}(\log n)^2).$$

Now set $X_i = X_{ni}$, $1 \leqslant i \leqslant n$, and $V_n^2 = V_{nn}^2$. Conditions (3.60) and (3.61) are satisfied with $M = 2^{1/2}$, $D = e$, and $C = 0$. Therefore the rate of convergence given in (3.62) is, in this case, the best possible.

(ii) Theorem 3.7 continues to hold for *any* sequence of σ-fields $\mathscr{F}_i$ such that $\{S_i, \mathscr{F}_i\}$ is a martingale, although with slightly altered constants. Observe that if $\{S_i, \mathscr{F}_i\}$ is a martingale, then so is $\{S_i, \mathscr{F}_i^*\}$, where $\mathscr{F}_i^*$ is the σ-field generated by $X_1, X_2, \ldots, X_i$. Carry out the proof in the same manner as below, obtaining the bound (3.66) with V_n^2 replaced by

$$V_n^{*^2} = \sum_1^n E(X_i^2 | \mathscr{F}_{i-1}^*).$$

Next observe that

$$P(|T_n - V_n^{*^2}| > \Delta) \leqslant P(|T_n - V_n^2| > \tfrac{1}{2}\Delta) + P(|V_n^2 - V_n^{*^2}| > \tfrac{1}{2}\Delta)$$

and

$$P(|V_n^{*^2} - 1| > \Delta) \leqslant P(|V_n^2 - 1| > \tfrac{1}{2}\Delta) + P(|V_n^2 - V_n^{*^2}| > \tfrac{1}{2}\Delta).$$

Since $\{V_i^2 - V_i^{*2}, \mathscr{F}_i^*\}$ is a martingale, inequalities like (3.68) and (3.69) give us the bound

$$P(|V_n^2 - V_n^{*^2}| > \tfrac{1}{2}\Delta) \leqslant A(Bp^2)^p p^{1/2} n^{p/2} \Delta^{-p} \max_{i \leqslant n} E|X_i|^{2p}$$

for absolute constants A and B. The proof is now completed as below.

Proof of Theorem 3.7. We begin with a lemma whose proof is due to Heyde and Brown (1970).

Lemma 3.2. *Let $W(t)$, $t \geqslant 0$, be a standard Brownian motion and let T be a nonnegative r.v. Then for all real x and all $\varepsilon > 0$,*

$$|P(W(T) \leqslant x) - \Phi(x)| \leqslant (2\varepsilon)^{1/2} + P(|T - 1| > \varepsilon).$$

Proof. First we note that if $0 < \varepsilon < \frac{1}{2}$,

$$P(W(T) \leqslant x) \leqslant P(W(T) \leqslant x;\, |T - 1| \leqslant \varepsilon) + P(|T - 1| > \varepsilon),$$

and

$$\begin{aligned} P(W(T) \leqslant x;\, |T - 1| \leqslant \varepsilon) &\leqslant P\Big(\inf_{|t-1| \leqslant \varepsilon} W(t) \leqslant x\Big) \\ &= P(W(1 - \varepsilon) + \inf_{|t-1| \leqslant \varepsilon} [W(t) - W(1 - \varepsilon)] \leqslant x) \\ &= \int_{-\infty}^0 P(W(1 - \varepsilon) \leqslant x - y)\, d\, P\Big(\inf_{0 \leqslant t \leqslant 2\varepsilon} W(t) \leqslant y\Big) \\ &= (\pi\varepsilon)^{-1/2} \int_0^\infty \Phi((1 - \varepsilon)^{-1/2}(x + y)) \exp(-y^2/4\varepsilon)\, dy \\ &= \pi^{-1/2} \int_0^\infty \Phi((1 - \varepsilon)^{-1/2}(x + \varepsilon^{1/2} z)) e^{-z^2/4}\, dz. \end{aligned}$$

Consequently,

$$\begin{aligned} P(W(T) \leqslant x) - \Phi(x) \leqslant \pi^{-1/2} \int_0^\infty \Phi((1-\varepsilon)^{-1/2}(x+\varepsilon^{1/2}z))e^{-z^2/4}\,dz \\ -\Phi(x) + P(|T-1| > \varepsilon) \\ \leqslant \pi^{-1/2} \int_0^\infty |\Phi((1-\varepsilon)^{-1/2}(x+\varepsilon^{1/2}z)) \\ - \Phi(x)|e^{-z^2/4}\,dz + P(|T-1| > \varepsilon). \end{aligned} \tag{3.63}$$

The term within modulus signs in the integrand is bounded by

$$|\Phi((1-\varepsilon)^{-1/2}(x+\varepsilon^{1/2}z)) - \Phi((1-\varepsilon)^{-1/2}x)| + |\Phi((1-\varepsilon)^{-1/2}x) - \Phi(x)|.$$

The first term here is not greater than $(2\pi)^{-1/2}(1-\varepsilon)^{-1/2}\varepsilon^{1/2}z$, and for $x \geqslant 0$ the second term is not greater than

$$(2\pi)^{-1/2}e^{-x^2/2}x[(1-\varepsilon)^{-1/2}-1] \leqslant (2\pi)^{-1/2}e^{-1/2}[(1-\varepsilon)^{-1/2}-1] < \pi^{-1/2}e^{-1/2}\varepsilon,$$

provided that $0 < \varepsilon < \frac{1}{2}$. By symmetry the same bound applies when $x < 0$, and combining these estimates we deduce that for $0 < \varepsilon < \frac{1}{2}$,

$$\begin{aligned} P(W(T) \leqslant x) - \Phi(x) - P(|T-1| > \varepsilon) \\ \leqslant \pi^{-1/2}e^{-1/2}\varepsilon + \pi^{-1/2}\varepsilon^{1/2} \int_0^\infty ze^{-z^2/4}\,dz \\ \leqslant \varepsilon^{1/2}((2\pi e)^{-1/2} + 2\pi^{-1/2}) \\ < (2\varepsilon)^{1/2}. \end{aligned}$$

That is,

$$P(W(T) \leqslant x) - \Phi(x) \leqslant (2\varepsilon)^{1/2} + P(|T-1| > \varepsilon). \tag{3.64}$$

This bound holds trivially if $\varepsilon \geqslant \frac{1}{2}$. Further, using a similar procedure we deduce that

$$\begin{aligned} P(W(T) \leqslant x) \geqslant P(W(T) \leqslant x;\, |T-1| \leqslant \varepsilon) - P(|T-1| > \varepsilon) \\ \geqslant P\Big(\sup_{|t-1|<\varepsilon} W(t) \leqslant x\Big) - P(|T-1| > \varepsilon) \\ = \pi^{-1/2} \int_0^\infty \Phi((1-\varepsilon)^{-1/2}(x-\varepsilon^{1/2}z))e^{-z^2/4}\,dz - P(|T-1| > \varepsilon) \\ \geqslant \Phi(x) - (2\varepsilon)^{1/2} - P(|T-1| > \varepsilon). \end{aligned} \tag{3.65}$$

Lemma 3.2 follows on combining (3.64) and (3.65). ∎

Returning to the proof of Theorem 3.7, we deduce from the Skorokhod representation (Theorem A.1 of Appendix I) that there exists a standard

Brownian motion W and nonnegative r.v. T_i, $1 \leqslant i \leqslant n$, such that (without loss of generality)

$$S_i = W(T_i), \qquad 1 \leqslant i \leqslant n.$$

Lemma 3.2 now asserts that for all n, x, and $\Delta > 0$,

$$|P(S_n \leqslant x) - \Phi(x)| \leqslant 2\Delta^{1/2} + P(|T_n - V_n^2| > \Delta) + P(|V_n^2 - 1| > \Delta). \quad (3.66)$$

Let $\tau_1 = T_1$ and $\tau_i = T_i - T_{i-1}$, $2 \leqslant i \leqslant n$. Theorem A.1 implies that each τ_i is $\mathscr{G}_i$-measurable and $E(\tau_i|\mathscr{G}_{i-1}) = E(X_i^2|\mathscr{F}_{i-1})$ a.s., where $\mathscr{G}_i$ is the σ-field generated by $S_1, \ldots, S_i$ and $W(t)$ for $t \leqslant T_i$. Therefore

$$T_n - V_n^2 = \sum_1^n (\tau_i - E(\tau_i|\mathscr{G}_{i-1}))$$

is a sum of martingale differences.

For any martingale with differences Z_i, $1 \leqslant i \leqslant n$, and any $p \geqslant 2$ we have from Hölder's and Burkholder's inequalities (see Theorem 2.10)

$$\begin{aligned} E\left|\sum_1^n Z_i\right|^p &\leqslant (18pq^{1/2})^p E\left|\sum_1^n Z_i^2\right|^{p/2} \leqslant (18pq^{1/2})^p n^{p/2-1} \sum_1^n E|Z_i|^p \\ &\leqslant (18pq^{1/2})^p n^{p/2} \max_{i \leqslant n} E|Z_i|^p, \end{aligned} \quad (3.67)$$

where $q = (1 - p^{-1})^{-1} \leqslant 2$ for $p \geqslant 2$. Applying these inequalities to the martingale with differences $Z_i = \tau_i - E(\tau_i|\mathscr{G}_{i-1})$ we deduce that

$$\begin{aligned} P(|T_n - V_n^2| > \Delta) &\leqslant \Delta^{-p} E\left|\sum_1^n Z_i\right|^p \\ &\leqslant \Delta^{-p}(18p2^{1/2})^p n^{p/2} \max_{i \leqslant n} E|Z_i|^p. \end{aligned} \quad (3.68)$$

Now, $|Z_i| \leqslant \max(\tau_i, E(\tau_i|\mathscr{G}_{i-1}))$ and $E[E(\tau_i|\mathscr{G}_{i-1})^p] \leqslant E(\tau_i^p)$, by Jensen's inequality. Hence by Theorem A.1,

$$\begin{aligned} E|Z_i|^p &\leqslant E[\tau_i^p + E(\tau_i|\mathscr{G}_{i-1})^p] \\ &\leqslant 2E(\tau_i^p) \leqslant 4\Gamma(p+1)E|X_i|^{2p}. \end{aligned} \quad (3.69)$$

Stirling's expansion of $\Gamma(p+1)$ implies that for $p \geqslant 2$,

$$\Gamma(p+1) \leqslant (2\pi)^{1/2} p^{p+1/2} e^{-p+1/24},$$

and combining this with both (3.68) and (3.69) we see that

$$P(|T_n - V_n^2| > \Delta) \leqslant 10.5(9.4p^2)^p p^{1/2} n^{p/2} \Delta^{-p} \max_{i \leqslant n} E|X_i|^{2p}. \quad (3.70)$$

From condition (3.60) we have

$$\max_{i \leqslant n} E|X_i|^{2p} \leqslant n^{-p} M^{2p},$$

and this together with (3.66) and (3.70) implies that for all $\Delta > 0$,

$$|P(S_n \leqslant x) - \Phi(x)| \leqslant 2\Delta^{1/2} + 10.5(9.4p^2M^2)^p p^{1/2} n^{-p/2}\Delta^{-p} + P(|V_n^2 - 1| > \Delta).$$

We now choose $\Delta = \Delta(n) \to 0$ and $p = p(n) \to \infty$ to minimize the sum of the first two terms above. Let

$$\Delta = 9.4M^2Dn^{-1/2}(\log n)^2 \qquad \text{and} \qquad p = \log n.$$

If $n > e^2$ then $p > 2$, and

$$\begin{aligned} &2\Delta^{1/2} + 10.5(9.4p^2M^2)^p p^{1/2} n^{-p/2}\Delta^{-p} \\ &\quad \leqslant 6.2MD^{1/2}n^{-1/4}\log n + 10.5D^{-\log n}(\log n)^{1/2} \\ &\quad \leqslant (7MD^{1/2} + 2)n^{-1/4}\log n, \end{aligned}$$

since we assumed that $D \geqslant e$. Consequently,

$$|P(S_n \leqslant x) - \Phi(x)| \leqslant (7MD^{1/2} + 2)n^{-1/4}\log n + P(|V_n^2 - 1| > \Delta),$$

and combined with (3.61) this implies (3.62). The bound in (3.62) applies trivially if $2 \leqslant n < e^2$. ∎

The uniform boundedness condition (3.60) will be too restrictive for some applications, but it can be replaced by a bound on the exponential moments.

Theorem 3.8. *In the notation of Theorem* 3.7, *suppose that for some* $\alpha > 0$ *and constants* M, C, *and* D,

$$\max_{i \leqslant n} E[\exp(|n^{1/2}X_i|^\alpha)] < M \tag{3.71}$$

and

$$P(|V_n^2 - 1| > Dn^{-1/2}(\log n)^{2+2/\alpha}) \leqslant Cn^{-1/4}(\log n)^{1+1/\alpha}. \tag{3.72}$$

Then for $n \geqslant 2$,

$$\sup_{-\infty < x < \infty} |P(S_n \leqslant x) - \Phi(x)| \leqslant An^{-1/4}(\log n)^{1+1/\alpha}, \tag{3.73}$$

where the constant A *depends only on* α, M, C, *and* D.

Remark. As in the case of Theorem 3.7, the result continues to hold for *any* sequence of σ-fields $\mathscr{F}_i$ such that $\{S_i, \mathscr{F}_i\}$ is a martingale.

Proof. Note first that

$$\sup_{x>0} x^p e^{-x} = p^p e^{-p},$$

and so for all $x > 0$,

$$x^p \leqslant p^p e^{-p} e^x.$$

Consequently,

$$n^p E|X_i|^{2p} = E[(|n^{1/2}X_i|^\alpha)^{2p/\alpha}] \leqslant (2p/\alpha)^{2p/\alpha} e^{-2p/\alpha} M,$$

using (3.71). Combining this with (3.70), we deduce that for positive constants a and b that do not depend on n, p, or Δ and may be taken arbitrarily large,

$$P(|T_n - V_n^2| > \Delta) \leqslant a(bp^{\alpha+1})^{2p/\alpha} p^{1/2} n^{-p/2} \Delta^{-p}.$$

From this and (3.66) we obtain

$$|P(S_n \leqslant x) - \Phi(x)| \leqslant 2\Delta^{1/2} + a(bp^{\alpha+1})^{2p/\alpha} p^{1/2} n^{-p/2} \Delta^{-p} + P(|V_n^2 - 1| > \Delta). \tag{3.74}$$

Again we choose Δ and p to minimize the sum of the first two terms. This time the optimal choice is

$$\Delta = b^{1+2/\alpha} n^{-1/2} (\log n)^{2+2/\alpha} \qquad \text{and} \qquad p = \log n.$$

Then

$$\begin{aligned}\Delta^{1/2} + (bp^{\alpha+1})^{2p/\alpha} p^{1/2} n^{-p/2} \Delta^{-p} &= b^{1/2+1/\alpha} n^{-1/4} (\log n)^{1+1/\alpha} + b^{-\log n} (\log n)^{1/2} \\ &\leqslant cn^{-1/4} (\log n)^{1+1/\alpha}\end{aligned}$$

for a constant c, if we choose $b \geqslant e$. Returning to (3.74) we find that

$$|P(S_n \leqslant x) - \Phi(x)| \leqslant dn^{-1/4} (\log n)^{1+1/\alpha} + P(|V_n^2 - 1| > \Delta)$$

for a constant d, and combined with (3.72) this implies (3.73). ∎

We now turn our attention to the more general form of the CLT in which the norming constant is $(\operatorname{var} S_n)^{1/2}$, not necessarily equal to $n^{1/2}$. Rates of convergence in this case have been obtained by Heyde and Brown (1970), Grams (1972), Basu (1976), and Erickson *et al.* (1978). Heyde and Brown's work is based on the Skorokhod representation, and we shall use similar techniques. The other authors rely on more classical methods and use the smoothing inequality for characteristic functions [see, e.g., Feller (1971, pp. 536–538)].

The results obtained in each case are very similar, and it is interesting to note that when they are specialized to an i.i.d. sequence they predict a rate of convergence of no better than $n^{-1/4}$. Indeed, the rate of $n^{-1/4}$ seems to be difficult to surpass for many types of dependent sequences. For example, Philipp (1969c) obtains a rate of $n^{-1/4}(\log n)^3$ for ϕ-mixing sequences, Landers and Rogge (1976) obtain a rate of $n^{-1/4}(\log n)^{1/4}$ for a class of Markov chains, and Sunklodas (1977) obtains a rate of $n^{-1/4}\log n$ for strong mixing sequences. The reason for the barrier at $n^{-1/4}$ may be related to the fact that many dependent sequences can be embedded in a Brownian motion, or at least approximated by embeddable sequences. Sawyer (1972) has proved a result which indicates that for many sequences of this type, the best rate of convergence that can be expected is $n^{-1/4}$.

The rates of convergence in Theorems 3.7 and 3.8 are of the uniform type, since they hold for all x. However, it is clear that if $|x|$ is very large, then the difference $|P(S_n \leqslant x) - \Phi(x)|$ will be much less than the uniform bound, and in fact a better estimate will be obtainable from crude moment bounds like

$$|P(S_n \leqslant x) - \Phi(x)| \leqslant P(|S_n| > |x|) + P(|N| > |x|) \leqslant x^{-2}(ES_n^2 + 1),$$

where N is distributed as $N(0,1)$. Clearly what is needed is a nonuniform bound for $|P(S_n \leqslant x) - \Phi(x)|$, depending on both x and n. In the independence case the theory of nonuniform rates of convergence is well developed, and a contemporary account is given in Petrov (1975, pp. 120–126). However, nonuniform rates are notably absent from the literature in the martingale case. We now prove a nonuniform version of Heyde and Brown's (1970) uniform rate of convergence.

Theorem 3.9. *Let* $\{S_i = \sum_{j=1}^{i} X_j, \mathscr{F}_i, 1 \leqslant i \leqslant n\}$ *be a zero-mean martingale. Set*

$$V_i^2 = \sum_{j=1}^{i} E(X_j^2 | \mathscr{F}_{j-1}) \qquad \text{and} \qquad U_i^2 = \sum_{j=1}^{i} X_i^2, \qquad 1 \leqslant i \leqslant n,$$

and suppose that $0 < \delta \leqslant 1$. *Define*

$$L_n = \sum_{i=1}^{n} E|X_i|^{2+2\delta} + E|V_n^2 - 1|^{1+\delta}$$

and

$$M_n = \sum_{i=1}^{n} E|X_i|^{2+2\delta} + E|U_n^2 - 1|^{1+\delta}.$$

There exist constants A_1 *and* A_2 *depending only on* δ, *such that whenever* $L_n \leqslant 1$,

$$|P(S_n \leqslant x) - \Phi(x)| \leqslant A_1 L_n^{1/(3+2\delta)}[1 + |x|^{4(1+\delta)^2/(3+2\delta)}]^{-1}, \tag{3.75}$$

and

$$|P(S_n \leqslant x) - \Phi(x)| \leqslant A_2 M_n^{1/(3+2\delta)}[1 + |x|^{4(1+\delta)^2/(3+2\delta)}]^{-1} \tag{3.76}$$

for all x.

As a preliminary to the proof of Theorem 3.9 we need a sharpening of Lemma 3.2.

Lemma 3.3. *Let $W(t)$, $t \geqslant 0$, be a standard Brownian motion and let T be a nonnegative variable. Then for all real x and all $0 < \varepsilon < \frac{1}{2}$,*

$$|P(W(T) \leqslant x) - \Phi(x)| \leqslant 16\varepsilon^{1/2}\exp(-x^2/4) + P(|T-1| > \varepsilon).$$

Proof. Combining (3.63) and (3.65) we deduce that

$$|P(W(T) \leqslant x) - \Phi(x)| \leqslant \pi^{-1/2}\int_0^\infty f(x,z)e^{-z^2/4}\,dz + P(|T-1| > \varepsilon), \tag{3.77}$$

where

$$f(x,z) = \Phi((1-\varepsilon)^{-1/2}(x+\varepsilon^{1/2}z)) - \Phi((1-\varepsilon)^{-1/2}(x-\varepsilon^{1/2}z)).$$

We shall assume that $x \geqslant 0$, the case $x < 0$ follows by symmetry. Now,

$$\int_0^\infty f(x,z)e^{-z^2/4}\,dz \leqslant \int_0^1 f(x,z)\,dz + \int_0^\infty f(x,z)ze^{-z^2/4}\,dz, \tag{3.78}$$

$$\int_0^1 f(x,z)\,dz \leqslant \sup_{0\leqslant z\leqslant 1} f(x,z) \leqslant (2\pi)^{-1/2}(1-\varepsilon)^{-1/2}2\varepsilon^{1/2}\exp(-u^2/2(1-\varepsilon)), \tag{3.79}$$

where $u = \inf_{|z|\leqslant 1}|x+\varepsilon^{1/2}z|$, and

$$\begin{aligned}
&\int_0^\infty f(x,z)ze^{-z^2/4}\,dz \\
&\quad= \int_0^\infty ze^{-z^2/4}\,dz\int_{(1-\varepsilon)^{-1/2}(x-\varepsilon^{1/2}z)}^{(1-\varepsilon)^{-1/2}(x+\varepsilon^{1/2}z)}(2\pi)^{-1/2}e^{-t^2/2}\,dt \\
&\quad= \int_{(1-\varepsilon)^{-1/2}x}^{\infty}(2\pi)^{-1/2}e^{-t^2/2}\,dt\int_{\varepsilon^{-1/2}[(1-\varepsilon)^{1/2}t-x]}^{\infty} ze^{-z^2/4}\,dz \\
&\qquad+ \int_{-\infty}^{(1-\varepsilon)^{-1/2}x}(2\pi)^{-1/2}e^{-t^2/2}\,dt\int_{-\varepsilon^{-1/2}[(1-\varepsilon)^{1/2}t-x]}^{\infty} ze^{-z^2/4}\,dz.
\end{aligned}\tag{3.80}$$

The first term on the right-hand side in (3.80) equals

$$\begin{aligned}
&(2/\pi)^{1/2}\int_{(1-\varepsilon)^{-1/2}x}^{\infty}\exp\{-\tfrac{1}{2}t^2 - [(1-\varepsilon)^{1/2}t-x]^2/4\varepsilon\}\,dt \\
&\quad= [2/\pi(1-\varepsilon)]^{1/2}\int_x^\infty \exp\{-\tfrac{1}{2}(1-\varepsilon)^{-1}u^2 - (u-x)^2/4\varepsilon\}\,du \\
&\quad\leqslant [2/\pi(1-\varepsilon)]^{1/2}\int_x^\infty \exp\{-\tfrac{1}{2}u^2 - (u-x)^2/4\varepsilon\}\,du \\
&\quad= [2/\pi(1-\varepsilon)]^{1/2}\int_x^\infty \exp\{-(1+2\varepsilon)[u-x/(1+2\varepsilon)]^2/4\varepsilon - x^2/2(1+2\varepsilon)\}\,du \\
&\quad\leqslant 2(1-\varepsilon)^{-1/2}\exp[-x^2/2(1+2\varepsilon)]\int_{-\infty}^{\infty}(2\pi)^{-1/2}\exp(-u^2/4\varepsilon)\,du \\
&\quad= 2^{3/2}\varepsilon^{1/2}(1-\varepsilon)^{-1/2}\exp[-x^2/2(1+2\varepsilon)].
\end{aligned}$$

The second term on the right-hand side in (3.80) equals

$$(2/\pi)^{1/2}\int_{-\infty}^{(1-\varepsilon)^{-1/2}x}\exp\{-\tfrac{1}{2}t^2-[(1-\varepsilon)^{1/2}t-x]^2/4\varepsilon\}\,dt$$

$$\leqslant [2/\pi(1-\varepsilon)]^{1/2}\int_{-\infty}^{x}\exp\{-(1+2\varepsilon)[u-x/(1+2\varepsilon)]^2/4\varepsilon-x^2/2(1+2\varepsilon)\}\,du$$

$$\leqslant 2^{3/2}\varepsilon^{1/2}(1-\varepsilon)^{-1/2}\exp[-x^2/2(1+2\varepsilon)],$$

using the same argument as before. Combining these two estimates and the inequalities (3.78)–(3.80) we deduce that for $0<\varepsilon<\frac{1}{2}$,

$$\int_0^\infty f(x,z)e^{-z^2/4}\,dz \leqslant 2\pi^{-1/2}\varepsilon^{1/2}\exp(-x^2/4+2^{3/2})+8\varepsilon^{1/2}\exp(-x^2/4)$$

$$\leqslant 28\varepsilon^{1/2}\exp(-x^2/4),$$

and together with (3.77) this completes the proof of Lemma 3.3. ∎

Proof of Theorem 3.9. We consider only the case where each $\mathscr{F}_i$ equals the σ-field generated by $X_1,\ldots,X_i$. The more general case may be proved easily from this specialisation; see remark (ii) following Theorem 3.7.

Let C, C', and C'' denote generic constants (not necessarily the same at each appearance). Applying the Skorokhod representation (Theorem A.1 of Appendix I) as in the proof of Theorem 3.7 we deduce that, without loss of generality, $S_n = W(T_n)$ a.s. for a nonnegative r.v. T_n, and

$$P(|T_n-1|>\varepsilon)\leqslant \varepsilon^{-1-\delta}E|T_n-1|^{1+\delta}$$

$$\leqslant \varepsilon^{-1-\delta}2^{\delta}E[|T_n-V_n^2|^{1+\delta}+|V_n^2-1|^{1+\delta}]. \tag{3.81}$$

Theorem 2.10 implies that

$$E|T_n-V_n^2|^{1+\delta}=E\left|\sum_1^n(\tau_i-E(\tau_i|\mathscr{G}_{i-1}))\right|^{1+\delta}$$

$$\leqslant CE\left|\sum_1^n(\tau_i-E(\tau_i|\mathscr{G}_{i-1}))^2\right|^{(1+\delta)/2}$$

$$\leqslant C\sum_1^n E|\tau_i-E(\tau_i|\mathscr{G}_{i-1})|^{1+\delta}\qquad (\text{since } \delta\leqslant 1)$$

$$\leqslant C'\sum_1^n\{E(\tau_i^{1+\delta})+E[E(\tau_i|\mathscr{G}_{i-1})^{1+\delta}]\}$$

$$\leqslant 2C'\sum_1^n E(\tau_i^{1+\delta}). \tag{3.82}$$

Since each $E(\tau_i^{1+\delta})\leqslant 2\Gamma(1+\delta)E|X_i|^{2+2\delta}$,

$$E|T_n-V_n^2|^{1+\delta}\leqslant C\sum_1^n E|X_i|^{2+2\delta}.$$

Returning to (3.81) and recalling the definition of L_n, we deduce that

$$P(|T_n - 1| > \varepsilon) \leqslant \varepsilon^{-1-\delta} C L_n.$$

This, together with Lemma 3.3, implies that

$$|P(W(T_n) \leqslant x) - \Phi(x)| \leqslant 16\varepsilon^{1/2}\exp(-x^2/4) + \varepsilon^{-1-\delta} C L_n. \tag{3.83}$$

We now choose $\varepsilon = \varepsilon(x,n)$ to minimize the right-hand side. Let

$$\varepsilon = L_n^{2/(3+2\delta)}(1 + |x|^{\alpha})^{1/(1+\delta)},$$

where $\alpha = \alpha(\delta) > 0$ will be chosen shortly. Suppose that $\varepsilon < \frac{1}{2}$. Then

$$\begin{aligned} &\varepsilon^{1/2}\exp(-x^2/4) + \varepsilon^{-1-\delta} L_n \\ &\quad = L_n^{1/(3+2\delta)}(1 + |x|^{\alpha})^{-1}[(1 + |x|^{\alpha})^{1+1/2(1+\delta)}\exp(-x^2/4) + 1] \\ &\quad \leqslant C L_n^{1/(3+2\delta)}(1 + |x|^{\alpha})^{-1}. \end{aligned} \tag{3.84}$$

From Markov's inequality we have for any x,

$$\begin{aligned} |P(S_n \leqslant x) - \Phi(x)| &\leqslant P(|S_n| > |x|) + P(|N| > |x|) \\ &\leqslant |x|^{-2-2\delta}(E|S_n|^{2+2\delta} + E|N|^{2+2\delta}), \end{aligned} \tag{3.85}$$

where N is distributed as $N(0,1)$. Now,

$$\begin{aligned} E|S_n|^{2+2\delta} &\leqslant CE\left[\left(\sum_1^n X_i^2\right)^{1+\delta}\right] \qquad \text{(by Theorem 2.10)} \\ &= CE(U_n^{2+2\delta}) \\ &\leqslant C'[E|U_n^2 - V_n^2|^{1+\delta} + E|V_n^2 - 1|^{1+\delta} + 1] \\ &\leqslant C''(L_n + 1) \end{aligned} \tag{3.86}$$

since, applying Theorem 2.10 to the martingale with differences $X_i^2 - E(X_i^2|\mathscr{F}_{i-1})$ and using the fact that $\delta \leqslant 1$, we derive

$$\begin{aligned} E|U_n^2 - V_n^2|^{1+\delta} &= E\left|\sum_1^n (X_i^2 - E(X_i^2|\mathscr{F}_{i-1}))\right|^{1+\delta} \\ &\leqslant CE\left|\sum_1^n (X_i^2 - E(X_i^2|\mathscr{F}_{i-1}))^2\right|^{(1+\delta)/2} \\ &\leqslant C\sum_1^n E|X_i^2 - E(X_i^2|\mathscr{F}_{i-1})|^{1+\delta} \\ &\leqslant C'\sum_1^n [E|X_i|^{2+2\delta} + E|E(X_i^2|\mathscr{F}_{i-1})|^{1+\delta}] \\ &\leqslant 2C'\sum_1^n E|X_i|^{2+2\delta}. \end{aligned} \tag{3.87}$$

Substituting (3.86) into (3.85) we see that if $L_n \leqslant 1$,

$$|P(S_n \leqslant x) - \Phi(x)| \leqslant C|x|^{-2-2\delta}(L_n + 1) \leqslant 2C|x|^{-2-2\delta}. \tag{3.88}$$

Suppose now that $L_n \leqslant 2^{-(3+2\delta)}$ and $L_n^{2/(3+2\delta)}(1 + |x|^{\alpha})^{1/(1+\delta)} > \frac{1}{2}$. Then $(1 + |x|^{\alpha})^{1/(1+\delta)} > 2$, and so $|x|^{\alpha} > 1$. Therefore

$$|x|^{-2-2\delta} \leqslant [\tfrac{1}{2}(1 + |x|^{\alpha})]^{-2(1+\delta)/\alpha} = C(1 + |x|^{\alpha})^{-2(1+\delta)/\alpha}. \tag{3.89}$$

Furthermore,

$$1 \leqslant 2^{1/2} L_n^{1/(3+2\delta)}(1 + |x|^{\alpha})^{1/2(1+\delta)}.$$

Substituting this inequality and (3.89) into (3.88) we deduce that

$$|P(S_n \leqslant x) - \Phi(x)| \leqslant C L_n^{1/(3+2\delta)}(1 + |x|^{\alpha})^{1/2(1+\delta) - 2(1+\delta)/\alpha}.$$

Comparing this with (3.84) we see that the best overall rate of convergence is obtained when α satisfies the equation

$$1/2(1 + \delta) - 2(1 + \delta)/\alpha = -1,$$

giving $\alpha = 4(1 + \delta)^2/(3 + 2\delta)$.

We have now proved (3.75) except when $1 \geqslant L_n > 2^{-(3+2\delta)}$. In this case we have for $|x| \leqslant 1$,

$$4L_n^{1/(3+2\delta)}(1 + |x|^{\alpha})^{-1} > 1 > |P(S_n \leqslant x) - \Phi(x)|,$$

and for $|x| > 1$,

$$\begin{aligned} 4L_n^{1/(3+2\delta)}(1 + |x|^{\alpha})^{-1} &\geqslant |x|^{-2-2\delta} \\ &= (C + E|N|^{2+2\delta})^{-1}(C + E|N|^{2+2\delta})|x|^{-2-2\delta} \\ &\qquad\qquad \text{(for any } C > 0\text{)} \\ &\geqslant C'(E|S_n|^{2+2\delta} + E|N|^{2+2\delta})|x|^{-2-2\delta} \qquad \text{(using (3.86))} \\ &\geqslant C'|P(S_n \leqslant x) - \Phi(x)| \qquad \text{(using (3.85))}. \end{aligned}$$

This completes the proof of (3.75).

Condition (3.76) follows from (3.75) and the inequalities (3.87) since

$$L_n \leqslant M_n + E|U_n^2 - V_n^2|^{1+\delta} \leqslant CM_n. \quad \blacksquare$$

The restriction $\delta \leqslant 1$ in Theorem 3.9 can be dropped if we require instead that the X_i have uniformly bounded moments.

Theorem 3.10. *In the notation of Theorem* 3.9, *suppose that* $0 < \delta < \infty$ *and*

$$\max_{i \leqslant n} E|X_i|^{2+2\delta} \leqslant n^{-1-\delta} M \tag{3.90}$$

for a constant M. Let

$$L_n = n^{-(1+\delta)/2} + E|V_n^2 - 1|^{1+\delta}$$

and

$$M_n = n^{-(1+\delta)/2} + E|U_n^2 - 1|^{1+\delta}.$$

There exist constants A_1 and A_2 depending only on M and δ such that whenever $L_n \leqslant 1$,

$$|P(S_n \leqslant x) - \Phi(x)| \leqslant A_1 L_n^{1/(3+2\delta)}[1 + |x|^{4(1+\delta)^2/(3+2\delta)}]^{-1} \tag{3.91}$$

and

$$|P(S_n \leqslant x) - \Phi(x)| \leqslant A_2 M_n^{1/(3+2\delta)}[1 + |x|^{4(1+\delta)^2/(3+2\delta)}]^{-1} \tag{3.92}$$

for all x.

Proof. The case $\delta \leqslant 1$ follows via Theorem 3.9, so we confine our attention to $\delta > 1$. Returning to the inequality (3.82) in the proof of Theorem 3.9, we deduce from the inequality

$$\left|\sum_1^n x_i\right|^p \leqslant n^{p-1}\sum_1^n |x_i|^p, \qquad p > 1,$$

that

$$\begin{aligned} E|T_n - V_n^2|^{1+\delta} &\leqslant CE\left|\sum_1^n (\tau_i - E(\tau_i|\mathcal{G}_{i-1}))^2\right|^{(1+\delta)/2} \\ &\leqslant Cn^{(1+\delta)/2-1}\sum_1^n E|\tau_i - E(\tau_i|\mathcal{G}_{i-1})|^{1+\delta} \\ &\leqslant C'n^{(1+\delta)/2-1}\sum_1^n E(\tau_i^{1+\delta}) \\ &\leqslant C''n^{(1+\delta)/2-1}\sum_1^n E|X_i|^{2+2\delta} \\ &\leqslant C''n^{-(1+\delta)/2}M, \end{aligned}$$

using (3.90). We now deduce from (3.81) that

$$P(|T_n - 1| > \varepsilon) \leqslant \varepsilon^{-1-\delta}CL_n,$$

and consequently

$$|P(S_n \leqslant x) - \Phi(x)| \leqslant 16\varepsilon^{1/2}\exp(-x^2/4) + \varepsilon^{-1-\delta}CL_n,$$

which is identical to (3.83). The proof of (3.91) and (3.92) can now be completed as before. ∎

These results are easily extended to reverse martingales by using the usual duality argument. [Of course, the reverse martingale results may be obtained directly; see Weber (1979).] For example, we have the following analog of Theorem 3.9.

Theorem 3.11. *Let* $\{S_n = \sum_{i=n}^{\infty} X_i, \mathscr{F}_n, 1 \leqslant n < \infty\}$ *be a zero-mean reverse martingale with moments of order* $2(1+\delta)$, $0 < \delta \leqslant 1$, *and suppose that* $S_n \to 0$ *a.s. Let* $s_n^2 = E(S_n^2)$,

$$V_n^2 = \sum_n^{\infty} E(X_i^2 | \mathscr{F}_{i+1}), \qquad U_n^2 = \sum_n^{\infty} X_i^2,$$

$$L_n = s_n^{-2-2\delta} \sum_n^{\infty} E|X_i|^{2+2\delta} + E|s_n^{-2} V_n^2 - 1|^{1+\delta},$$

and

$$M_n = s_n^{-2-2\delta} \sum_n^{\infty} E|X_i|^{2+2\delta} + E|s_n^{-2} U_n^2 - 1|^{1+\delta}.$$

There exist constants A_1 *and* A_2 *depending only on* δ *such that whenever* $L_n \leqslant 1$,

$$|P(s_n^{-1} S_n \leqslant x) - \Phi(x)| \leqslant A_1 L_n^{1/(3+2\delta)} [1 + |x|^{4(1+\delta)^2/(3+2\delta)}]^{-1} \tag{3.93}$$

and

$$|P(s_n^{-1} S_n \leqslant x) - \Phi(x)| \leqslant A_2 M_n^{1/(3+2\delta)} [1 + |x|^{4(1+\delta)^2/(3+2\delta)}]^{-1} \tag{3.94}$$

for all x.

Proof. Fix m and $n \geqslant 1$, and let $S_i' = s_n^{-1}(S_{m+n-i} - S_{m+n})$ and $\mathscr{F}_i' = \mathscr{F}_{m+n-i}$, $1 \leqslant i \leqslant m$. Then $\{S_i', \mathscr{F}_i', 1 \leqslant i \leqslant m\}$ is a (forward) martingale. Applying Theorem 3.9 to this martingale we deduce that

$$\begin{aligned}
&|P(s_n^{-1}(S_n - S_{m+n}) \leqslant x) - \Phi(x)| [1 + |x|^{4(1+\delta)^2/(3+2\delta)}] \\
&\leqslant A_1 \Bigg[\sum_1^m E|s_n^{-1} X_{m+n-i}|^{2+2\delta} \\
&\qquad + E\Big|s_n^{-2} \sum_1^m E(X_{m+n-i}^2 | \mathscr{F}_{m+n-i+1}) - 1\Big|^{1+\delta} \Bigg]^{1/(3+2\delta)} \\
&\to A_1 L_n^{1/(3+2\delta)}
\end{aligned}$$

as $m \to \infty$. This proves (3.93), and (3.94) follows in the same way.

4

Invariance Principles in the Central Limit Theorem and Law of the Iterated Logarithm

4.1. Introduction

In this chapter we present invariance principles associated with the central limit theorem (CLT) and law of the iterated logarithm (LIL) for martingales. For the most part these may be regarded as generalizations of the more familiar 1-dimensional results, such as those proved in Chapter 3 for the CLT. Since we do not consider the ordinary LIL *per se*, we show how to derive it as a corollary to its invariance principle (see Theorem 4.8). Section 4.2 contains invariance principles in the CLT, and new results on the rate of convergence are given in Section 4.3. Section 4.4 is devoted to invariance principles in the LIL.

The invariance principle for sums of independent r.v. grew from a basic idea of Erdös and Kac (1946, 1947), who saw that the limiting behavior of a function of sums could often be obtained by working out the limit when the sums had a certain special distribution. Later developments showed that the most efficient way of solving Erdös and Kac's problem was to consider the convergence of probability measures on a function space (see Appendix II). The name "invariance principle" is now frequently taken to mean a limit theorem in a function space, although the term "functional limit theorem" might be more apt.

4.2. Invariance Principles in the CLT

Donsker (1951, 1952) provided the first general invariance principle for sums of independent r.v. Billingsley (1968) derived a principle for martingales with stationary and ergodic differences, extending earlier work of Rosén

(1967a). These martingale results were simply generalizations of Donsker's theorem. The random functions were constructed like Donsker's: if $\{S_n,\mathscr{F}_n\}$ is a zero-mean, square-integrable martingale whose differences X_n form a stationary and ergodic sequence with zero mean and finite variance σ^2, let $\xi_n(t)$, $t \in [0,1]$, be the random function obtained by linearly interpolating between the points (0,0), $(n^{-1},(n\sigma^2)^{-1/2}S_1)$, $(2n^{-1},(n\sigma^2)^{-1/2}S_2), \ldots, (1,(n\sigma^2)^{-1/2}S_n)$. That is,

$$\xi_n(t) = (n\sigma^2)^{-1/2}\{S_{[nt]} + (nt - [nt])X_{[nt]+1}\},$$

where $[nt]$ denotes the integer part of nt. Then $\xi_n \xrightarrow{d} W$ in the sense (C,ρ), where W is standard Brownian motion on $[0,1]$. It follows that for any continuous functional h: $C[0,1] \to \mathbb{R}$, $h(\xi_n) \xrightarrow{d} h(W)$. (See Theorem A.3, Appendix II). Here $C = C[0,1]$ is the space of real, continuous functions on $[0,1]$ with the uniform metric ρ. In particular, if $h(x) = x(1)$, then we deduce that $(n\sigma^2)^{-1/2}S_n \xrightarrow{d} W(1) \stackrel{d}{=} N(0,1)$, and if $h(x) = \sup_{t\in[0,1]} x(t)$, then

$$(n\sigma^2)^{-1/2} \sup_{i\leqslant n} S_i \xrightarrow{d} \sup_{t\in[0,1]} W(t),$$

thereby answering one of Erdös and Kac's problems.

Billingsley's invariance principle is a straightforward extension of the 1-dimensional martingale CLT. However, there are many other ways of constructing invariance principles, and we shall indicate the method of Drogin (1972). If $\{S_n,\mathscr{F}_n\}$ is a martingale, let $X_n = S_n - S_{n-1}$ $(S_0 = 0)$ and $V_n^2 = \sum_{i=1}^n E(X_i^2|\mathscr{F}_{i-1})$, and define the continuous function S on $[0,\infty)$ by interpolating between the points (0,0), (V_1^2,S_1), $(V_2^2,S_2), \ldots$, so that

$$S(t) = S_n + E(X_{n+1}^2|\mathscr{F}_n)^{-1}(t - V_n^2)X_{n+1} \qquad \text{if} \quad V_n^2 \leqslant t < V_{n+1}^2.$$

Now let $\xi_n(t) = S(nt)/\sqrt{n}$, $t \in [0,1]$. Define $T_n = \inf\{m | V_m^2 \geqslant n\}$ and suppose that $E|n^{-1}V_{T_n}^2 - 1| \to 0$. Under this assumption, Drogin found necessary and sufficient conditions for $\xi_n \xrightarrow{d} W$ in the sense (C,ρ). Letting $h(x) = x(1)$ and using the fact that $h(\xi_n) \xrightarrow{d} h(W)$, we deduce that $n^{-1/2}S_{T_n} \xrightarrow{d} N(0,1)$, and letting $h(x) = \sup_{t\in[0,1]} x(t)$, we see that $n^{-1/2}\sup_{i\leqslant T_n} S_i \xrightarrow{d} \sup_{t\in[0,1]} W(t)$.

Rootzén (1977a, 1980) and Gänssler and Häusler (1980) have given similar necessary and sufficient conditions for invariance principles. Results like these have the advantage that they hold without the need for an assumption such as $(n\sigma^2)^{-1}V_n^2 \xrightarrow{p} 1$. However, from the point of view of applications the most useful invariance principle involves a fixed rather than a random sample size, and should provide a direct extension of a CLT for the martingale $\{S_n\}$. Drogin's invariance principle involves a random time change from n to T_n, which is often inconvenient. Of course, it may be possible subsequently to remove the time change by using a random norming, but this is rarely a straightforward matter.

In Chapter 3 we considered a CLT of the form $S_n/U_n \xrightarrow{d} N(0,1)$. We shall establish an invariance principle which allows us to extend this result—for example, to prove that $U_n^{-1} \max_{i \leqslant n} S_i \xrightarrow{d} \sup_{t \in [0,1]} W(t)$. Let $\{S_n, \mathscr{F}_n, n \geqslant 1\}$ be a zero-mean, square-integrable martingale with differences X_n, let $U_n^2 = \sum_1^n X_i^2$ and set $s_n^2 = E(S_n^2) = E(U_n^2)$. Define ξ_n to be the random element of $C[0,1]$ obtained by interpolating between the points $(0,0), (U_n^{-2}U_1^2, U_n^{-1}S_1), \ldots, (1, U_n^{-1}S_n)$, namely,

$$\xi_n(t) = U_n^{-1}\{S_i + X_{i+1}^{-2}(tU_n^2 - U_i^2)X_{i+1}\} \qquad \text{if} \quad U_i^2 \leqslant tU_n^2 < U_{i+1}^2.$$

Theorem 4.1. *If the Lindeberg condition holds, namely*

$$\text{for all} \quad \varepsilon > 0, \qquad s_n^{-2} \sum_{i=1}^{n} E[X_i^2 I(|X_i| > \varepsilon s_n)] \to 0 \tag{4.1}$$

as $n \to \infty$, and if

$$s_n^{-2} U_n^2 \xrightarrow{p} \eta^2 > 0 \quad a.s., \tag{4.2}$$

then $\xi_n \xrightarrow{d} W$ in the sense (C,ρ).

If the limit r.v. η^2 is equal to 1 a.s., it is more usual to use a different definition of the random functions ξ_n. Brown (1971) used

$$\xi_n^*(t) = s_n^{-1}\{S_i + (s_{i+1}^2 - s_i^2)^{-1}(ts_n^2 - s_i^2)X_{i+1}\} \qquad \text{if} \quad s_i^2 \leqslant ts_n^2 < s_{i+1}^2.$$

He showed that when (4.1) and (4.2) hold with $\eta^2 = 1$ a.s., $\xi_n^* \xrightarrow{d} W$. When η^2 is not constant we should multiply ξ_n^* by the r.v. $s_n U_n^{-1}$ in order to cancel the effect of η. Let $\eta_n = s_n U_n^{-1} \xi_n^*$. It turns out that if the convergence in (4.2) is strengthened to a.s. convergence (which is often true in applications), then the functions ξ_n and η_n are uniformly close.

Theorem 4.2. *Suppose that* (4.1) *holds. If*

$$s_n^{-2} U_n^2 \xrightarrow{\text{a.s.}} \eta^2 > 0 \qquad \text{and} \qquad s_n^2 \to \infty, \tag{4.3}$$

then

$$\sup_{t \in [0,1]} |\xi_n(t) - \eta_n(t)| \xrightarrow{p} 0. \tag{4.4}$$

(If $\xi_n \xrightarrow{d} W$ and (4.4) holds, then $\eta_n \xrightarrow{d} W$.)

A variety of methods have been used to establish the convergence of random functions. It is usually necessary to prove both (a) the convergence of finite-dimensional distributions, and (b) tightness (see Appendix II). Part

(a) is usually accomplished relatively easily, but tightness is often more difficult. The Skorokhod representation allows us to establish the invariance principle directly, without using a two-part proof and we shall use it to prove Theorem 4.1. It permits a very simple criterion for tightness (Hall, 1979a).

Theorem 4.3. *If* (4.1) *holds and*

$$\liminf_{n\to\infty} P(\delta < s_n^{-2}U_n^2 < \lambda) \to 1 \qquad \text{as} \quad \delta \to 0 \text{ and } \lambda \to \infty \tag{4.5}$$

(that is, the sequences $\{s_n^{-2}U_n^2\}$ *and* $\{U_n^{-2}s_n^2\}$ *are tight), then the sequence of random functions* $\{\xi_n\}$ *is tight in* $C[0,1]$.

The examples in Section 4.3 show that (4.1) and (4.5) alone are not sufficient to establish an invariance principle. On the other hand, condition (4.2) clearly implies (4.5). A sufficient criterion for tightness of a general sequence of stochastic processes has been obtained by Aldous (1978). It too is applicable in the martingale context (see Aldous' Theorem 3) as are the results of Loynes (1976).

In Chapter 3 we showed that norming by the sequence $\{U_n\}$ in the CLT is usually equivalent to norming by $\{V_n\}$, where $V_n^2 = \sum_1^n E(X_i^2|\mathscr{F}_{i-1})$. The same is true in the invariance principle. The analog of ξ_n, using V_n instead of U_n, is

$$\zeta_n(t) = V_n^{-1}\{S_i + (V_{i+1}^2 - V_i^2)^{-1}(tV_n^2 - V_i^2)X_{i+1}\} \qquad \text{if} \quad V_i^2 \leqslant tV_n^2 < V_{i+1}^2,$$

for $t \in [0,1]$ and $0 \leqslant i \leqslant n-1$.

Theorem 4.4. *If* (4.1) *and* (4.5) *hold, then*

$$\sup_{t\in[0,1]} |\xi_n(t) - \zeta_n(t)| \xrightarrow{\text{p}} 0.$$

Therefore under (4.1) *and* (4.2)*—or equivalently, if* (4.1) *holds and* $s_n^{-2}V_n^2 \xrightarrow{\text{p}} \eta^2 > 0$ *a.s. (see Theorem 2.23)—*$\zeta_n \xrightarrow{\text{d}} W$ *in the sense* (C,ρ).

Proof of Theorems 4.1 and 4.3. Our proof of Theorem 4.1 is based on the following limit theorem for Brownian motion.

Theorem A. *Let* $W(t)$, $t \geqslant 0$, *be standard Brownian motion and* T_n, $n \geqslant 1$, *be positive r.v. Let* W_1 *denote the restriction of* W *to* $[0,1]$ *and write* $\zeta_n(t) = W(tT_n)/\sqrt{T_n}$, $t \in [0,1]$. *If there exist constants* c_n *such that*

$$T_n/c_n \xrightarrow{\text{p}} \eta^2 > 0 \quad \text{a.s.}, \tag{4.6}$$

then $\zeta_n \xrightarrow{\text{d}} W_1$ *in the sense* (C,ρ).

Proof. Let W' be a copy of W which is independent of η^2 (that is, the σ-field generated by $\{W'(t), t > 0\}$ is independent of η^2) and let W'_λ denote the restriction of W' to $[0,\lambda]$. Put $\alpha_n(t) = W(tT_n)/\sqrt{c_n}$ and $\alpha(t) = W'(\eta t)$, $t \in [0,1]$. Since α/η has the same distribution as W_1, it suffices to prove that

$$(\alpha_n, T_n/c_n) \xrightarrow{\text{d}} (\alpha, \eta^2) \tag{4.7}$$

(convergence in distribution relative to the product topology). Let $\lambda > 0$ and define $\beta_n(t) = W(tc_n)/\sqrt{c_n}$, $t \in [0,\lambda]$; β_n is a random element of $C[0,\lambda]$. Suppose we can prove that

$$(\beta_n, T_n/c_n) \xrightarrow{\text{d}} (W'_\lambda, \eta^2). \tag{4.8}$$

Let $h: \{C[0,\lambda]\} \times [0,\lambda] \to \{C[0,1]\} \times [0,\lambda]$ be the continuous functional defined by $h(x(\cdot),r) = (x(\cdot r),r)$. If

$$P(T_n > \lambda c_n) = 0 \qquad \text{for all} \quad n, \tag{4.9}$$

then in view of Theorem A.4 of Appendix II,

$$h(\beta_n, T_n/c_n) \xrightarrow{\text{d}} h(W'_\lambda, \eta^2),$$

which is (4.7).

We have to check (4.8). Let $\varepsilon, \delta > 0$,

$$\beta'_n(t) = [W((t+\delta)c_n) - W(\delta c_n)]/\sqrt{c_n}, \qquad t \in [0,\lambda],$$

and write $\mathscr{G}_n$ for the σ-field generated by $\{W(t) - W(\delta c_n), t > \delta c_n\}$. If A is a continuity set of W'_λ, let

$$A_\varepsilon = \left\{ x \in C[0,\lambda] \,\middle|\, \sup_{t \leqslant \lambda} |x(t) - y(t)| < \varepsilon \text{ for some } y \in A \right\}.$$

Let x be a continuity point of η^2. Then

$$\begin{aligned} d_n &= P(\beta_n \in A,\, T_n/c_n \leqslant x) - P(W'_\lambda \in A,\, \eta^2 \leqslant x) \\ &\leqslant P\left(\sup_{t \leqslant \lambda} |\beta_n(t) - \beta'_n(t)| > \varepsilon \right) + P(W'_\lambda \in A_\varepsilon - A) \\ &\quad + P(\beta'_n \in A_\varepsilon,\, T_n/c_n \leqslant x) - P(W'_\lambda \in A_\varepsilon,\, \eta^2 \leqslant x). \end{aligned}$$

The second term on the right-hand side of the above inequality can be made arbitrarily small by choosing ε sufficiently small. The first term equals

$$P\left(\sup_{t \leqslant \lambda} |W(t) - W(t+\delta) + W(\delta)| > \varepsilon \right),$$

which can be made small by choosing δ small. Since W'_λ and β'_n have the same distribution, the difference of the last two terms equals

$$\begin{aligned}
E[I(\beta'_n \in A_\varepsilon)I(T_n/c_n \leqslant x) - I(\beta'_n \in A_\varepsilon)P(\eta^2 \leqslant x)] \\
= E[I(\beta'_n \in A_\varepsilon)\{E[I(T_n/c_n \leqslant x)|\mathscr{G}_n] - P(\eta^2 \leqslant x)\}] \\
\leqslant E|E[I(T_n/c_n \leqslant x)|\mathscr{G}_n] - P(\eta^2 \leqslant x)| \\
\leqslant E|E[I(T_n/c_n \leqslant x) - I(\eta^2 \leqslant x)|\mathscr{G}_n]| \\
+ E|E[I(\eta^2 \leqslant x)|\mathscr{G}_n] - P(\eta^2 \leqslant x)|.
\end{aligned}$$

The first term on the right-hand side of the above inequality is dominated by

$$E|I(T_n/c_n \leqslant x) - I(\eta^2 \leqslant x)| \to 0$$

as $n \to \infty$. Since the process W has independent increments, $\mathscr{G}_n$ converges to a σ-field which contains only events of probability zero or one, and so the last term of the above inequality converges to zero. It follows that $\limsup d_n \leqslant 0$; similarly, $\liminf d_n \geqslant 0$, and so $d_n \to 0$. This proves (4.8).

Finally we must remove condition (4.9). Given $\varepsilon > 0$, choose a continuity point λ of η^2 such that for all n, $P(T_n > \lambda c_n) < \varepsilon$. Define $T'_n = T_n I(T_n \leqslant \lambda c_n) + \lambda c_n I(T_n > \lambda c_n)$ and $\alpha'_n(t) = W(tT'_n)/\sqrt{c_n}$, $t \in [0,1]$. Since $T'_n/c_n \xrightarrow{\text{p}} \eta'^2 = \eta^2 I(\eta^2 \leqslant \lambda) + \lambda I(\eta^2 > \lambda)$, and since $P(T'_n > \lambda c_n) = 0$ for all n, from the result just proved above we deduce that for each continuity point $\lambda > 0$, $(\alpha'_n, T'_n/c_n) \xrightarrow{\text{d}} (\alpha', \eta'^2)$, where $\alpha'(t) = W'(\eta' t)$, $t \in [0,1]$. But

$$P(\alpha'_n \neq \alpha_n \text{ or } T'_n \neq T_n) + P(\alpha' \neq \alpha \text{ or } \eta' \neq \eta) \leqslant 2\varepsilon,$$

and so (4.7) must hold. ∎

Let T_n, ζ_n and W be the stochastic processes defined in Theorem A. If (4.6) holds, then of course the sequence $\{\zeta_n\}$ is tight. However, tightness holds under more general conditions.

Lemma 4.1. *If*

$$\liminf_{n\to\infty} P(\Delta < T_n/c_n < \lambda) \to 1 \qquad \textit{as} \quad \Delta \to 0 \textit{ and } \lambda \to \infty \tag{4.10}$$

for some sequence of constants c_n, then for all $\varepsilon > 0$,

$$\lim_{\delta\to 0} \limsup_{n\to\infty} P\left(\sup_{|u-v|<\delta} |\zeta_n(u) - \zeta_n(v)| > \varepsilon\right) = 0.$$

Proof. The probability above is dominated by

$$P\left(\sup_{|u-v|<\lambda\delta,\ 0\leqslant u\leqslant\lambda} |W(u) - W(v)| > \varepsilon\sqrt{\Delta}\right) + P(T_n/c_n < \Delta \text{ or } > \lambda),$$

and the required result follows by letting $n \to \infty$, $\delta \to 0$, $\Delta \to 0$, and $\lambda \to \infty$ in that order, and using the tightness of W. ∎

Let Z_n, $n \geqslant 1$, be r.v. such that $0 \leqslant Z_1 \leqslant Z_2 \leqslant \cdots$. Define random elements ζ_n' of $C[0,1]$ by interpolating between the points

$$(0,0),\ (Z_n^{-1}Z_1, Z_n^{-1/2}W(T_1)), \ldots, (1, Z_n^{-1/2}W(T_n)).$$

If the Z are close to the T, then ζ_n is close to ζ_n', as is shown by the following result.

Lemma 4.2. *If* (4.10) *holds,* $T_n/Z_n \xrightarrow{\text{p}} 1$,

$$\max_{i \leqslant n}(Z_i - Z_{i-1})/Z_n \xrightarrow{\text{p}} 0, \qquad \textit{and} \qquad \max_{i \leqslant n}|T_n^{-1}T_i - Z_n^{-1}Z_i| \xrightarrow{\text{p}} 0,$$

then $\rho(\zeta_n,\zeta_n') \xrightarrow{\text{p}} 0$, *where* $\rho(x,y) = \sup_{t \in [0,1]}|x(t) - y(t)|$ *denotes the uniform metric.*

Proof. Let ζ_n'' be the random function obtained by interpolating between the points $(0,0), (Z_n^{-1}Z_1, T_n^{-1/2}W(T_1)), \ldots, (1, T_n^{-1/2}W(T_n))$. Now,

$$\rho(\zeta_n,\zeta_n') \leqslant \rho(\zeta_n,\zeta_n'') + \rho(\zeta_n'',\zeta_n'),$$

and

$$\begin{aligned}\rho(\zeta_n'',\zeta_n') &\leqslant T_n^{-1/2} \sup_{t \leqslant T_n}|W(t)|\,|1 - (T_n/Z_n)^{1/2}| \\ &= \sup_{t \in [0,1]} |\zeta_n(t)|\,|1 - (T_n/Z_n)^{1/2}| \xrightarrow{\text{p}} 0,\end{aligned}$$

since $\{\zeta_n\}$ is tight. Hence it suffices to prove that $\rho(\zeta_n,\zeta_n'') \xrightarrow{\text{p}} 0$. Let ε, Δ, and $\delta > 0$, and define

$$E_n = \left\{ \sup_{|u-v| \leqslant 2\delta} |\zeta_n(u) - \zeta_n(v)| > \varepsilon \right\}$$

and

$$F_n = \left\{ \max_{i \leqslant n}(Z_i - Z_{i-1})/Z_n \leqslant \delta; \max_{i \leqslant n}|T_n^{-1}T_i - Z_n^{-1}Z_i| \leqslant \delta \right\}.$$

Since $\{\zeta_n\}$ is tight, we can choose δ and N_1 such that for all $n \geqslant N_1$, $P(E_n) \leqslant \Delta$. Now choose $N \geqslant N_1$ so large that for all $n \geqslant N$, $P(F_n^c) \leqslant \Delta$. Suppose that $n \geqslant N$. Then

$$\begin{aligned}&P\left(\sup_{z \in [0,1]} |\zeta_n(z) - \zeta_n''(z)| > \varepsilon \right) \\ &\quad \leqslant P\left(\max_{i \leqslant n} \sup_{Z_{i-1}/Z_n < z \leqslant Z_i/Z_n} |W(T_i) - W(zT_n)|/\sqrt{T_n} > \varepsilon \right) \\ &\qquad + P\left(\max_{i \leqslant n} \sup_{Z_{i-1}/Z_n < z \leqslant Z_i/Z_n} |W(T_{i-1}) - W(zT_n)|/\sqrt{T_n} > \varepsilon \right).\end{aligned}$$

The first term on the right-hand side does not exceed

$$P\left(\max_{i\leqslant n}\sup_{Z_{i-1}/Z_n<z\leqslant Z_i/Z_n}|W(T_i)-W(zT_n)|/\sqrt{T_n}>\varepsilon;\,F_n\right)+P(F_n^c)$$
$$\leqslant P(E_n)+P(F_n^c)\leqslant 2\Delta.$$

The second term can be handled in a similar way, proving that the left-hand side is dominated by 4Δ. This establishes that $\rho(\zeta_n,\zeta_n'')\xrightarrow{\mathrm{p}}0$. ∎

Now we introduce the martingale theory. We approximate to the martingale $\{(S_n,\mathscr{F}_n)\}$ by a truncated martingale $\{(S_n^*,\mathscr{F}_n^*)\}$ and show that the approximation is uniformly close. Then we apply the Skorokhod representation (Appendix I), which gives $S_n^*=W(T_n)$ a.s., $n\geqslant 1$, for an increasing sequence of positive r.v. $\{T_n\}$. If (4.6) holds, then the T_n satisfy $T_n/s_n^2\xrightarrow{\mathrm{p}}\eta^2$ and so by Theorem A, $\zeta_n\xrightarrow{\mathrm{d}}W_1$. ζ_n is uniformly close to ξ_n and so $\xi_n\xrightarrow{\mathrm{d}}W_1$. Some of the techniques of our approximation are drawn from Scott (1973).

The Lindeberg condition (4.1) is equivalent to the apparently stronger condition

$$\text{for all}\quad\varepsilon>0,\qquad s_n^{-2}\sum_1^n E[X_i^2I(|X_i|>\varepsilon s_i)]\to 0. \tag{4.11}$$

To see this, let $0<\delta<\varepsilon$ and $k_n=\max\{i\leqslant n|\varepsilon s_i\leqslant\delta s_n\}$. If (4.1) holds, then the left-hand side of (4.11) does not exceed

$$s_n^{-2}\sum_1^n E[X_i^2I(|X_i|>\delta s_n)]+s_n^{-2}\sum_1^{k_n}E(X_i^2)\leqslant o(1)+\delta^2/\varepsilon^2,$$

and (4.11) follows on letting $n\to\infty$ and then $\delta\to 0$.

Fix $\varepsilon>0$ and define $X_0^*=0$ and

$$X_i^*=X_iI(|X_i|\leqslant\varepsilon s_i)-E[X_iI(|X_i|\leqslant\varepsilon s_i)|\mathscr{F}_{i-1}^*],\qquad 1\leqslant i\leqslant n,$$

where $\mathscr{F}_{i-1}^*$ is the σ-field generated by $X_1^*,X_2^*,\ldots,X_{i-1}^*$. Let $S_n^*=\sum_1^n X_i^*$, $U_n^{*2}=\sum_1^n X_i^{*2}$, and let ξ_n^* be the process obtained by interpolating between the points $(0,0)$, $(U_n^{-2}U_1^2,U_n^{-1}S_1^*),\ldots,(1,U_n^{-1}S_n^*)$.

Lemma 4.3. *Under condition* (4.11),

$$s_n^{-1}\sum_1^n E[|X_i|I(|X_i|>\varepsilon s_i)]\to 0, \tag{4.12}$$

$$s_n^{-4}\sum_1^n E[X_i^4I(|X_i|\leqslant\varepsilon s_i)]\to 0, \tag{4.13}$$

and

$$s_n^{-2}\max_{i\leqslant n}|U_i^2-U_i^{*2}|\xrightarrow{\mathrm{p}}0. \tag{4.14}$$

Under conditions (4.5) *and* (4.11),

$$\rho(\xi_n, \xi_n^*) \xrightarrow{\mathrm{p}} 0. \tag{4.15}$$

Proof. Let δ and k_n be as above. Then

$$\begin{aligned} s_n^{-1} \sum_1^n E[|X_i| I(|X_i| > \varepsilon s_i)] &\leqslant s_n^{-1} \sum_1^n \varepsilon^{-1} s_i^{-1} E[X_i^2 I(|X_i| > \varepsilon s_i)] \\ &\leqslant \delta \varepsilon^{-2} s_{k_n}^{-1} \sum_1^{k_n} s_i^{-1} E(X_i^2) + \delta^{-1} s_n^{-2} \sum_1^n E[X_i^2 I(|X_i| > \varepsilon s_i)]. \end{aligned} \tag{4.16}$$

The last term is $o(1)$ and the first converges to $2\delta\varepsilon^{-2}$ since

$$s_n^{-1} \sum_1^n s_i^{-1} E(X_i^2) = s_n^{-1} \sum_1^n (s_i - s_{i-1})(1 + s_i^{-1} s_{i-1}) \sim 2 s_n^{-1} \sum_1^n (s_i - s_{i-1}) = 2.$$

Condition (4.12) follows on letting $n \to \infty$ and then $\delta \to 0$ in (4.16).

Again suppose that $0 < \delta < \varepsilon$. The left side of (4.13) equals

$$\begin{aligned} s_n^{-4} \sum_1^n E[X_i^4 I(|X_i| \leqslant \delta s_i)] &+ s_n^{-4} \sum_1^n E[X_i^4 I(\delta s_i < |X_i| \leqslant \varepsilon s_i)] \\ &\leqslant \delta^2 + \varepsilon^2 s_n^{-2} \sum_1^n E[X_i^2 I(|X_i| > \delta s_i)], \end{aligned}$$

and this tends to zero as $n \to \infty$ and then $\delta \to 0$.

Condition (4.14) follows from (4.11), (4.12), and the inequalities

$$\begin{aligned} E\left[s_n^{-2} \max_{i \leqslant n} |U_i^2 - U_i^{*2}| \right] &\leqslant s_n^{-2} \sum_1^n E|X_i^2 - X_i^{*2}| \\ &\leqslant s_n^{-2} \sum_1^n E[X_i^2 I(|X_i| > \varepsilon s_i)] \\ &\quad + 3\varepsilon s_n^{-1} \sum_1^n E[|X_i| I(|X_i| > \varepsilon s_i)]. \end{aligned}$$

(Note that $E(X_i | \mathscr{F}_{i-1}) = E(X_i | \mathscr{F}_{i-1}^*) = 0$.)

Since $\rho(\xi_n, \xi_n^*) = U_n^{-1} \max_{i \leqslant n} |S_i - S_i^*|$, to prove (4.15) it suffices to show that $s_n^{-1} \max_{i \leqslant n} |S_i - S_i^*| \xrightarrow{\mathrm{p}} 0$. But

$$\begin{aligned} s_n^{-1} \max_{j \leqslant n} |S_j - S_j^*| &\leqslant s_n^{-1} \max_{j \leqslant n} \left| \sum_1^j \{ X_i I(|X_i| > \varepsilon s_i) - E[X_i I(|X_i| > \varepsilon s_i) | \mathscr{F}_{i-1}] \} \right| \\ &\quad + s_n^{-1} \sum_1^n E[|X_i| I(|X_i| > \varepsilon s_i) | \mathscr{F}_{i-1}] \\ &\quad + s_n^{-1} \sum_1^n E[|X_i| I(|X_i| > \varepsilon s_i) | \mathscr{F}_{i-1}^*]. \end{aligned}$$

Condition (4.12) implies that the last two terms converge to zero in L^1, and Kolmogorov's inequality (Corollary 2.1) implies that

$$P\left(s_n^{-1}\max_{j\leqslant n}\left|\sum_1^j\{X_iI(|X_i|>\varepsilon s_i)-E[X_iI(|X_i|>\varepsilon s_i)|\mathscr{F}_{i-1}]\}\right|>\delta\right)$$

$$\leqslant \delta^{-2}s_n^{-2}\sum_1^n E[X_i^2I(|X_i|>\varepsilon s_i)]\to 0 \qquad \text{as} \quad n\to\infty.$$

Hence $s_n^{-1}\max_{j\leqslant n}|S_j-S_j^*|\xrightarrow{\text{p}}0$. ∎

Now we introduce the Skorokhod representation (Appendix I). Without loss of generality there exists a Brownian motion W and an increasing sequence of nonnegative r.v. T_n such that $S_n^*=W(T_n)$ a.s., $n\geqslant 1$. Put $\tau_n=T_n-T_{n-1}$, $n\geqslant 1$ ($T_0=0$), let $\mathscr{G}_n$ be the σ-field generated by $S_1^*, S_2^*,\ldots,S_n^*$, and $W(t)$ for $0\leqslant t\leqslant T_n$ ($n\geqslant 1$), and let $\mathscr{G}_0$ and $\mathscr{F}_0^*$ denote the trivial σ-field. Theorem A.1 of Appendix I tells us that the T_n can be chosen such that τ_n is $\mathscr{G}_n$-measurable, $E(\tau_n|\mathscr{G}_{n-1})=E(X_n^{*2}|\mathscr{F}_{n-1}^*)$ a.s. ($n\geqslant 1$), and for a constant $L>0$, $E(\tau_n^2|\mathscr{G}_{n-1})\leqslant LE(X_n^{*4}|\mathscr{F}_{n-1}^*)$ a.s. ($n\geqslant 1$).

Lemma 4.4. *Under condition* (4.11),

$$s_n^{-2}\max_{j\leqslant n}\left|T_j-\sum_1^j E(\tau_i|\mathscr{G}_{i-1})\right|\xrightarrow{\text{p}}0 \tag{4.17}$$

and

$$s_n^{-2}\max_{j\leqslant n}\left|\sum_1^j E(\tau_i|\mathscr{G}_{i-1})-U_j^{*2}\right|\xrightarrow{\text{p}}0. \tag{4.18}$$

Proof. Applying Kolmogorov's inequality to the martingale with differences $\tau_i-E(\tau_i|\mathscr{G}_{i-1})$ and σ-fields $\mathscr{G}_i$ gives

$$P\left(s_n^{-2}\max_{j\leqslant n}\left|T_j-\sum_1^j E(\tau_i|\mathscr{G}_{i-1})\right|>\delta\right)\leqslant\delta^{-2}s_n^{-4}\sum_1^n E(\tau_i^2)\leqslant L\delta^{-2}s_n^{-4}\sum_1^n E(X_i^{*4})$$

$$\leqslant L\delta^{-2}s_n^{-4}\sum_1^n\{E[X_i^4I(|X_i|\leqslant\varepsilon s_i)]+15\varepsilon^3s_i^3E[|X_i|I(|X_i|>\varepsilon s_i)]\}$$

$$\to 0 \qquad \text{as} \quad n\to\infty,$$

using Lemma 4.3. This proves (4.17); (4.18) is proved in the same way by applying Kolmogorov's inequality to the martingale with differences $X_i^{*2}-E(X_i^{*2}|\mathscr{F}_{i-1}^*)$. ∎

We are now in a position to prove Theorems 4.1 and 4.3. Suppose that (4.1) and (4.5) hold. Then so does (4.11), and hence $\rho(\xi_n,\xi_n^*)\xrightarrow{\text{p}}0$ in view of

Lemma 4.3. Furthermore, in view of (4.5), (4.14), (4.17) and (4.18),

$$U_n^{-2} \max_{i \leqslant n} |U_i^2 - T_i| \xrightarrow{\text{p}} 0.$$

For $T_n > 0$,

$$\max_{i \leqslant n} |T_n^{-1}T_i - U_n^{-2}U_i^2| \leqslant U_n^{-2} \max_{i \leqslant n} |U_i^2 - T_i| + |1 - U_n^{-2}T_n|,$$

and so $\max_{i \leqslant n} |T_n^{-1}T_i - U_n^{-2}U_i^2| \xrightarrow{\text{p}} 0$. Since

$$s_n^{-2} \max_{i \leqslant n} X_i^2 \leqslant \varepsilon^2 + s_n^{-2} \sum_1^n X_i^2 I(|X_i| > \varepsilon s_n) \xrightarrow{\text{p}} \varepsilon^2,$$

$U_n^{-2} \max_{i \leqslant n} X_i^2 \xrightarrow{\text{p}} 0$. It follows from Lemma 4.2 that $\rho(\zeta_n, \xi_n^*) \xrightarrow{\text{p}} 0$. Theorem 4.3 now follows from Lemma 4.1. If (4.1) and (4.2) hold, then $s_n^{-2}T_n \xrightarrow{\text{p}} \eta^2$, and so (4.6) holds with $c_n = s_n^2$. Since $\rho(\xi_n, \zeta_n^*) \xrightarrow{\text{p}} 0$ and $\rho(\zeta_n, \xi_n^*) \xrightarrow{\text{p}} 0$, Theorem 4.1 follows from Theorem A. ∎

Proof of Theorem 4.2. Conditions (4.1) and (4.3) imply that $\xi_n \xrightarrow{\text{d}} W$, and so the sequence of r.v. $\{U_n^{-1} \max_{i \leqslant n} |S_i|\}$ is tight. As in the proof above we deduce that $U_n^2/T_n \xrightarrow{\text{p}} 1$, $\rho(\xi_n, \zeta_n^*) \xrightarrow{\text{p}} 0$, and $\rho(\zeta_n, \xi_n^*) \xrightarrow{\text{p}} 0$. Hence to establish (4.4) we need only prove that $\rho(\zeta_n, \eta_n) \xrightarrow{\text{p}} 0$. (Here ζ_n is defined as in Theorem A.)

The random function η_n is obtained by interpolating between the points $(0,0)$, $(s_n^{-2}s_1^2, U_n^{-1}S_1), \ldots, (1, U_n^{-1}S_n)$. Let η_n' and η_n'' be the functions defined by interpolating between $(0,0)$, $(s_n^{-2}s_1^2, T_n^{-1/2}S_1), \ldots, (1, T_n^{-1/2}S_n)$ and between $(0,0)$, $(s_n^{-2}s_1^2, T_n^{-1/2}W(T_1)), \ldots, (1, T_n^{-1/2}W(T_n))$, respectively. Then

$$\rho(\eta_n, \eta_n') \leqslant U_n^{-1} \max_{i \leqslant n} |S_i| \, |1 - (U_n^2/T_n)^{1/2}| \xrightarrow{\text{p}} 0$$

and

$$\begin{aligned} \rho(\eta_n', \eta_n'') &= T_n^{-1/2} \sup_{i \leqslant n} |S_i - S_i^*| \\ &= (U_n^2/T_n)^{1/2} \rho(\zeta_n, \xi_n^*) \xrightarrow{\text{p}} 0. \end{aligned}$$

Hence it suffices to prove that $\rho(\zeta_n, \eta_n'') \xrightarrow{\text{p}} 0$.

As in the proof of Theorems 4.1 and 4.3, we can show that the sequence $\{\zeta_n\}$ is tight and that

$$\max_{i \leqslant n} |T_n^{-1}T_i - U_n^{-2}U_i^2| \xrightarrow{\text{p}} 0.$$

Condition (4.3) implies that

$$\max_{i \leqslant n} s_n^{-2}s_i^2 |s_i^{-2}U_i^2 - \eta^2| \xrightarrow{\text{a.s.}} 0,$$

and consequently

$$\max_{i\leqslant n}|U_n^{-2}U_i^2 - s_n^{-2}s_i^2| \leqslant \max_{i\leqslant n} U_n^{-2}s_i^2|s_i^{-2}U_i^2 - \eta^2| + |s_n^2 U_n^{-2}\eta^2 - 1| \xrightarrow{\text{a.s.}} 0.$$

Therefore $\max_{i\leqslant n}|T_n^{-1}T_i - s_n^{-2}s_i^2| \xrightarrow{\text{p}} 0$. Let ε, Δ, and $\delta > 0$ and define

$$E_n = \left\{ \sup_{|u-v|\leqslant 2\delta} |\zeta_n(u) - \zeta_n(v)| > \varepsilon \right\}$$

and

$$F_n = \left\{ \max_{i\leqslant n} |T_n^{-1}T_i - s_n^{-2}s_i^2| \leqslant \delta \right\}.$$

Since $\{\zeta_n\}$ is tight we can choose δ and N_1 such that for all $n \geqslant N_1$, $P(E_n) \leqslant \Delta$. Now choose $N \geqslant N_1$ so large that for all $n \geqslant N$, $P(F_n^c) \leqslant \Delta$ and

$$s_n^{-2} \sup_{i\leqslant n} (s_i^2 - s_{i-1}^2) \leqslant \delta.$$

If $n \geqslant N$,

$$\begin{aligned} P&\left(\sup_{z\in[0,1]} |\zeta_n(z) - \eta_n''(z)| > \varepsilon \right) \\ &\leqslant P\left(\max_{i\leqslant n} \sup_{s_{i-1}^2/s_n^2 \leqslant z \leqslant s_i^2/s_n^2} |W(T_i) - W(zT_n)|/\sqrt{T_n} > \varepsilon \right) \\ &\quad + P\left(\max_{i\leqslant n} \sup_{s_{i-1}^2/s_n^2 \leqslant z \leqslant s_i^2/s_n^2} |W(T_{i-1}) - W(zT_n)|/\sqrt{T_n} > \varepsilon \right). \end{aligned}$$

The first term on the right does not exceed

$$\begin{aligned} P&\left(\max_{i\leqslant n} \sup_{s_{i-1}^2/s_n^2 \leqslant z \leqslant s_i^2/s_n^2} |W(T_i) - W(zT_n)|/\sqrt{T_n} > \varepsilon;\, F_n \right) + P(F_n^c) \\ &\leqslant P(E_n) + P(F_n^c) \leqslant 2\Delta. \end{aligned}$$

The second term can be handled in the same way, and so for all $n \geqslant N$,

$$P(\rho(\zeta_n, \eta_n'') > \varepsilon) \leqslant 4\Delta,$$

proving that $\rho(\zeta_n, \eta_n'') \xrightarrow{\text{p}} 0$. ∎

Proof of Theorem 4.4. In view of Theorem 2.23, conditions (4.1) and (4.5) imply that $V_n^{-1}U_n \xrightarrow{\text{p}} 1$. It also follows from Theorem 4.3 that the sequence $\{\xi_n\}$ is tight. Let $\alpha_n(t) = V_n^{-1}U_n\xi_n(t)$, $t \in [0,1]$. Then

$$P\left(\sup_{t\in[0,1]} |\alpha_n(t) - \xi_n(t)| > \varepsilon \right) = P\left(\sup_{t\in[0,1]} |\xi_n(t)|\,|1 - V_n^{-1}U_n| > \varepsilon \right) \to 0$$

as $n \to \infty$. Therefore it suffices to prove that for each $\varepsilon > 0$,

$$P\left(\sup_{t\in[0,1]} |\alpha_n(t) - \zeta_n(t)| > \varepsilon\right) \to 0. \tag{4.19}$$

(Here ζ_n is defined as in Theorem 4.4.) For any $\delta > 0$,

$$P\left(\sup_{t\in[0,1]} |\alpha_n(t) - \zeta_n(t)| > \varepsilon\right) \leqslant P\left(\max_{j\leqslant n} |U_n^{-2}U_j^2 - V_n^{-2}V_j^2| > \delta\right)$$
$$+ P\left(\sup_{|s-t|\leqslant 2\delta} |\alpha_n(s) - \alpha_n(t)| > \varepsilon\right).$$

The second term on the right-hand side can be made arbitrarily small by choosing δ sufficiently small and then n sufficiently large, since the sequence $\{\alpha_n\}$ is tight. The first term is dominated by

$$P\left(U_n^{-2}\max_{j\leqslant n}|U_j^2 - V_j^2| > \delta/2\right) + P(|U_n^{-2}V_n^2 - 1| > \delta/2),$$

which converges to zero as $n \to \infty$, since from Theorem 2.23,

$$s_n^{-2}\max_{j\leqslant n}|U_j^2 - V_j^2| \xrightarrow{\text{p}} 0.$$

This proves (4.19). ∎

4.3. Rates of Convergence for the Invariance Principle in the CLT

Suppose that conditions (4.1) and (4.2) hold with $\eta^2 = 1$, and let α_n be the random element defined by

$$\alpha_n(t) = s_n^{-1}\{S_i + (V_{i+1}^2 - V_i^2)^{-1}(tV_n^2 - V_i^2)X_{i+1}\} \qquad \text{if} \quad V_i^2 \leqslant tV_n^2 < V_{i+1}^2,$$

for $t \in [0,1]$ and $0 \leqslant i \leqslant n-1$. In the notation of Theorem 4.4 we have $\alpha_n(t) = s_n^{-1}V_n\zeta_n(t)$, and so according to that theorem, $\alpha_n \xrightarrow{\text{d}} W$ in the sense (C,ρ). It follows that $h(\alpha_n) \xrightarrow{\text{d}} h(W)$ for all continuous functionals $h\colon C[0,1] \to \mathbb{R}$ (see Theorem A.3 of Appendix II). We shall obtain a rate of convergence in the invariance principle for α_n by studying the rate of convergence of $h(\alpha_n)$ to $h(W)$.

It simplifies the notation if we replace $s_n^{-1}S_i$ by S_i, $1 \leqslant i \leqslant n$. Then

$$\alpha_n(t) = S_i + (V_{i+1}^2 - V_i^2)^{-1}(tV_n^2 - V_i^2)X_{i+1} \quad \text{if} \quad V_i^2 \leqslant tV_n^2 < V_{i+1}^2, \tag{4.20}$$

for $t \in [0,1]$ and $0 \leqslant i \leqslant n-1$. If x and $y \in C[0,1]$, set

$$\rho(x,y) = \sup_{t\in[0,1]} |x(t) - y(t)|.$$

Theorem 4.5. *Let $\{S_i = \sum_1^i X_j, \mathcal{F}_i, 1 \leqslant i \leqslant n\}$ be a zero-mean martingale. Define α_n as in (4.20), and*

$$V_i^2 = \sum_1^i E(X_j^2|\mathcal{F}_{j-1}) \qquad \text{and} \qquad U_i^2 = \sum_1^i X_j^2.$$

Let $h: C[0,1] \to \mathbb{R}$ be a functional with the properties that for constants K and L,

$$|h(x) - h(y)| \leqslant K\rho(x,y) \tag{4.21}$$

and

$$\sup_{-\infty<z<\infty} |P(h(W) \leqslant z) - P(h(W) \leqslant z + \varepsilon)| \leqslant L\varepsilon \tag{4.22}$$

for all $x, y \in C[0,1]$ and $\varepsilon > 0$. Let $0 < \delta < 1$ and define

$$L_n = \sum_{i=1}^n E|X_i|^{2+2\delta} + E|V_n^2 - 1|^{1+\delta}$$

and

$$M_n = \sum_{i=1}^n E|X_i|^{2+2\delta} + E|U_n^2 - 1|^{1+\delta}.$$

There exist constants A_1 and A_2 depending only on K, L and δ such that

$$\Delta_n = \sup_{-\infty<z<\infty} |P(h(\alpha_n) \leqslant z) - P(h(W) \leqslant z)|$$

$$\leqslant A_1 L_n^{1/(3+2\delta)} |\log L_n|^{(1+\delta)/(3+2\delta)} \qquad \text{whenever} \quad L_n \leqslant \tfrac{1}{2} \tag{4.23}$$

$$\leqslant A_2 M_n^{1/(3+2\delta)} |\log M_n|^{(1+\delta)/(3+2\delta)} \qquad \text{whenever} \quad M_n \leqslant \tfrac{1}{2}. \tag{4.24}$$

Setting $h(x) = \sup_{t\in[0,1]} x(t)$ we obtain a rate of convergence in the limit theorem for the maximum of partial sums. If $h(x) = x(1)$, we have a rate of convergence in the central limit theorem. In this case the rate of convergence given by Theorem 3.9 is a little faster, and implies that

$$\Delta_n \leqslant A_1 L_n^{1/(3+2\delta)} \leqslant A_2 M_n^{1/(3+2\delta)}.$$

The same techniques may be used to obtain similar results for random functions different from α_n, and under alternative conditions to (4.21) and (4.22). For example, (4.21) might be replaced by the more general Lipshitz-type condition

$$|h(x) - h(y)| \leqslant K\rho(x,y)^\alpha.$$

Rates of convergence may also be proved under conditions like those in Theorems 3.7, 3.8, and 3.10. For example, we have

Theorem 4.6. *In the notation of Theorem 4.4, suppose that*

$$\max_{i\leqslant n} |X_i| \leqslant n^{-1/2} M \quad \text{a.s.} \tag{4.25}$$

and

$$P(|V_n^2 - 1| > 20M^2 n^{-1/2}(\log n)^2) \leqslant Cn^{-1/4}(\log n)^{3/2} \tag{4.26}$$

for constants M and C. Assume also that (4.21) *and* (4.22) *hold. There exists a constant A depending only on M, C, K, and L such that whenever* $n \geqslant 2$,

$$\sup_{-\infty < z < \infty} |P(h(\alpha_n) \leqslant z) - P(h(W) \leqslant z)| \leqslant An^{-1/4}(\log n)^{3/2}.$$

Results like Theorems 4.5 and 4.6 for sums of independent r.v. have been obtained by Rosenkrantz (1967), Heyde (1969), Sawyer (1967, 1968, 1972), and Fraser (1973).

Proof of Theorems 4.5 and 4.6. As in the proofs given in Section 3.6, there is no real loss of generality in assuming that each $\mathscr{F}_i$ equals the σ-field generated by $X_1, \ldots, X_i$. To obtain the general result from this specialization, use the techniques of remark (ii) following Theorem 3.7.

Extend the domain of W to the positive half-line, so that W denotes standard Brownian motion on $[0,\infty)$. We shall let W_λ denote the restriction of W to $[0,\lambda]$. By defining the martingale $\{S_n\}$ on a new probability space and using the Skorokhod representation (Theorem A.1 of Appendix I) we may assume without loss of generality that there exist nonnegative variables $\tau_1, \ldots, \tau_n$ such that for $T_i = \sum_1^i \tau_j$, $1 \leqslant i \leqslant n$, we have $S_i = W(T_i)$, $1 \leqslant i \leqslant n$. In this case

$$\alpha_n(t) = W(T_i) + (V_{i+1}^2 - V_i^2)^{-1}(tV_n^2 - V_i^2)[W(T_{i+1}) - W(T_i)] \quad \text{if} \quad V_i^2 \leqslant tV_n^2 < V_{i+1}^2.$$

We first prove an analog of Lemma 3.2.

Lemma 4.5. *For all* $\varepsilon > 0$, $0 < \Delta < \frac{1}{2}$ *and* $p \geqslant 1$,

$$P(\rho(\alpha_n, W_1) > \varepsilon) \leqslant 28\varepsilon^{-1}\Delta^{-1/2}\exp(-\varepsilon^2/144\Delta) + \Delta^{-p}\left(E|T_n - V_n^2|^p + \sum_1^n E|X_i|^{2p}\right) + P(|V_n^2 - 1| > \Delta). \tag{4.27}$$

Proof. If $V_i^2 \leqslant V_n^2 t \leqslant V_{i+1}^2$, then

$$|\alpha_n(t) - W(t)| \leqslant \max(|W(T_i) - W(t)|, |W(T_{i+1}) - W(t)|), \tag{4.28}$$

and

$$|W(T_i) - W(t)| \leqslant |W(T_i) - W(V_n^{-2}V_i^2)| + |W(V_n^{-2}V_i^2) - W(t)|.$$

If also $|T_i - V_i^2| \leqslant \Delta$, $|V_{i+1}^2 - V_i^2| \leqslant \Delta$, and $|V_n^2 - 1| \leqslant \Delta$, then

$$|T_i - V_n^{-2}V_i^2| \leqslant |T_i - V_i^2| + |V_i^2 - V_n^{-2}V_i^2| \leqslant |T_i - V_i^2| + |V_n^2 - 1| \leqslant 2\Delta$$

and

$$|V_n^{-2}V_i^2 - t| \leqslant V_n^{-2}\Delta \leqslant 2\Delta.$$

Therefore, on the set

$$\left\{\max_{i\leqslant n}|T_i - V_i^2| \leqslant \Delta, \max_{i\leqslant n}|V_{i+1}^2 - V_i^2| \leqslant \Delta, \text{ and } |V_n^2 - 1| \leqslant \Delta\right\},$$

it follows that

$$|W(T_i) - W(t)| \leqslant 2 \sup_{|s-t|\leqslant 2\Delta} |W(s) - W(t)|.$$

The same inequality holds for $|W(T_{i+1}) - W(t)|$, and combining this with (4.28) we see that

$$\begin{aligned} P(\rho(\alpha_n, W_1) > \varepsilon) &\leqslant P\left(\sup_{|s-t|\leqslant 2\Delta} |W_2(s) - W_2(t)| > \varepsilon/2\right) \\ &\quad + P\left(\max_{i\leqslant n}|T_i - V_i^2| > \Delta\right) + P\left(\max_{i\leqslant n} E(X_i^2|\mathscr{F}_{i-1}) > \Delta\right) \\ &\quad + P(|V_n^2 - 1| > \Delta) \\ &= A_n + B_n + C_n + P(|V_n^2 - 1| > \Delta), \end{aligned} \tag{4.29}$$

say. Now, Theorem A.1 of Appendix I implies that

$$T_i - V_i^2 = \sum_1^i (\tau_j - E(\tau_j|\mathscr{G}_{j-1})) \quad \text{a.s.,} \qquad 1 \leqslant i \leqslant n,$$

where $\mathscr{G}_j$ is the σ-field generated by $W(T_1), \ldots, W(T_j)$ and by $W(t)$ for $0 \leqslant t \leqslant T_j$. Therefore $\{T_i - V_i^2, \mathscr{G}_i, 1 \leqslant i \leqslant n\}$ is a martingale, and by Kolmogorov's inequality (see Corollary 2.1),

$$B_n \leqslant \Delta^{-p} E|T_n - V_n^2|^p \tag{4.30}$$

for any $p \geqslant 1$. Furthermore,

$$\begin{aligned} C_n &\leqslant \Delta^{-p} E\left[\max_{i\leqslant n} E(X_i^2|\mathscr{F}_{i-1})^p\right] \\ &\leqslant \Delta^{-p} \sum_1^n E[E(X_i^2|\mathscr{F}_{i-1})^p] \\ &\leqslant \Delta^{-p} \sum_1^n E|X_i|^{2p}, \end{aligned} \tag{4.31}$$

using Jensen's inequality. It remains to estimate A_n. Now,

$$\begin{aligned}
A_n &= P\left(\sup_{|s-t|\leqslant 2\Delta} |W_2(s) - W_2(t)| > \varepsilon/2\right) \\
&\leqslant P\left(\bigcup_{0\leqslant i\Delta<1}\left\{\sup_{2i\Delta<t\leqslant 2(i+1)\Delta} |W(2i\Delta) - W(t)| > \varepsilon/6\right\}\right) \\
&\leqslant \sum_{0\leqslant i\Delta<1} P\left(\sup_{2i\Delta<t\leqslant 2(i+1)\Delta} |W(2i\Delta) - W(t)| > \varepsilon/6\right) \\
&\leqslant 2\Delta^{-1}P\left(\sup_{0<t\leqslant 2\Delta} |W(t)| > \varepsilon/6\right) \\
&= 2\Delta^{-1}P\left(\sup_{0<t\leqslant 1} |W(t)| > \varepsilon/6(2\Delta)^{1/2}\right) \\
&\leqslant 8\Delta^{-1}P(N > \varepsilon/6(2\Delta)^{1/2}) \qquad [\text{where } N \text{ is } N(0,1)] \\
&\leqslant 8\Delta^{-1}(2\pi)^{-1/2}[\varepsilon/6(2\Delta)^{1/2}]^{-1}\exp[-\tfrac{1}{2}(\varepsilon/6(2\Delta)^{1/2})^2] \\
&\leqslant 28\varepsilon^{-1}\Delta^{-1/2}\exp(-\varepsilon^2/144\Delta). \qquad (4.32)
\end{aligned}$$

Substituting the estimates (4.30)–(4.32) into (4.29), we obtain (4.27). ∎

We have from (4.21) and (4.22) that for all real z and any $\varepsilon > 0$,

$$\begin{aligned}
P(h(\alpha_n) \leqslant z) &\leqslant P(h(\alpha_n) \leqslant z;\, \rho(\alpha_n, W_1) \leqslant \varepsilon) + P(\rho(\alpha_n, W_1) > \varepsilon) \\
&\leqslant P(h(W_1) \leqslant z + K\varepsilon) + P(\rho(\alpha_n, W_1) > \varepsilon) \\
&\leqslant P(h(W_1) \leqslant z) + KL\varepsilon + P(\rho(\alpha_n, W_1) > \varepsilon).
\end{aligned}$$

Similarly, $P(h(\alpha_n) \leqslant z) \geqslant P(h(W_1) \leqslant z) - KL\varepsilon - P(\rho(\alpha_n, W_1) > \varepsilon)$, and from Lemma 4.5 we see that if $0 < \Delta < \frac{1}{2}$,

$$\begin{aligned}
\Delta_n = &\sup_{-\infty<z<\infty} |P(h(\alpha_n) \leqslant z) - P(h(W_1) \leqslant z)| \\
&\leqslant KL\varepsilon + 28\varepsilon^{-1}\Delta^{-1/2}\exp(-\varepsilon^2/144\Delta) \\
&+ \Delta^{-p}\left(E|T_n - V_n^2|^p + \sum_1^n E|X_i|^{2p}\right) + P(|V_n^2 - 1| > \Delta). \qquad (4.33)
\end{aligned}$$

We first prove Theorem 4.5, where $p = 1 + \delta$, $0 < \delta \leqslant 1$. As in the inequalities (3.82) we have

$$E|T_n - V_n^2|^{1+\delta} \leqslant C\sum_1^n E(\tau_i^{1+\delta}) \leqslant C'\sum_1^n E|X_i|^{2+2\delta},$$

where C and C' depend only on δ. Furthermore,

$$P(|V_n^2 - 1| > \Delta) \leqslant \Delta^{-1-\delta}E|V_n^2 - 1|^{1+\delta},$$

and substituting these estimates into (4.33) we see that for a constant C depending only on K, L, and δ,

$$\Delta_n \leqslant C(\varepsilon + \varepsilon^{-1}\Delta^{-1/2}\exp(-\varepsilon^2/144\Delta) + \Delta^{-1-\delta}L_n).$$

The best rate of convergence is obtained by setting

$$\varepsilon = L_n^{1/(3+2\delta)}|\log L_n|^{(1+\delta)/(3+2\delta)}$$

and

$$\Delta = \tfrac{1}{144}L_n^{2/(3+2\delta)}|\log L_n|^{-1/(3+2\delta)}.$$

Then $\Delta \leqslant \frac{1}{2}$ if $L_n \leqslant \frac{1}{2}$, $\varepsilon^2/144\Delta = |\log L_n|$, and

$$\begin{aligned}\Delta_n &\leqslant C(1 + 12L_n^{2\delta/(3+2\delta)}|\log L_n|^{-(3+4\delta)/2(3+2\delta)} + 144^{1+\delta})\\ &\quad\times L_n^{1/(3+2\delta)}|\log L_n|^{(1+\delta)/(3+2\delta)}\\ &\leqslant C'L_n^{1/(3+2\delta)}|\log L_n|^{(1+\delta)/(3+2\delta)}\end{aligned}$$

if $L_n \leqslant \frac{1}{2}$, where C' depends only on K, L, and δ. This proves (4.23), and (4.24) follows on noting that by Theorem 2.10,

$$\begin{aligned}E|U_n^2 - V_n^2|^{1+\delta} &= E\left|\sum_1^n (X_i^2 - E(X_i^2|\mathscr{F}_{i-1}))\right|^{1+\delta}\\ &\leqslant CE\left|\sum_1^n (X_i^2 - E(X_i^2|\mathscr{F}_{i-1}))^2\right|^{(1+\delta)/2}\\ &\leqslant C\sum_1^n E|X_i^2 - E(X_i^2|\mathscr{F}_{i-1})|^{1+\delta} \qquad \text{(since } \delta \leqslant 1)\\ &\leqslant C'\sum_1^n E\{|X_i|^{2+2\delta} + E(X_i^2|\mathscr{F}_{i-1})^{1+\delta}\}\\ &\leqslant 2C'\sum_1^n E|X_i|^{2+2\delta},\end{aligned}$$

where C and C' depend only on δ.

Now we prove Theorem 4.6. From the inequalities (3.67) and (3.69) we see that, setting $q = (1 - p^{-1})^{-1}$,

$$\begin{aligned}E|T_n - V_n^2|^p &= E\left|\sum_1^n (\tau_i - E(\tau_i|\mathscr{G}_{i-1}))\right|^p\\ &\leqslant (18pq^{1/2})^p n^{p/2} \max_{i\leqslant n} E|\tau_i - E(\tau_i|\mathscr{G}_{i-1})|^p\\ &\leqslant 4(18pq^{1/2})^p n^{p/2}\Gamma(p+1) \max_{i\leqslant n} E|X_i|^p,\end{aligned}$$

using result (iii) from Theorem A.1 of Appendix I. Condition (4.25) implies that

$$E|X_i|^{2p} \leqslant n^{-p}M^{2p},$$

and since $q \leqslant 2$ for $p \geqslant 2$, and

$$\Gamma(p+1) \leqslant (2\pi)^{1/2}p^{p+1/2}e^{-p+1/24},$$

it follows that

$$E|T_n - V_n^2|^p \leqslant 10.5(9.4p^2M^2)^p p^{1/2} n^{-p/2}$$

for all $p \geqslant 2$. Furthermore,

$$\sum_1^n E|X_i|^{2p} \leqslant n^{-p+1}M^{2p} \leqslant (0.1)(9.4p^2M^2)^p p^{1/2} n^{-p/2}$$

for $p \geqslant 2$, and using these estimates in (4.33) we see that

$$\begin{aligned}\Delta_n \leqslant KL\varepsilon &+ 28\varepsilon^{-1}\Delta^{-1/2}\exp(-\varepsilon^2/144\Delta) \\ &+ \Delta^{-p}10.6(9.4p^2M^2)^p p^{1/2} n^{-p/2} + P(|V_n^2 - 1| > \Delta). \qquad (4.34)\end{aligned}$$

Let

$$\Delta = 20M^2n^{-1/2}(\log n)^2 \qquad \text{and} \qquad \varepsilon = 12(20M^2)^{1/2}n^{-1/4}(\log n)^{3/2},$$

where n is sufficiently large for $\Delta \leqslant \frac{1}{2}$. Then

$$\begin{aligned}\varepsilon^{-1}\Delta^{-1/2}\exp(-\varepsilon^2/144\Delta) &= (240M^2)^{-1}n^{-1/2}(\log n)^{-5/2} \\ &= o(n^{-1/4}(\log n)^{3/2})\end{aligned}$$

as $n \to \infty$, and with $p = \log n$,

$$\begin{aligned}\Delta^{-p}10.6(9.4p^2M^2)^p p^{1/2} n^{-p/2} &= 10.6(0.47)^{\log n}(\log n)^{1/2} \\ &\leqslant 10.6n^{-0.7}(\log n)^{1/2} \\ &= o(n^{-1/4}(\log n)^{3/2}).\end{aligned}$$

Substituting in (4.34) we see that

$$\Delta_n \leqslant Bn^{-1/4}(\log n)^{3/2} + P(|V_n^2 - 1| > \Delta),$$

where B depends only on M, C, K, and L. Theorem 4.6 follows from this and (4.26). ∎

4.4. The Law of the Iterated Logarithm and its Invariance Principle

The LIL for sums of independent r.v. grew from the early efforts of Hausdorff (1913) and Hardy and Littlewood (1914) to determine the rate of convergence in the law of large numbers for normal numbers. The law for Bernoulli variables was obtained by Khintchine (1924), and for more general sequences by Kolmogorov (1929). The LIL's principal importance in applications is still as a rate of convergence. Let $X_1, X_2, \ldots$ be i.i.d. r.v. with zero means and unit variances, and let $S_n = \sum_1^n X_i$. Lindeberg's CLT implies that $n^{-1/2}S_n \xrightarrow{d} N(0,1)$, although it is well known that the sequence $\{n^{-1/2}S_n\}$ has as its set of a.s. limit points the whole real line. Hartman and Wintner's (1941) LIL describes this behavior in much more detail, asserting that the lim sup and lim inf of the sequence $\{(2n\log\log n)^{-1/2}S_n\}$ are, respectively, ± 1, with probability 1.

Donsker's (1951) extension of the CLT implies that the random function ξ_n defined by

$$\xi_n(t) = n^{-1/2}\{S_i + (nt - i)X_{i+1}\} \qquad \text{for} \quad i \leqslant nt \leqslant i + 1, \quad t \in [0,1],$$

converges in distribution to standard Brownian motion on $[0,1]$. Strassen (1964) extended Hartman and Wintner's LIL to obtain a rate of convergence in Donsker's invariance principle. Let K be the set of absolutely continuous functions x in $C[0,1]$ with $x(0) = 0$ and whose derivatives $\dot{x}$ are such that

$$\int_0^1 \dot{x}^2(t)\,dt \leqslant 1.$$

Then K is compact. The sequence $\{(2\log\log n)^{-1/2}\xi_n\}$ is relatively compact with a.s. limit set K.

An early LIL for martingales was stated by Lévy (1954), but it was some time before martingale results were developed which approach the generality of those for sums of independent r.v. Stout (1970a) obtained the martingale analog of Kolmogorov's LIL and (1970b) an analog of the Hartman–Wintner law. Heyde and Scott (1973) extended Strassen's (1964) invariance principle in the LIL to the martingale case. A further extension was provided by Hall and Heyde (1976).

We have seen in Chapter 3 that the most general CLTs involve a random norming, and so it is to be expected that their associated LILs will also use such a norming. This approach has been adopted by a number of workers. For example, Jain *et al.* (1975) used a norming based on the conditional variances to obtain sharp estimates of tail probabilities for martingales. They applied these to give an integral test for the LIL. Freedman (1975) obtained an LIL by constructing bounds on the distributions of upcrossing times.

Our aim in this section is to prove an invariance principle for the LIL, extending Strassen's result. We use the same approach as Hall and Heyde (1976), applying the Skorokhod representation. This method provides sharp results without the technicalities involved in estimating tail probabilities.

Let $\{S_n, \mathscr{F}_n\}$ be a zero-mean, square-integrable martingale and suppose that the σ-field $\mathscr{F}_n$ is generated by $S_1, S_2, \ldots, S_n$. Let ξ_n be the random function considered in Theorem 4.1, namely,

$$\xi_n(t) = U_n^{-1}\{S_i + X_{i+1}^{-2}(tU_n^2 - U_i^2)X_{i+1}\} \quad \text{for } U_i^2 \leqslant tU_n^2 < U_{i+1}^2,\ i \leqslant n-1,$$

where $U_i^2 = \sum_1^i X_j^2$. Define $\zeta_n = (2\log\log U_n)^{-1/2}\xi_n$, $n \geqslant 1$. (In order to avoid difficulties in specification we will suppose that $X_1^2 > 0$ a.s. and adopt the convention that $\log\log x = 1$ if $0 < x \leqslant e^e$.) By analogy with Strassen's invariance principle we would expect that under certain conditions the sequence $\{\zeta_n, n \geqslant 1\}$ would be relatively compact with limit set K.

However, $\{U_n\}$ is just one possible norming sequence. Another would be the sequence $\{V_n\}$. We shall accordingly formulate an invariance principle for a general norming sequence. Let $\{W_n, n \geqslant 1\}$ be r.v. such that $0 < W_1 \leqslant W_2 \leqslant \cdots$, and define the continuous function ζ_n on $[0,1]$ by

$$\zeta_n(t) = [\phi(W_n^2)]^{-1}\{S_i + (W_{i+1}^2 - W_i^2)^{-1}(tW_n^2 - W_i^2)X_{i+1}\}$$
$$\text{for } \quad W_i^2 \leqslant tW_n^2 < W_{i+1}^2, \quad i \leqslant n-1,$$

where $\phi(t) = (2t\log\log t)^{1/2}$.

Theorem 4.7. *Let $\{Z_n, n \geqslant 1\}$ be a sequence of nonnegative r.v. and suppose that Z_n and W_n are $\mathscr{F}_{n-1}$-measurable. If*

$$\lim_{n\to\infty} [\phi(W_n^2)]^{-1}\sum_1^n \{X_i I(|X_i| > Z_i) - E[X_i I(|X_i| > Z_i)|\mathscr{F}_{i-1}]\} \xrightarrow{\text{a.s.}} 0, \tag{4.35}$$

$$W_n^{-2}\sum_1^n \{E[X_i^2 I(|X_i| \leqslant Z_i)|\mathscr{F}_{i-1}] - [E(X_i I(|X_i| \leqslant Z_i)|\mathscr{F}_{i-1})]^2\} \xrightarrow{\text{a.s.}} 1, \tag{4.36}$$

$$\sum_1^\infty W_i^{-4} E[X_i^4 I(|X_i| \leqslant Z_i)|\mathscr{F}_{i-1}] < \infty \quad a.s., \tag{4.37}$$

and

$$W_{n+1}^{-1}W_n \xrightarrow{\text{a.s.}} 1 \quad \text{and} \quad W_n \xrightarrow{\text{a.s.}} \infty, \tag{4.38}$$

then

with probability 1 the sequence $\{\zeta_n\}$ is relatively compact in $C[0,1]$ and its set of a.s. limit points coincides with K. (4.39)

(The restriction that W_n be $\mathscr{F}_{n-1}$-measurable does not exclude the norming sequence U_n. If $W_n = U_{n-1}$, then W_n is $\mathscr{F}_{n-1}$-measurable, and (4.38) ensures that norming by U_n is asymptotically equivalent to norming by U_{n-1}.)

Corollary 4.1. *If $\{S_n, \mathscr{F}_n\}$ is a martingale with uniformly bounded differences, then* (4.39) *holds with $W_n = V_n$ on the set $\{V_n \to \infty\}$.*

(This extends Neveu's (1975) Proposition VII.2.7. A variant of Example 1 of Chapter 3 shows that it is possible for a martingale to have uniformly bounded differences and satisfy the condition $V_n \to \infty$ a.s. but not satisfy the CLT $S_n/V_n \xrightarrow{d} N(0,1)$.)

Corollary 4.2. *If*

$$s_n^{-2}U_n^2 \xrightarrow{\text{a.s.}} \eta^2 > 0 \quad a.s., \tag{4.40}$$

$$\textit{for all} \quad \varepsilon > 0, \qquad \sum_1^\infty s_i^{-1} E[|X_i| I(|X_i| > \varepsilon s_i)] < \infty, \tag{4.41}$$

and for some $\delta > 0$,

$$\sum_1^\infty s_i^{-4} E[X_i^4 I(|X_i| \leqslant \delta s_i)] < \infty, \tag{4.42}$$

then (4.39) *holds with* $W_n = U_n$.

Before proving Theorem 4.7 and the corollaries above, let us verify that the functional form of the LIL, condition (4.39), implies the classical form.

Theorem 4.8. *If* (4.39) *holds, then*

$$\limsup_{n\to\infty} [\phi(W_n^2)]^{-1} S_n = +1 \quad a.s.$$

and

$$\liminf_{n\to\infty} [\phi(W_n^2)]^{-1} S_n = -1 \quad a.s.$$

Proof. Note first that for any $x \in K$,

$$x(t)^2 = \left[\int_0^t \dot{x}(s)\, ds\right]^2 \leqslant \left[\int_0^t 1^2\, ds\right]\left[\int_0^t \dot{x}^2(s)\, ds\right] \leqslant t,$$

using the Cauchy–Schwarz inequality, so that $|x(t)| \leqslant t^{1/2}$. It follows that $\sup_{t\in[0,1]}|x(t)| \leqslant 1$. Hence from (4.39),

$$\limsup_{n\to\infty} \sup_{t\in[0,1]} |\zeta_n(t)| \leqslant 1 \quad \text{a.s.}$$

and, letting $t = 1$,

$$\limsup_{n\to\infty} [\phi(W_n^2)]^{-1} |S_n| \leqslant 1 \quad \text{a.s.}$$

If we show that

$$\limsup_{n\to\infty} [\phi(W_n^2)]^{-1} S_n \geqslant 1 \quad \text{a.s.}, \tag{4.43}$$

then the LIL will follow by symmetry. Let $x(t) \equiv t$, $t \in [0,1]$. Then $x \in K$ and so for almost all ω there is a subsequence $n_k = n_k(\omega)$ such that

$$\zeta_{n_k}(\cdot)(\omega) \to x(\cdot),$$

the convergence being in the uniform metric on $[0,1]$. In particular, since $x(1) = 1$, $\zeta_{n_k}(1)(\omega) \to 1$. That is,

$$[\phi(W_{n_k}^2(\omega))]^{-1} S_{n_k}(\omega) \to 1,$$

which implies (4.43). ∎

Proof of Theorem 4.7. Strassen (1964) based his proof on a limit law for Brownian motion, which we state here in the interests of clarity. Let $u > e$ and define

$$\alpha_u(t) = (2u \log\log u)^{-1/2} W(ut), \qquad t \in [0,1],$$

where $W(t)$, $t \geqslant 0$, is standard Brownian motion on $[0,\infty)$.

Lemma 4.6. *With probability* 1 *the sequence* $\{\alpha_u\}$ *is relatively compact, and the set of its a.s. limit points coincides with K.*

As in the proof of Theorem 4.1, our approach is via a limit theorem for Brownian motion.

Theorem B. *Let* $W(t)$, $t \geqslant 0$, *be standard Brownian motion and let* $\{T_n, n \geqslant 1\}$ *and* $\{W_n, n \geqslant 1\}$ *be nondecreasing sequences of positive r.v. Set* $S_n^* = W(T_n)$ *and* $X_n^* = S_n^* - S_{n-1}^*$, *and let* ζ_n^* *be the random element of* $C[0,1]$ *defined by*

$$\zeta_n^*(t) = [\phi(W_n^2)]^{-1}\{S_i^* + (W_{i+1}^2 - W_i^2)^{-1}(tW_n^2 - W_i^2)X_{i+1}^*\}$$
$$\text{for} \quad W_i^2 \leqslant tW_n^2 < W_{i+1}^2, \quad i \leqslant n-1.$$

If

$$T_n \xrightarrow{\text{a.s.}} \infty, \qquad T_{n+1}^{-1} T_n \xrightarrow{\text{a.s.}} 1, \qquad \text{and} \qquad T_n^{-1} W_n^2 \xrightarrow{\text{a.s.}} 1, \quad (4.44)$$

then with probability 1 *the sequence* $\{\zeta_n^*\}$ *is relatively compact and the set of its a.s. limit points coincides with K.*

Proof. Define $\beta(t)$ for $t \in [0,\infty)$ by

$$\beta(t) = S_p^* + (W_{p+1}^2 - W_p^2)^{-1}(t - W_p^2)X_{p+1}^*,$$

where

$$p = p(t) = \max\{i | W_i^2 \leqslant t\}.$$

Then

$$\zeta_n^*(t) = [\phi(W_n^2)]^{-1}\beta(W_n^2 t),$$

and so in view of Lemma 4.6 it suffices to prove that under (4.44),

$$\lim_{t\to\infty}[\phi(t)]^{-1}|\beta(t) - W(t)| = 0 \quad \text{a.s.} \tag{4.45}$$

Suppose that $|1 - t^{-1}s| \leqslant \varepsilon < \frac{1}{2}$ and let $u = t(1+\varepsilon)$. Then for large t,

$$\begin{aligned}[\phi(t)]^{-1}|W(s) - W(t)| &\leqslant 2[\phi(u)]^{-1}|W(s) - W(t)| \\ &= 2|\alpha_u(u^{-1}s) - \alpha_u(u^{-1}t)| \\ &\leqslant 4 \sup_{1-2\varepsilon\leqslant r\leqslant 1} |\alpha_u(r) - \alpha_u(1)|,\end{aligned}$$

where α_u is as in Lemma 4.6. Let K_ε denote the set of functions distant $<\varepsilon$ from K. Lemma 4.6 implies that if λ is sufficiently large,

$$P(\alpha_{t(1+\varepsilon)} \in K_\varepsilon \text{ for all } t > \lambda) \geqslant 1 - \varepsilon,$$

and if $\alpha_u \in K_\varepsilon$, then for some $x \in K$,

$$\sup_{1-2\varepsilon\leqslant r\leqslant 1} |\alpha_u(r) - \alpha_u(1)| \leqslant 2\varepsilon + \sup_{1-2\varepsilon\leqslant r\leqslant 1} |x(r) - x(1)|.$$

But

$$|x(r) - x(1)| = \left|\int_r^1 \dot{x}(t)\,dt\right| \leqslant \left[\left(\int_r^1 1^2\,dt\right)\left(\int_r^1 \dot{x}^2(t)\,dt\right)\right]^{1/2} \leqslant (1-r)^{1/2},$$

and consequently

$$\sup_{1-2\varepsilon\leqslant r\leqslant 1} |\alpha_u(r) - \alpha_u(1)| \leqslant 2\varepsilon + (2\varepsilon)^{1/2}$$

if $\alpha_u \in K_\varepsilon$. Since

$$|\beta(t) - W(t)| \leqslant \max\{|W(T_{p(t)}) - W(t)|, |W(T_{p(t)+1}) - W(t)|\},$$

then combining the results above we deduce that if λ is sufficiently large,

$$\begin{aligned}&P([\phi(t)]^{-1}|\beta(t) - W(t)| > 4(2\varepsilon + (2\varepsilon)^{1/2}) \text{ for some } t > \lambda) \\ &\quad \geqslant 1 - \varepsilon - P(|1 - t^{-1}T_{p(t)}| > \varepsilon \text{ for some } t > \lambda) \\ &\qquad - P(|1 - t^{-1}T_{p(t)+1}| > \varepsilon \text{ for some } t > \lambda).\end{aligned}$$

Therefore (4.45) will follow if we show that

$$t^{-1}T_{p(t)} \xrightarrow{\text{a.s.}} 1 \quad \text{and} \quad t^{-1}T_{p(t)+1} \xrightarrow{\text{a.s.}} 1 \quad \text{as} \quad t \to \infty.$$

Now, (4.44) implies that $W_{n+1}^{-2}W_n^2 \xrightarrow{\text{a.s.}} 1$, and hence

$$1 \geqslant t^{-1}W_{p(t)}^2 \geqslant W_{p(t)+1}^{-2}W_{p(t)}^2 \xrightarrow{\text{a.s.}} 1,$$

so that

$$t^{-1}W^2_{p(t)} \xrightarrow{\text{a.s.}} 1.$$

Similarly,

$$t^{-1}W^2_{p(t)+1} \xrightarrow{\text{a.s.}} 1,$$

and these combined with (4.44) give the required result. ∎

To establish Theorem 4.7 we first define

$$\begin{aligned}\tilde{X}_i = X_i I(c_i < |X_i| \leqslant Z_i) + \tfrac{1}{2}X_i I(|X_i| \leqslant c_i) \\ + \tfrac{1}{2}\operatorname{sgn}(X_i)c_i(1 + Z_i|X_i|^{-1})I(|X_i| > Z_i),\end{aligned}$$

where $\{c_i, i \geqslant 1\}$ is a monotone sequence of positive constants with $c_i \to 0$ as $i \to \infty$ so fast that

$$\sum_1^\infty c_i < \infty \qquad \text{and} \qquad \sum_1^\infty c_i Z_i W_i^{-2} < \infty \quad \text{a.s.}$$

(If $Z_i(\omega) < c_i$, let $I(c_i < |X_i| \leqslant Z_i)(\omega) = 0$.) Set also

$$X_i^* = \tilde{X}_i - E(\tilde{X}_i|\mathscr{F}_{i-1}).$$

It is easily checked that $\tilde{X}_1, \ldots, \tilde{X}_n$ and hence $X_1^*, \ldots, X_n^*$ also generate the σ-field $\mathscr{F}_n$.

Set $S_n^* = \sum_1^n X_i^*$ and $V_n^{*2} = \sum_1^n E(X_i^{*2}|\mathscr{F}_{i-1})$, and define ζ_n^* as in the statement of Theorem B. From the definition of X_i^*,

$$|X_i - X_i^* - \{X_i I(|X_i| > Z_i) - E[X_i I(|X_i| > Z_i)|\mathscr{F}_{i-1}]\}| \leqslant 3c_i,$$

and so

$$\begin{aligned}&\sup_{0\leqslant t\leqslant 1} |\zeta_n(t) - \zeta_n^*(t)| \\ &\quad \leqslant |\phi(W_n^2)|^{-1} \sup_{1\leqslant j\leqslant n} \left|\sum_{i=1}^j (X_i - X_i^*)\right| \\ &\quad \leqslant |\phi(W_n^2)|^{-1} \sup_{1\leqslant j\leqslant n} \left|\sum_1^j \{X_i I(|X_i| > Z_i) - E[X_i I(|X_i| > Z_i)|\mathscr{F}_{i-1}]\}\right| \\ &\qquad + |\phi(W_n^2)|^{-1} \sum_1^n 3c_i \xrightarrow{\text{a.s.}} 0 \qquad (4.46)\end{aligned}$$

as $n \to \infty$, in view of (4.35).

Next we introduce the Skorokhod representation. By extending the original probability space if necessary, we may suppose that there exists a Brownian motion W and a sequence $\{T_n, n \geqslant 1\}$ of nonnegative r.v. defined on our probability space such that $S_n^* = W(T_n)$ almost surely for all n. Let $\tau_n = T_n - T_{n-1}$, $n \geqslant 1$ ($T_0 = 0$). If $\mathscr{G}_n$ is the σ-field generated by $X_1, \ldots, X_n$

and $W(u)$ for $u < T_n$, then τ_n is $\mathscr{G}_n$-measurable,

$$E(\tau_n|\mathscr{G}_{n-1}) = E(X_n^{*2}|\mathscr{G}_{n-1}) = E(X_n^{*2}|\mathscr{F}_{n-1}) \quad \text{a.s.},$$

and for some constant L,

$$E(\tau_n^2|\mathscr{G}_{n-1}) \leqslant LE(X_n^{*4}|\mathscr{G}_{n-1}) = LE(X_n^{*4}|\mathscr{F}_{n-1}) \quad \text{a.s.}$$

In view of (4.38), (4.46) and Theorem B, it suffices to prove that

$$W_n^{-2}T_n \xrightarrow{\text{a.s.}} 1. \tag{4.47}$$

To this end we will first show that

$$T_n - V_n^{*2} = o(W_n^2) \quad \text{a.s.} \qquad \text{as} \quad n \to \infty. \tag{4.48}$$

Since

$$\begin{aligned} E(X_i^{*4}|\mathscr{F}_{i-1}) &= E(\tilde{X}_i^4|\mathscr{F}_{i-1}) - 4E(\tilde{X}_i^3|\mathscr{F}_{i-1})E(\tilde{X}_i|\mathscr{F}_{i-1}) \\ &\quad + 6E(\tilde{X}_i^2|\mathscr{F}_{i-1})[E(\tilde{X}_i|\mathscr{F}_{i-1})]^2 - 3[E(\tilde{X}_i|\mathscr{F}_{i-1})]^4 \\ &\leqslant 11E(\tilde{X}_i^4|\mathscr{F}_{i-1}) \\ &\leqslant 11E[X_i^4 I(|X_i| \leqslant Z_i)|\mathscr{F}_{i-1}] + 11c_i^4, \end{aligned}$$

we have from (4.37) that

$$\sum_1^\infty W_i^{-4}E(X_i^{*4}|\mathscr{F}_{i-1}) < \infty \quad \text{a.s.}$$

and hence, using Theorem 2.18,

$$\sum_1^n [\tau_i - E(\tau_i|\mathscr{G}_{i-1})] = o(W_n^2) \quad \text{a.s.},$$

which is equivalent to (4.48).

Next, we have

$$|E(\tilde{X}_i^2|\mathscr{F}_{i-1}) - E[X_i^2 I(|X_i| \leqslant Z_i)|\mathscr{F}_{i-1}]| \leqslant 2c_i^2, \tag{4.49}$$

$$E(X_i^{*2}|\mathscr{F}_{i-1}) = E(\tilde{X}_i^2|\mathscr{F}_{i-1}) - [E(\tilde{X}_i|\mathscr{F}_{i-1})]^2, \tag{4.50}$$

and

$$|E(\tilde{X}_i|\mathscr{F}_{i-1}) - E[X_i I(|X_i| \leqslant Z_i)|\mathscr{F}_{i-1}]| \leqslant 2c_i. \tag{4.51}$$

Condition (4.51) implies that

$$\sum_1^n \{[E(\tilde{X}_i|\mathscr{F}_{i-1})]^2 - [E(X_i I(|X_i| \leqslant Z_i)|\mathscr{F}_{i-1})]^2\} = o(W_n^2) \quad \text{a.s.} \tag{4.52}$$

since

$$\sum_1^n c_i|E[X_i I(|X_i| \leqslant Z_i)|\mathscr{F}_{i-1}]| = o(W_n^2) \quad \text{a.s.},$$

by virtue of Kronecker's lemma and the fact that $\sum_1^\infty c_i Z_i W_i^{-2} < \infty$ a.s. Conditions (4.36), (4.49), (4.50), and (4.52) now imply that

$$V_n^{*2} - W_n^2 = o(W_n^2) \quad \text{a.s.}$$

Combined with (4.48) this establishes (4.47) and completes the proof. ▮

Proof of Corollary 4.1. Suppose that $|X_n| \leqslant C < \infty$ for all n. Let $Z_n = C + 1$ and

$$W_n^2 = V_n^2 = \sum_1^n E(X_i^2 | \mathscr{F}_{i-1}).$$

Conditions (4.35) and (4.36) hold trivially while (4.38) holds on the set $\{V_n \to \infty\}$. The series in (4.37) reduces to

$$\sum_1^\infty V_i^{-4} E(X_i^4 | \mathscr{F}_{i-1}) \leqslant C^2 \sum_1^\infty V_i^{-4} E(X_i^2 | \mathscr{F}_{i-1}),$$

which is finite on the set $\{V_n \to \infty\}$. To see this, let a_n, $n \geqslant 1$, be any sequence of nonnegative numbers with $a_1 > 0$ and such that $a_n / \sum_1^n a_i \to 0$ as $n \to \infty$. Set $b_n = \sum_1^n a_i$ and $b_0 = 0$. Then

$$\sum_1^n b_i^{-2} a_i = \sum_1^n b_i^{-2}(b_i - b_{i-1}) = \sum_1^{n-1} b_i(b_i^{-2} - b_{i+1}^{-2}) + b_n^{-1}$$

$$\leqslant 2 \sum_1^{n-1} (b_i^{-1} - b_{i+1}^{-1}) + b_n^{-1} = 2b_1^{-1} - b_n^{-1}. \quad ▮$$

Proof of Corollary 4.2. First we shall establish (4.39) for $W_n = U_{n-1}$. Let $Z_j = \delta s_j$ in Theorem 4.7. In view of (4.40) it suffices to verify (4.35)–(4.38) with W_n replaced by s_n (except that in (4.36), the limit then has to be η^2 rather than 1).

Condition (4.41) implies that

$$\sum_1^\infty s_i^{-1} |X_i| I(|X_i| > \varepsilon s_i) < \infty \quad \text{a.s.},$$

and so by Kronecker's lemma,

$$s_n^{-1} \max_{i \leqslant n} |X_i| I(|X_i| > \varepsilon s_i) \leqslant s_n^{-1} \sum_1^n |X_i| I(|X_i| > \varepsilon s_i) \xrightarrow{\text{a.s.}} 0.$$

It follows that for any $\varepsilon > 0$,

$$s_n^{-1} \max_{i \leqslant n} |X_i| \leqslant \varepsilon + s_n^{-1} \max_{i \leqslant n} |X_i| I(|X_i| > \varepsilon s_i) \xrightarrow{\text{a.s.}} \varepsilon.$$

Hence

$$s_n^{-2} \max_{i \leqslant n} X_i^2 \xrightarrow{\text{a.s.}} 0. \tag{4.53}$$

Combined with (4.40) this implies that

$$1 - s_{n+1}^{-2}s_n^2 = U_n^{-2}s_n^2(s_{n+1}^{-2}X_{n+1}^2 + s_n^{-2}U_n^2 - s_{n+1}^{-2}U_{n+1}^2) \xrightarrow{\text{a.s.}} 0,$$

proving (4.38).

Condition (4.35) follows from (4.40), (4.41), and Kronecker's lemma, while (4.37) follows from (4.40) and (4.42). To prove (4.36) note that

$$\begin{aligned} s_n^{-2}\sum_1^n\{E[X_i^2I(|X_i| \leqslant \delta s_i)|\mathscr{F}_{i-1}] - [E(X_iI(|X_i| \leqslant \delta s_i)|\mathscr{F}_{i-1})]^2\} \\ = s_n^{-2}U_n^2 - s_n^{-2}\sum_1^n X_i^2I(|X_i| > \delta s_i) - s_n^{-2}\sum_1^n[E(X_iI(|X_i| \leqslant \delta s_i)|\mathscr{F}_{i-1})]^2 \\ + s_n^{-2}\sum_1^n\{E[X_i^2I(|X_i| \leqslant \delta s_i)|\mathscr{F}_{i-1}] - X_i^2I(|X_i| \leqslant \delta s_i)\}. \end{aligned}$$

The first term on the right-hand side converges to η^2 a.s. The second is dominated by

$$\left(s_n^{-1}\max_{i \leqslant n}|X_i|\right)s_n^{-1}\sum_1^n|X_i|I(|X_i| > \delta s_i) \xrightarrow{\text{a.s.}} 0,$$

and the absolute value of the third equals

$$\begin{aligned} s_n^{-2}\sum_1^n|E[X_iI(|X_i| \leqslant \delta s_i)|\mathscr{F}_{i-1}]E[X_iI(|X_i| > \delta s_i)|\mathscr{F}_{i-1}]| \\ \leqslant \delta s_n^{-1}\sum_1^n E[|X_i|I(|X_i| > \delta s_i)|\mathscr{F}_{i-1}] \xrightarrow{\text{a.s.}} 0 \end{aligned}$$

(using (4.41) and Kronecker's lemma). Hence to establish (4.36) it suffices to show that

$$\sum_1^\infty s_i^{-2}\{X_i^2I(|X_i| \leqslant \delta s_i) - E[X_i^2I(|X_i| \leqslant \delta s_i)|\mathscr{F}_{i-1}]\}$$

converges a.s. The terms in the series are martingale differences and so the series' mean square equals

$$\begin{aligned} \sum_1^\infty s_i^{-4}E\{X_i^2I(|X_i| \leqslant \delta s_i) - E[X_i^2I(|X_i| \leqslant \delta s_i)|\mathscr{F}_{i-1}]\}^2 \\ \leqslant \sum_1^\infty s_i^{-4}E[X_i^4I(|X_i| \leqslant \delta s_i)] < \infty, \quad \text{(use Theorem 2.5)} \end{aligned}$$

which proves (4.36) and completes the proof in the case $W_n = U_{n-1}$. Condition (4.53) ensures that the result remains true if we set $W_n = U_n$. ∎

We close this chapter by mentioning a tail version of the LIL due to Heyde (1977a), which forms a supplement to the tail CLT (Corollary 3.5). It has an invariance principle analog along the lines of Corollary 4.2, but the more familiar 1-dimensional version will suffice for the applications we present in subsequent chapters.

Let $\{S_n = \sum_1^n X_i, \mathscr{F}_n\}$ be a zero-mean, square-integrable martingale. Suppose $E(S_n^2)$ is bounded as $n \to \infty$, implying that $W_n^2 = \sum_n^\infty X_j^2$ and $T_n = \sum_n^\infty X_j$ converge a.s. for each n (see Theorem 2.5). Let $t_n^2 = E(T_n^2)$; then $t_n \to 0$ as $n \to \infty$.

Theorem 4.9. *If*

$$t_n^{-2} W_n^2 \xrightarrow{\text{a.s.}} \eta^2 > 0 \quad \textit{a.s.},$$

$$\sum_1^\infty t_j^{-1} E[|X_j| I(|X_j| > \varepsilon t_j)] < \infty \qquad \textit{for all} \quad \varepsilon > 0,$$

and (4.54)

$$\sum_1^\infty t_j^{-4} E[X_j^4 I(|X_j| \leqslant \delta t_j)] < \infty \qquad \textit{for some} \quad \delta > 0,$$

then

$$\limsup_{n\to\infty} [\psi(W_n^2)]^{-1} T_n = +1 \quad \textit{a.s.}$$

and (4.55)

$$\liminf_{n\to\infty} [\psi(W_n^2)]^{-1} T_n = -1 \quad \textit{a.s.},$$

where $\psi(t) = (2t \log|\log t|)^{1/2}$.

We remark that under the conditions of the theorem and if

$$\sum_1^\infty t_n^{-2}[X_n^2 - E(X_n^2|\mathscr{F}_{n-1})] \qquad \text{converges a.s.,} \tag{4.56}$$

then (4.55) holds with W_n^2 replaced by the conditional variance $Z_n^2 = \sum_n^\infty E(X_j^2|\mathscr{F}_{j-1})$. For then

$$1 - Z_n^2/W_n^2 = W_n^{-2} t_n^2 \{t_n^{-2} \sum_n^\infty [X_j^2 - E(X_j^2|\mathscr{F}_{j-1})]\} \xrightarrow{\text{a.s.}} 0$$

by (4.54) and Kronecker's lemma. Of course, (4.56) holds if

$$\sum_1^\infty t_n^{-4} E[X_n^2 - E(X_n^2|\mathscr{F}_{n-1})]^2 < \infty.$$

5

Limit Theory for Stationary Processes via Corresponding Results for Approximating Martingales

5.1. Introduction

Many random phenomena are approximately in steady-state operation. Since these can reasonably be modeled by stationary processes, such processes have considerable practical importance in a wide variety of contexts. The use of these processes inevitably involves large sample (limit) theory. The strong law behavior is well known through the classical ergodic theorem and our concern in this chapter will be the central limit theorem (CLT) and law of the iterated logarithm (LIL). A review of the concepts of stationarity and ergodicity, together with a listing of pertinent references, can be found in Appendix IV.

Let $\{X_k, -\infty < k < \infty\}$ be a stationary ergodic sequence of random variables with $EX_0 = 0$, $EX_0^2 < \infty$, and write $S_n = \sum_{k=1}^{n} X_k$, $n \geqslant 1$. Suppose it happens that $\lim_{n\to\infty} n^{-1}ES_n^2 = \sigma^2$, $0 < \sigma^2 < \infty$. Then, under fairly mild additional conditions, S_n behaves asymptotically like a sum of n martingale differences each with variance σ^2. In fact it is often possible to approximate S_n by a naturally related martingale with stationary ergodic differences each having variance σ^2. Since the CLT and LIL hold for martingales with stationary ergodic increments, such approximation provides a possible approach to proving the corresponding limit results for $\{S_n\}$. We shall explore this approach, which was first utilized by Gordin (1969), in the present chapter. It turns out that this method leads to very general theorems which can be used to prove or improve many of the standard CLT and LIL results originally obtained by other methods.

There is a very large related literature. Classical methods have mostly involved the use of conditions of strong or uniform mixing [introduced by

Rosenblatt (1956) and by Cogburn (1960) and Ibragimov (1962), respectively]. The technique is to approximate the distribution of S_n by an appropriate sum of independent random variables. The proof of Theorem 5.6 follows this pattern. The best results based on the use of classical methods appear to be those in Oodaira and Yoshihara (1971a,b, 1972), and Ibragimov (1975).

Martingale approximations in the style of this chapter have also been used by Philipp and Stout (1975) and McLeish (1975c, 1977). Their concern has been mostly with the general case rather than the stationary one. Philipp and Stout are concerned with establishing results of the kind

$$S_n - W(n) = O(n^{1/2-\lambda}) \quad \text{a.s,}$$

where $\{W(t)\}$ is standard Brownian motion and $\lambda > 0$ depends on the process considered. McLeish is concerned with functional central limit theorems for mixingales (see Section 2.3) and in the stationary case these reduce to results akin to our Theorems 5.4 and 5.5. As far as the 1-dimensional CLT is concerned they are certainly weaker than Theorem 5.4.

5.2. The Probabilistic Framework

We start from a probability space $(\Omega,\mathscr{F},P)$ with ergodic one-to-one measure-preserving transformation T. Write L^2 for the Hilbert space of random variables with finite second moment and define the unitary operator U on L^2 by $UX(\omega) = X(T\omega)$ for $X \in L^2$, $\omega \in \Omega$. Let $\mathscr{M}_0$ be a sub-σ-field of $\mathscr{F}$ such that $\mathscr{M}_0 \subseteq T^{-1}(\mathscr{M}_0)$ and put $\mathscr{M}_k = T^{-k}(\mathscr{M}_0)$. We shall fix attention on a particular X_0 with $EX_0 = 0$ and $EX_0^2 < \infty$, and deal with sums $S_n = \sum_{k=1}^n X_k$ from the stationary ergodic sequence $\{X_k = U^k X_0, -\infty < k < \infty\}$. (If $T^{-1}(\mathscr{M}_0) \subseteq \mathscr{M}_0$, the framework is essentially the same, amounting to a reversal of the numbering in the $\{X_k\}$ process.)

5.3. The Central Limit Theorem

In order to develop our results we need some preliminary notation. If $\mathscr{L}$ is a sub-σ-field of $\mathscr{F}$, we denote the Hilbert space of those functions in L^2 which are measurable with respect to $\mathscr{L}$ by $H(\mathscr{L})$. Then set $H_k = H(\mathscr{M}_k)$ and $J_k = H_k \ominus H_{k-1}$ (meaning that $H_k = J_k \oplus H_{k-1}$; each element in H_k can be represented uniquely as a sum of an element of J_k and an element of H_{k-1}). Denote by Q the linear space of elements $g \in L^2$ such that $g \in H_k \ominus H_j$ for some finite k and j $(-\infty < j \leqslant k < \infty)$. [Simmons (1963, Chapter 10), for example, gives an adequate theoretical background for all the results we shall need concerning Hilbert spaces.]

Our first important result is given in the following theorem which is due to Gordin (1969).

Theorem 5.1. *Let $f \in L_2$ and suppose that*

$$\inf_{g \in Q} \lim_{n \to \infty} n^{-1} E\left[\sum_{k=0}^{n-1} U^k(f - g)\right]^2 = 0. \tag{5.1}$$

Then

$$\lim_{n \to \infty} n^{-1} E\left[\sum_{k=0}^{n-1} U^k f\right]^2 = \sigma^2, \qquad 0 \leqslant \sigma^2 < \infty, \tag{5.2}$$

and if $\sigma > 0$, $n^{-1/2} \sum_{k=0}^{n-1} U^k f$ converges in distribution to the normal law with mean zero and variance σ^2.

Proof. Given a sequence $\{\varepsilon_j\}$ of positive constants with $\varepsilon_j \to 0$ as $j \to \infty$ we choose $f_j \in Q$ such that

$$\limsup_{n \to \infty} n^{-1} E\left[\sum_{k=0}^{n-1} U^k(f - f_j)\right]^2 < \varepsilon_j. \tag{5.3}$$

This can be accomplished in view of (5.1).

Now for $h \in H(\mathscr{L})$ write

$$P_k h = E(h|\mathscr{M}_k) - E(h|\mathscr{M}_{k-1}).$$

We consider the chain of equations

$$\begin{aligned}
f &= f_j + f - f_j \\
&= \sum_{l=-\infty}^{\infty} P_l f_j + f - f_j \\
&= \sum_{l=-\infty}^{\infty} U^{-l} P_l f_j + U\left[\sum_{l>0} \sum_{m=-l}^{-1} U^m P_l f_j - \sum_{l<0} \sum_{m=0}^{-l-1} U^m P_l f_j\right] \\
&\quad - \left[\sum_{l>0} \sum_{m=-l}^{-1} U^m P_l f_j - \sum_{l<0} \sum_{m=0}^{-l-1} U^m P_l f_j\right] + f - f_j \\
&= h_j + U g_j - g_j + f - f_j,
\end{aligned} \tag{5.4}$$

say, the infinite sums containing only finitely many nonzero terms since $f_j \in Q$. Then

$$\begin{aligned}
\limsup_{n \to \infty} n^{-1} E\left[\sum_{k=0}^{n-1} U^k(f - h_j)\right]^2 &= \limsup_{n \to \infty} n^{-1} E\left[(U^n g_j - g_j) + \sum_{k=0}^{n-1} U^k(f - f_j)\right]^2 \\
&\leqslant 2 \limsup_{n \to \infty} n^{-1} E\left[\sum_{k=0}^{n-1} U^k(f - f_j)\right]^2 \leqslant 2\varepsilon_j,
\end{aligned} \tag{5.5}$$

using (5.3) and the fact that $g_j \in L^2$ together with the inequality $(a+b)^2 \leqslant 2(a^2+b^2)$. Notice that $h_j = \sum_{l=-\infty}^{\infty} U^{-l} P_l f_j \in J_0$, while $U^k h_j \in J_k$. It follows that the sequence $\{U^k h_j, -\infty < k < \infty\}$ is a stationary ergodic sequence of martingale differences and hence the martingale CLT (Theorem 3.2) and the ensuing remarks ensure that

$$n^{-1/2} \sum_{k=0}^{n-1} U^k h_j \xrightarrow{\mathrm{d}} N(0,\sigma_j^2), \tag{5.6}$$

where $\sigma_j^2 = Eh_j^2$ ($N(0,0)$ is understood to be the law which is degenerate at zero).

Now, σ_j converges to some limit σ as $j \to \infty$ since

$$\begin{aligned}
(\sigma_j - \sigma_{j'})^2 &= [(Eh_j^2)^{1/2} - (Eh_{j'}^2)^{1/2}]^2 \\
&\leqslant E(h_j - h_{j'})^2 \\
&= n^{-1} E\left[\sum_{k=0}^{n-1} U^k (h_j - h_{j'})\right]^2 \\
&\leqslant 2 \limsup_{n\to\infty} n^{-1} E\left[\sum_{k=0}^{n-1} U^k (f - h_j)\right]^2 \\
&\quad + 2 \limsup_{n\to\infty} n^{-1} E\left[\sum_{k=0}^{n-1} U^k (f - h_{j'})\right]^2 \\
&\leqslant 2(\varepsilon_j + \varepsilon_{j'}).
\end{aligned} \tag{5.7}$$

The result (5.2) then follows using (5.5) together with an application of Schwarz's inequality since $\sigma_j \to \sigma$ as $j \to \infty$ while convergence in distribution of $n^{-1/2} \sum_{k=0}^{n-1} U^k f$ to $N(0,\sigma^2)$ follows from (5.3), (5.4), and (5.6) since $\sigma_j \to \sigma$ as $j \to \infty$. This completes the proof. ∎

Theorem 5.1 is in fact a very general result but suffers from the major shortcoming that condition (5.1) is difficult to check. A variety of more concrete results, such as those of Theorems 5.2 and 5.3 below, can be established, however, via Theorem 5.1. An extension of Theorem 5.1 to the case of general stationary processes (i.e., not necessarily ergodic) has been provided by Eagleson (1975b); see also Theorem 3.2. In the general case the corresponding limit law is a mixture of normals.

Theorem 5.2. *Let $\{X_k\}$ denote a stationary and ergodic sequence with $EX_0 = 0$, $EX_0^2 < \infty$, and put $S_n = \sum_{k=1}^{n} X_k$. Suppose that X_0 is $\mathscr{M}_0$-measurable. If*

$$\sum_{k=1}^{\infty} E(X_k E(X_N | \mathscr{M}_0)) \qquad \textit{converges for each } N \geqslant 0 \tag{5.8}$$

and

$$\lim_{N\to\infty} \sum_{k=K}^{\infty} E(X_k E(X_N|\mathcal{M}_0)) = 0 \qquad \textit{uniformly in } K \geqslant 1, \tag{5.9}$$

then $\lim n^{-1}ES_n^2 = \sigma^2, 0 \leqslant \sigma^2 < \infty$. *If* $\sigma^2 > 0$, *then* $S_n/\sigma\sqrt{n}$ *converges in distribution to the standard normal law.*

Proof. We shall show that, when $X_0 = f$, condition (5.1) is implied by the conditions of Theorem 5.2.

From (5.9) it follows immediately that

$$E[X_N E(X_N|\mathcal{M}_0)] = E[E(X_0|\mathcal{M}_{-N})]^2 \to 0$$

as $N \to \infty$. Given $\varepsilon > 0$, choose N such that $E[E(X_0|\mathcal{M}_{-N})]^2 < \varepsilon$. Set $g = X_0 - E(X_0|\mathcal{M}_{-N})$. Then, because X_0 is $\mathcal{M}_0$-measurable, $g \in H_0 \ominus H_{-N}$ and hence $g \in Q$.

We have

$$\begin{aligned} n^{-1}E\left[\sum_{k=0}^{n-1} U^k(X_0 - g)\right]^2 &= n^{-1}E\left[\sum_{k=0}^{n-1} E(X_k|\mathcal{M}_{-N+k})\right]^2 \\ &= E[E(X_0|\mathcal{M}_{-N})]^2 \\ &\quad + 2n^{-1}\sum_{j<k}\sum E[E(X_j|\mathcal{M}_{-N+j})E(X_k|\mathcal{M}_{-N+k})]. \end{aligned}$$

The first term has been chosen $<\varepsilon$ and the second term is equal to

$$\begin{aligned} &2n^{-1}\sum_{j<k}\sum E[E(X_0|\mathcal{M}_{-N})E(X_{k-j}|\mathcal{M}_{-N+k-j})] \\ &= 2n^{-1}\sum_{k=1}^{n-1}(n-k)E[E(X_0|\mathcal{M}_{-N})E(X_k|\mathcal{M}_{-N+k})] \\ &= 2n^{-1}\sum_{k=1}^{n-1}(n-k)E[X_k E(X_0|\mathcal{M}_{-N})] \\ &= 2n^{-1}\sum_{k=1}^{n-1}(n-k)E[X_{k+N}E(X_N|\mathcal{M}_0)] \\ &= 2\sum_{k=N+1}^{N+n-1} E[X_k E(X_N|\mathcal{M}_0)] - 2n^{-1}\sum_{k=1}^{n-1} kE[X_{k+N}E(X_N|\mathcal{M}_0)]. \end{aligned}$$

The first term on the right-hand side of this last expression tends to zero uniformly in n as $N \to \infty$ by (5.9), while for fixed N the second term tends to zero as $n \to \infty$ by (5.3), Kronecker's lemma, and the fact that $\sum_{k=1}^{\infty} E(X_{k+N}E(X_N|\mathcal{M}_0))$ differs by at most N terms from $\sum_{k=1}^{\infty} E(X_k E(X_N|\mathcal{M}_0)) < \infty$. Thus, by choosing N large enough, the result (5.1) follows with $f = X_0$, and Theorem 5.1 now implies Theorem 5.2. ∎

It is interesting to remark on an alternative but less concise route that can be used to establish Theorem 5.2. Let

$$x_l = E(X_l|\mathcal{M}_0) - E(X_l|\mathcal{M}_{-1}), \qquad 0 \leqslant l < \infty.$$

Then, under the conditions of the theorem, $Y_0 = \sum_{l=0}^{\infty} x_l \in L^2$ and $EY_0^2 = \sigma^2$. The variables Y_k defined by $Y_k = U^k Y_0$ are martingale differences and $T_n = \sum_{k=1}^{n} Y_k$ closely approximates S_n. In fact, $n^{-1}E(S_n - T_n)^2 \to 0$ as $n \to \infty$ (which amounts to checking condition (5.1) with $f = X_0$, $g = Y_0$), so that $n^{-1/2}(S_n - T_n)$ converges in probability to zero. Of course the central limit theorem for martingales with stationary ergodic increments ensures that $\sigma^{-1/2}n^{-1/2}T_n$ converges in distribution to the standard normal law and the required result then follows. This proof is interesting in view of the explicit construction of an approximating martingale. We shall say more about such explicit approximations later in the chapter.

For a simple application of Theorem 5.2, suppose that $\{X_k\}$ is a stationary process with $EX_0 = 0$, $EX_0^2 < \infty$, satisfying the *strong mixing* condition

$$\sup_{A \in \mathscr{F}_{-\infty}^{k},\, B \in \mathscr{F}_{k+n}^{\infty}} |P(A \cap B) - P(A)P(B)| = \alpha(n) \downarrow 0 \qquad \text{as} \quad n \to \infty,$$

where $\mathscr{F}_{-\infty}^{a}$ denotes the σ-field generated by $\{X_k, -\infty < k \leqslant a\}$, and $\mathscr{F}_{b}^{\infty}$ the σ-field generated by $\{X_k, b \leqslant k < \infty\}$. Strong mixing was introduced by Rosenblatt (1956) as a condition under which a central limit result for stationary processes could be obtained. We shall establish the following result.

Corollary 5.1. *Let $0 < \delta \leqslant \infty$ be fixed. Suppose that the stationary sequence $\{X_k\}$ satisfies the strong mixing condition with $EX_0 = 0$ and $E|X_0|^{2+\delta} < \infty$, in case $0 < \delta < \infty$, or $|X_0| \leqslant c < \infty$ if $\delta = \infty$, while $\sum_{n=1}^{\infty} [\alpha(n)]^{\delta/(2+\delta)} < \infty$. Put $S_n = \sum_{k=1}^{n} X_k$. Then $\lim_{n\to\infty} n^{-1}\, ES_n^2 = \sigma^2$, $0 \leqslant \sigma^2 < \infty$. If $\sigma^2 > 0$, $S_n/\sigma\sqrt{n}$ converges in distribution to the standard normal law.*

Proof. The strong mixing condition implies ergodicity of the process and hence there exists an ergodic one-to-one bimeasurable measure-preserving transformation T on the σ-field generated by X_k, $-\infty < k < \infty$, such that $X_k(\omega) = X_0(T^k\omega)$, $\omega \in \Omega$ [see, e.g., Ibragimov and Linnik (1971, Chapter 17), or Doob (1953, Chapter X)]. We choose $\mathscr{F}_{-\infty}^{0}$ as our $\mathcal{M}_0$.

For the case $\delta = \infty$ we have, using Theorem 5.1 and Corollary A.1 of Appendix III,

$$\begin{aligned}
\sum_{k=1}^{\infty} |E(X_k E(X_N|\mathcal{M}_0))| &= \sum_{k=1}^{N} |E(X_k E(X_N|\mathcal{M}_0))| + \sum_{k=1}^{\infty} |E(X_{k+N} E(X_N|\mathcal{M}_0))| \\
&\leqslant 6c \sum_{k=1}^{N} [E(E(X_0|\mathcal{M}_{-N}))^2]^{1/2}\alpha^{1/2}(k) + 4c^2 \sum_{k=1}^{\infty} \alpha(k+N) \\
&\leqslant 36c^2\alpha^{1/2}(N) \sum_{k=1}^{N} \alpha^{1/2}(k) + 4c^2 \sum_{k=1}^{\infty} \alpha(k+N),
\end{aligned}$$

which is finite and tends to zero as $N \to \infty$ since, in view of the monotonicity of the α's, $k\alpha(k) \to 0$ as $k \to \infty$ and hence $k\alpha(k) < A < \infty$ for all $k \geqslant 1$, so that

$$\alpha^{1/2}(N) \sum_{k=1}^{N} \alpha^{1/2}(k) < A^{1/2}\alpha^{1/2}(N) \sum_{k=1}^{N} k^{-1/2}$$
$$\sim 2A^{1/2}[N\alpha(N)]^{1/2} \to 0$$

as $N \to \infty$. Theorem 5.2 thus applies.

For the case $0 < \delta < \infty$ we have from Corollary A.2 of Appendix III,

$$\sum_{k=1}^{\infty} |E(X_k E(X_N|\mathscr{M}_0))|$$
$$\leqslant 12(E|X_0|^{2+\delta})^{1/(2+\delta)}(E|E(X_0|\mathscr{M}_{-N})|^{2+\delta})^{1/(2+\delta)} \sum_{k=1}^{\infty} [\alpha(k)]^{\delta/(2+\delta)},$$

which is finite and tends to zero as $N \to \infty$ since $E|E(X_0|\mathscr{M}_{-N})|^{2+\delta} \to 0$ as $N \to \infty$. This last result follows from the dominated convergence theorem since, by Jensen's inequality,

$$|E(X_0|\mathscr{M}_{-N})|^{2+\delta} \leqslant E(|X_0|^{2+\delta}|\mathscr{M}_{-N})$$

and

$$E(E(|X_0|^{2+\delta}|\mathscr{M}_{-N})) = E|X_0|^{2+\delta} < \infty.$$

Then the martingale convergence theorem (Theorem 2.5) gives

$$E(X_0|\mathscr{M}_{-N}) \xrightarrow{\text{a.s.}} 0 \quad \text{as} \quad N \to \infty$$

(since $\{E(X_0|\mathscr{M}_{-N}), \mathscr{M}_{-N}, N \geqslant 1\}$ is a martingale and the strong mixing condition implies that $\mathscr{M}_{-\infty} = \bigcap_{k=-\infty}^{\infty} \mathscr{M}_k$ is the trivial σ-field). Theorem 5.2 again applies and this completes the proof. ∎

The results of Corollary 5.1 are due to Ibragimov (1962), who used very different methods. See also Ibragimov and Linnik (1971, Theorem 18.5.3). The present proof, together with Theorems 5.2 and 5.3 and Corollary 5.2, comes from Heyde (1974a).

For some applications it is desirable to have a central limit result for the case in which X_0 is not necessarily $\mathscr{M}_0$-measurable. We shall treat this case in Theorem 5.3. It is possible to approach the problem in a similar way to that of Theorem 5.2, but we shall directly use approximating martingales. Conditions analogous to those of Theorem 5.2 then provide sufficient conditions for the validity of Theorem 5.3.

Theorem 5.3. *Let $\{X_k\}$ denote a stationary and ergodic sequence with $EX_0 = 0$, $EX_0^2 < \infty$, and put $S_n = \sum_{k=1}^{n} X_k$. Suppose that $\mathscr{M}_0$ is a sub-σ-field* *of* $\mathscr{F}$ *and* $\mathscr{M}_0 \subseteq T^{-1}(\mathscr{M}_0)$, *and put* $\mathscr{M}_k = T^{-k}(\mathscr{M}_0)$ *and*

$$x_l = E(X_l|\mathscr{M}_0) - E(X_l|\mathscr{M}_{-1}), \qquad -\infty < l < \infty.$$

If $\sum_{l=-\infty}^{\infty} x_l = Y_0 \in L^2$, $EY_0^2 = \sigma^2 > 0$, *and* $n^{-1}ES_n^2 \to \sigma^2$ *as* $n \to \infty$, *then* $S_n/\sigma\sqrt{n}$ *converges in distribution to the standard normal law.*

Remark. Theorem 5.2 provides sufficient conditions for the validity of the conditions of Theorem 5.3 in the case where X_0 is $\mathscr{M}_0$-measurable (and hence $x_l = 0, l < 0$).

Proof of Theorem 5.3. We have $Y_0 \in L^2$ and $Y_0 \in J_0 = H_0 \ominus H_{-1}$, and we apply Theorem 5.1 with $g = Y_0$, $f = X_0$.

Write $Y_k = U^k Y_0$ and note that $\{Y_k\}$ forms a stationary ergodic sequence of martingale differences. Put $T_n = \sum_{k=1}^{n} Y_k$, $n \geqslant 1$. We have to check that $n^{-1}E(S_n - T_n)^2 \to 0$ as $n \to \infty$. However,

$$E(S_n - T_n)^2 = ES_n^2 + ET_n^2 - 2ES_nT_n$$

and $n^{-1}ES_n^2 \to \sigma^2$, $n^{-1}ET_n^2 = \sigma^2$, so in order to complete the proof it just suffices to show that $n^{-1}ES_nT_n \to \sigma^2$ as $n \to \infty$.

Now,

$$n^{-1}ES_nT_n = n^{-1}\sum_{i=1}^{n}\sum_{j=1}^{n} E(X_iY_j) = \sum_{j=-(n-1)}^{n-1} (1 - |j|n^{-1})E(Y_0X_j).$$

Thus from Kronecker's lemma, $n^{-1}ES_nT_n \to \sum_{j=-\infty}^{\infty} E(Y_0X_j)$ provided that $\sum_{j=-\infty}^{\infty} E(Y_0X_j)$ converges.

However,

$$E(Y_0X_j) = E(Y_0E(X_j|\mathscr{M}_0)) = E(Y_0x_j)$$

since $Y_0 \in J_0 = H_0 \ominus H_{-1}$ while $E(X_j|\mathscr{M}_{-1}) \in H_{-1}$. Also, since $Y_0 \in L^2$ we easily see that $\sum_{j=-\infty}^{\infty} E(x_jY_0) = EY_0^2 = \sigma^2$. ∎

Theorem 5.3 provides a powerful tool for some applications. To illustrate we introduce the stationary and ergodic linear process $\{x(n)\}$ given by

$$x(n) - \mu = \sum_{j=-\infty}^{\infty} \beta(j)\varepsilon(n-j), \qquad \sum_{j=-\infty}^{\infty} \beta^2(j) < \infty, \tag{5.10}$$

where the $\{\varepsilon(n)\}$ are independent and identically distributed with mean zero and variance σ^2. This process is widely used in time-series analysis.

Suppose that $\{x(j), 1 \leqslant j \leqslant n\}$ is a sample of n consecutive observations on the process (5.10). Anderson (1970, Theorem 7.7.8) shows that

$n^{-1/2}\sum_{k=1}^{n}(x(k)-\mu)$ converges in distribution to the normal law with mean zero and variance $\sigma^2[\sum_{j=-\infty}^{\infty}\beta(j)]^2$, provided $\sum_{j=-\infty}^{\infty}|\beta(j)|<\infty$. Hannan (1970, Theorem 11, p. 221) weakens this to show that the limit result continues to hold provided only that the spectral density

$$f(\lambda)=\sigma^2(2\pi)^{-1}\left[\sum_{j=-\infty}^{\infty}\beta(j)e^{i\lambda j}\right]^2 \tag{5.11}$$

is uniformly bounded and is continuous at $\lambda=0$ with $f(0)>0$. We shall observe here that the result in fact continues to hold without the uniform boundedness of the spectral density.

Corollary 5.2. *Suppose that the linear process* (5.10) *has spectral density* $f(\lambda)$ *(given by* (5.11)*) which is continuous at* $\lambda=0$. *Then* $n^{-1/2}\sum_{k=1}^{n}(x(k)-\mu)$ *converges in distribution to the normal law with mean zero and variance* $2\pi f(0)$.

Proof. We put $X_k=x(k)-\mu$ and can take $\mathscr{M}_k$ as the σ-field generated by $\varepsilon(j), j\leqslant k$. Then,

$$x_l=E(X_l|\mathscr{M}_0)-E(X_l|\mathscr{M}_{-1})=\beta(l)\varepsilon(0),\qquad -\infty<l<\infty,$$

and

$$Y_0=\sum_{l=-\infty}^{\infty}x_l=\varepsilon(0)\sum_{l=-\infty}^{\infty}\beta(l)\in L^2$$

with

$$EY_0^2=\sigma^2\left[\sum_{l=-\infty}^{\infty}\beta(l)\right]^2=2\pi f(0).$$

Also, under the assumption of continuity of $f(\lambda)$ at $\lambda=0$, $n^{-1}ES_n^2\to 2\pi f(0)$ as $n\to\infty$ [see, e.g., Ibragimov and Linnik (1971, Theorem 18.2.1)]. The required result then follows from Theorem 5.3. ∎

The result of Corollary 5.2 is extended to consideration of the periodogram in Rootzén (1976).

If instead of (5.10), $x(n)-\mu$ is a stationary purely nondeterministic process [e.g., Hannan (1970, Chapter III)], then it may be represented in the basic form

$$x(n)-\mu=\sum_{j=0}^{\infty}\beta(j)\varepsilon(n-j),\qquad \sum_{j=0}^{\infty}\beta^2(j)<\infty,\qquad \beta(0)=1,$$

$$E\varepsilon(n)=0,\qquad E(\varepsilon(m)\varepsilon(n))=0,\qquad m\neq n,$$

with the $\varepsilon(n)$ as the linear prediction errors, each having variance σ^2. The condition that the best linear predictor is the best predictor (both in the least squares sense) for the $\{x(n)\}$ process is precisely the condition that $E(\varepsilon(n)|\mathscr{F}_{n-1}) = 0$ a.s., for all n, where $\mathscr{F}_n$ is the σ-field generated by the $\varepsilon(m)$, $m \leqslant n$ [Hannan and Heyde (1972)]. That is, the $\varepsilon(n)$ are martingale differences. Under this condition it should be observed that the results of Corollary 5.2 continue to apply. Estimation theory for this process is considered in Chapter 6.

All the theory so far has been based on the assumption $EX_0^2 < \infty$, which is usual in the context of the central limit theorem. There is, however, no formal need to assume anything further than $E|X_0| < \infty$, as we shall demonstrate in the following theorem. This result was announced by Gordin at a conference in Vilnius in 1975 but it does not appear to have been published.

Theorem 5.4. *Let $\{X_k(\omega) = X_0(T^k\omega)\}$ be a stationary and ergodic sequence with $E|X_0| < \infty$ and $EX_0 = 0$. Put $S_n = \sum_{k=1}^n X_k$. Suppose that $\mathscr{M}_0$ is a sub-σ-field of $\mathscr{F}$ with $\mathscr{M}_0 \subseteq T^{-1}(\mathscr{M}_0)$, and set $\mathscr{M}_k = T^{-k}(\mathscr{F}_0)$. If*

$$\sum_{k=0}^{\infty} \{E|E(X_0|\mathscr{M}_{-k})| + E|X_0 - E(X_0|\mathscr{M}_k)|\} < \infty \tag{5.12}$$

and

$$\limsup_{n\to\infty} n^{-1/2} E|S_n| < \infty, \tag{5.13}$$

then

$$\lim_{n\to\infty} n^{-1/2} E|S_n| = \lambda \tag{5.14}$$

for some λ, $0 \leqslant \lambda < \infty$. If $\lambda > 0$, then $n^{-1/2}S_n$ converges in distribution to the normal law with zero mean and variance $\pi\lambda^2/2$.

Proof. We begin by pursuing the idea expressed in equation (5.4) and showing that we can represent X_k in the form

$$X_k = Y_k + Z_k - Z_{k+1}, \tag{5.15}$$

where $Y_k(\omega) = Y_0(T^k\omega)$, $-\infty < k < \infty$, are stationary ergodic margingale differences, $Z_k(\omega) = Z_0(T^k\omega)$, $-\infty < k < \infty$, and $E|Z_0| < \infty$.

For $m, n > 0$,

$$\begin{aligned} X_0 &= E(X_0|\mathscr{M}_n) - E(X_0|\mathscr{M}_{-m}) + X_0 - E(X_0|\mathscr{M}_n) + E(X_0|\mathscr{M}_{-m}) \\ &= \sum_{l=-m+1}^{n} [E(X_0|\mathscr{M}_l) - E(X_0|\mathscr{M}_{l-1})] + X_0 - E(X_0|\mathscr{M}_n) + E(X_0|\mathscr{M}_{-m}). \end{aligned}$$

Furthermore, from (5.12).

$$X_0 - E(X_0|\mathscr{M}_n) \xrightarrow{L^1} 0, \qquad E(X_0|\mathscr{M}_{-m}) \xrightarrow{L^1} 0,$$

as $n, m \to \infty$, and hence

$$X_0 = \sum_{l=-\infty}^{\infty} [E(X_0|\mathscr{M}_l) - E(X_0|\mathscr{M}_{l-1})]. \tag{5.16}$$

Now set

$$Y_0 = \sum_{l=-\infty}^{\infty} [E(X_l|\mathscr{M}_0) - E(X_l|\mathscr{M}_{-1})], \tag{5.17}$$

observing that the sum is well defined in view of (5.12) and that $\{Y_k(\omega) = Y_0(T^k\omega), \mathscr{M}_k, -\infty < k < \infty\}$ is a martingale difference sequence. Then we note, using (5.16) and (5.17), that (5.15) is formally satisfied for

$$Z_0 = \sum_{k=0}^{\infty} E(X_k|\mathscr{M}_{-1}) - \sum_{k=-\infty}^{-1} \{X_k - E(X_k|\mathscr{M}_{-1})\}. \tag{5.18}$$

That (5.18) is well defined and $E|Z_0| < \infty$ follows from (5.12), and this establishes (5.15).

The next step in the proof is to show that $EY_0^2 < \infty$ under (5.13). For $u > 0$, define

$$X_{ku} = \begin{cases} X_k & \text{if } |X_k| \leqslant u \\ 0 & \text{if } |X_k| > u, \end{cases}$$

and put

$$Y_{ku} = X_{ku} - E(X_{ku}|\mathscr{G}_{k-1}),$$

where $\mathscr{G}_k$ is the σ-field generated by $X_j, j \leqslant k$. The $\{Y_{ku}, -\infty < k < \infty\}$ are stationary martingale differences and hence $n^{-1/2}\sum_{k=1}^n Y_{ku}$ converges in distribution to the normal law with zero mean and variance EY_{0u}^2 as $n \to \infty$. The sequence $\{n^{-1/2}\sum_{k=1}^n Y_{ku}, n \geqslant 1\}$ is, furthermore, uniformly integrable since

$$E\left(n^{-1/2}\sum_{k=1}^n Y_{ku}\right)^2 = EY_{0u}^2 \leqslant 4u^2,$$

and hence

$$\begin{aligned} n^{-1/2}E\left|\sum_{k=1}^n Y_{ku}\right| &\to (EY_{0u}^2)^{1/2} 2(2\pi)^{-1/2} \int_0^\infty xe^{-x^2/2}\,dx \\ &= (2\pi^{-1}EY_{0u}^2)^{1/2}. \end{aligned} \tag{5.19}$$

Also, (5.13) and the representation (5.15) with $E|Z_0| < \infty$ ensure that we can find an $A < \infty$ such that

$$n^{-1/2}E\left|\sum_1^n Y_k\right| \leqslant A \qquad \text{for all} \quad n. \tag{5.20}$$

Now suppose that $EY_0^2 = \infty$. In view of (5.19) we can find a $U(A)$ which is so large that for $u > U(A)$,

$$n^{-1/2}E\left|\sum_{k=1}^n Y_{ku}\right| > 2A \qquad \text{for} \quad n > n_0\{U(A)\}.$$

Then, for fixed $n > n_0\{U(A)\}$,

$$\lim_{n\to\infty} n^{-1/2}E\left|\sum_{k=1}^n Y_{ku}\right| = n^{-1/2}E\left|\sum_{k=1}^n Y_k\right| \geqslant 2A,$$

which contradicts (5.20). It follows that $EY_0^2 < \infty$.

From (5.15) together with $E|Z_0| < \infty$ and $EY_0^2 < \infty$, we have from the central limit theorem for martingales with stationary ergodic differences that

$$n^{-1/2}\sum_1^n X_k = n^{-1/2}\sum_1^n Y_k + n^{-1/2}(UZ_0 - U^{n+1}Z_0)$$
$$\xrightarrow{d} N(0, EY_0^2),$$

with the obvious interpretation if $EY_0^2 = 0$. Then, for $EY_0^2 > 0$,

$$\lim_{n\to\infty} n^{-1/2}E\left|\sum_1^n X_k\right| = \lim_{n\to\infty} n^{-1/2}E\left|\sum_1^n Y_k\right| = (2\pi^{-1}EY_0^2)^{1/2},$$

using an argument similar to that leading up to (5.19) to obtain the right-hand equality, and the proof of the theorem is completed by identifying λ as $(2\pi^{-1}EY_0^2)^{1/2}$. ∎

The conditions of Theorem 5.4 make an interesting contrast with those of other results in this chapter where the conditions typically ensure that $n^{-1}ES_n^2 \to \sigma^2$. This contrast is most effectively seen in the following Corollary 5.3, which is concerned with stationary mixing sequences.

In Corollary 5.1 we dealt with stationary strong mixing processes. Now we introduce a more stringent condition of asymptotic independence called uniform mixing. The stationary process $\{X_k, -\infty < k < \infty\}$ is said to satisfy the *uniform mixing* condition if

$$\sup_{A\in\mathscr{F}_{-\infty}^k,\, B\in\mathscr{F}_{k+n}^\infty} \{P(A)\}^{-1}|P(A\cap B) - P(A)P(B)| = \phi(n) \downarrow 0$$

as $n \to \infty$, where $\mathscr{F}_{-\infty}^a$ and $\mathscr{F}_b^\infty$ denote the σ-fields generated by $\{X_k, -\infty < k \leqslant a\}$ and $\{X_k, b \leqslant k < \infty\}$, respectively. It should be noted that this condition is not symmetric in the past and future. Time reversal of the process may be necessary under some circumstances and indeed is illustrated in the following result.

Corollary 5.3. *Suppose that the stationary process* $\{X_k\}$ *satisfies* $EX_0 = 0$, $E|X_0|^{2+\delta} < \infty$, *for some* $0 \leqslant \delta < \infty$, *and that*

$$n^{-1}ES_n^2 \to \sigma^2 > 0$$

as $n \to \infty$, *where* $S_n = \sum_{k=1}^n X_k$. *If either*

(*i*) *the uniform mixing condition is satisfied by the time reversed process* $\{X_{-k}\}$ *with* $\sum_{n=1}^\infty \phi(n) < \infty$, *or*
(*ii*) *the strong mixing condition is satisfied with* $\sum_{n=1}^\infty [\alpha(n)]^{(1+\delta)/(2+\delta)} < \infty$,

then $S_n/\sigma n^{1/2}$ *converges in distribution to the standard normal law.*

Proof. In the case of either (i) or (ii) the process $\{X_k\}$ is strong mixing and ergodic. As noted in the proof of Corollary 5.1, there exists an ergodic transformation T on the σ-field generated by $\{X_k, -\infty < k < \infty\}$ such that $X_k(\omega) = X_0(T^k\omega)$, $\omega \in \Omega$, and our present situation fits within the framework of Theorem 5.4 with $\mathscr{F}_{-\infty}^0$ as the $\mathscr{M}_0$ therein.

To check the condition (5.12) we first note that for $k \geqslant 0$, X_0 is $\mathscr{M}_k$-measurable, so that $X_0 = E(X_0|\mathscr{M}_k)$ a.s. To deal with $E|E(X_0|\mathscr{M}_{-k})|$ we use the results of Appendix III.

Setting $\xi = \operatorname{sgn} E(X_0|\mathscr{M}_{-k})$, we have

$$E|E(X_0|\mathscr{M}_{-k})| = E\xi E(X_0|\mathscr{M}_{-k}) = E\xi X_0 \leqslant |E\xi X_0 - E\xi EX_0|, \quad (5.21)$$

where ξ and X_0 are $\mathscr{M}_{-k}$ and $\mathscr{M}_0$-measurable, respectively. In the case of (i) we use Theorem A.6 of Appendix III and obtain

$$|E\xi X_0 - E\xi EX_0| \leqslant 2\phi(k)E|X_0|, \quad (5.22)$$

where ϕ refers to the time reversed process $\{X_{-k}\}$. For (ii) we use Corollary A.1 of Appendix III and obtain

$$|E\xi X_0 - E\xi EX_0| \leqslant 6[\alpha(k)]^{(1+\delta)/(2+\delta)}\|X_0\|_{2+\delta}. \quad (5.23)$$

The condition (5.12) then follows from (5.21) and (5.22) or (5.23) and this completes the proof since $n^{-1}ES_n^2 \to \sigma^2$ implies (5.13). ∎

The result of Corollary 5.3 should be contrasted with those of Corollary 5.1 and the subsequent Corollary 5.5, where stronger mixing rate conditions are imposed. In these results the explicit use of the condition $n^{-1}ES_n^2 \to \sigma^2$ is avoided, but the mixing rate conditions are strong enough to ensure its validity.

5.4. Functional Forms of the Central Limit Theorem and Law of the Iterated Logarithm

In the proof of Theorem 5.4 it was noted that it is possible, under certain circumstances, to represent the increments of a stationary ergodic process $\{X_k, -\infty < k < \infty\}$ in the form

$$X_k = Y_k + Z_k - Z_{k+1},$$

where $\{Y_k, -\infty < k < \infty\}$ is a stationary ergodic sequence of martingale differences and $\{Z_k, -\infty < k < \infty\}$ is a stationary sequence whose sum of differences telescopes and disappears under suitable norming. The asymptotic behavior of $\{\sum_{k=1}^n X_k\}$ can therefore be conveniently deduced from that of the approximating martingale. In Theorem 5.4 the conditions imposed were only sufficient to ensure that $E|Z_0| < \infty$, but for functional forms of the central limit and iterated logarithm laws we need $EZ_0^2 < \infty$.

Set $S_0 = 0$, $S_n = \sum_{k=1}^n X_k$ for $n \geqslant 1$, and $\sigma_n^2 = ES_n^2$. Let $\{\theta_n(\cdot)\}$ and $\{\eta_n(\cdot)\}$ be sequences of random function on $[0,1]$ defined, respectively, by

$$\theta_n(t) = \sigma_n^{-1}(S_k + (nt-k)X_{k+1}), \qquad k \leqslant nt \leqslant k+1, \quad k = 0, 1, \ldots, n-1,$$

and

$$\eta_n(t) = [\phi(\sigma_n^2)]^{-1}(S_k + (nt-k)X_{k+1}), \quad k \leqslant nt \leqslant k+1, \quad k = 0, 1, \ldots, n-1,$$

where

$$\phi(t) = (2t \log\log t)^{1/2}, \qquad e < t < \infty.$$

Note that $\theta_n, \eta_n \in C = C[0,1]$, the space of continuous functions on $[0,1]$.

Let K be the set of absolutely continuous $x \in C$ such that

$$x(0) = 0$$

and

$$\int_0^1 [\dot{x}(t)]^2 \, dt \leqslant 1,$$

where $\dot{x}$ denotes the derivative of x determined almost everywhere with respect to Lebesgue measure. Also define

$$g = \sup\{n : \sigma_n^2 \leqslant e\}.$$

We write $\mathscr{M}_{-\infty} = \bigcap_{k=-\infty}^{\infty} \mathscr{M}_k$ and $\mathscr{M}_\infty$ for the σ-field generated by $\bigcup_{k=-\infty}^{\infty} \mathscr{M}_k$. Our object is the following theorem. This is from Heyde (1975a) and refines Scott (1973) and Heyde and Scott (1973).

Theorem 5.5. *Suppose that $\mathscr{M}_0$ is a sub-σ-field of $\mathscr{F}$ with $\mathscr{M}_0 \subset T^{-1}(\mathscr{M}_0)$, and put $\mathscr{M}_k = T^{-k}(\mathscr{M}_0)$. Let $EX_0^2 < \infty$ and $EX_0 = 0$. If*

$$\sum_{m=1}^{\infty} \left\{ \limsup_{n\to\infty} E\left(\sum_{l=m}^{n} x_l \right)^2 + \limsup_{n\to\infty} E\left(\sum_{l=m}^{n} x_{-l} \right)^2 \right\} < \infty, \tag{5.24}$$

where

$$x_l = E(X_{-l}|\mathscr{M}_0) - E(X_{-l}|\mathscr{M}_{-1}), \qquad -\infty < l < \infty,$$

and

$$E(X_0|\mathscr{M}_\infty) = X_0 \quad a.s., \qquad E(X_0|\mathscr{M}_{-\infty}) = 0 \quad a.s., \tag{5.25}$$

then $\lim_{n\to\infty} \sigma_n/n^{1/2} = \sigma$ exists for $0 \leqslant \sigma < \infty$. If $\sigma > 0$, then $\theta_n \xrightarrow{d} W$ in the sense (C, ρ), where W is a standard Brownian motion on $[0,1]$. Also, $g < \infty$, $\{\eta_n, n > g\}$ is relatively compact, and the set of its limit points coincides with K.

Proof. Following the proof of Theorem 5.4 we have

$$X_0 = \sum_{l=-\infty}^{\infty} U^l x_l, \tag{5.26}$$

U being the unitary operator associated with T, and we construct a solution of the equation

$$X_0 = Y_0 + Z_0 - UZ_0 \tag{5.27}$$

where $Y_0 \in J_0$ but now $Z_0 \in L^2$. This will provide the desired martingale approximation. Note that the condition (5.24) gives

$$\lim_{m\to\infty} \limsup_{n\to\infty} E\left(\sum_{l=0}^{n} x_l \right)^2 = \lim_{m\to\infty} \limsup_{n\to\infty} E\left(\sum_{l=m}^{n} x_{-l} \right)^2 = 0$$

so that $\sum_{l=0}^{\infty} x_l$ and $\sum_{l=0}^{\infty} x_{-l}$ converge in L^2 and

$$\limsup_{n\to\infty} E\left(\sum_{l=m}^{n} x_l \right)^2 + \limsup_{n\to\infty} E\left(\sum_{l=m}^{n} x_{-l} \right)^2 = E\left(\sum_{l=m}^{\infty} x_l \right)^2 + E\left(\sum_{l=m}^{\infty} x_{-l} \right)^2.$$

Observe now that there exists a solution to the set of equations

$$x_l = Y_0 \delta_{0l} + z_l - z_{l-1}$$

where δ_{0l} is the Kronecker delta and $z_l \in J_0$. Indeed,

$$z_l = -\sum_{k=l+1}^{\infty} x_k, \quad l \geqslant 0, \qquad z_l = \sum_{k=-\infty}^{l} x_k, \quad l < 0.$$

Thus, in view of (5.26), (5.27) holds with

$$Z_0 = \sum_{l=-\infty}^{\infty} U^l z_l \in L^2, \qquad Y_0 = \sum_{l=-\infty}^{\infty} x_l \in L^2.$$

In fact,

$$EZ_0^2 = \sum_{m=1}^{\infty} E\left(\left(\sum_{l=m}^{\infty} x_l\right)^2 + \left(\sum_{l=m}^{\infty} x_{-l}\right)^2\right).$$

The above construction for Z_0 is different from the one provided in Theorem 5.4. However, the result is the same a.s. since if (5.27) holds for $Z_0^{(1)}$ and $Z_0^{(2)}$, then

$$U(Z_0^{(1)} - Z_0^{(2)}) = Z_0^{(1)} - Z_0^{(2)},$$

which yields $Z_0^{(1)} = Z_0^{(2)}$ a.s. because $EZ_0^{(1)} = EZ_0^{(2)} = 0$.

It should be remarked that $\limsup_{n\to\infty} E(\sum_{k=1}^{n} U^k(X_0 - Y_0))^2 < \infty$ for $Y_0 \in L^2$ with $E(Y_0|\mathcal{M}_{-1}) = 0$ a.s. if and only if $X_0 - Y_0$ is representable in the form (5.27) [using Theorem 18.2.2 of Ibragimov and Linnik (1971)].

Set $\sigma^2 = EY_0^2$. In view of the representation (5.27) and the fact that $EZ_0^2 < \infty$, we obtain $\lim_{n\to\infty} n^{-1}\sigma_n^2 = \sigma^2$ and $\sigma = 0$ only if $X_n = U^n Z_0 - U^{n+1} Z_0$ for some $Z_0 \in L^2$. Henceforth we shall assume that $\sigma > 0$.

Now we are in a position to deal with the CLT and LIL. We take the case of the CLT first and define

$$\xi_n(t) = \sigma^{-1} n^{-1/2}\left(\sum_{j=1}^{k} Y_j + (nt - k)Y_{k+1}\right), \quad k \leqslant nt \leqslant k+1, \ k = 0, 1, \ldots, n-1,$$

where $Y_j = U^j Y_0$. Let $\mathcal{N}_0$ be the σ-field generated by $Y_j, j \leqslant 0$, and set $\mathcal{N}_n = T^{-n}(\mathcal{N}_0)$, the σ-field generated by $Y_j, j \leqslant n$. Then $\mathcal{N}_n \subseteq \mathcal{M}_n$ for $n \geqslant 0$, and so

$$E(Y_n|\mathcal{N}_{n-1}) = E[E(Y_n|\mathcal{M}_{n-1})|\mathcal{N}_{n-1}] = 0 \quad \text{a.s.}$$

It follows that $\{\sum_{j=1}^{n} Y_j, \mathcal{N}_n, n \geqslant 1\}$ is a martingale with stationary ergodic differences, and so by Theorem 4.1,

$$\xi_n \xrightarrow{\text{d}} W, \tag{5.28}$$

where W is a standard Brownian motion. Furthermore, using the continuous mapping theorem (see Theorem A.3, Appendix II) and (5.28),

$$\lim_{n\to\infty} P\left(\sup_{0\leqslant t\leqslant 1} |\xi_n(t)| > K\right) = \lim_{n\to\infty} P\left(\sup_{1\leqslant k\leqslant n} \sigma^{-1} n^{-1/2}\left|\sum_{j=1}^{k} Y_j\right| > K\right)$$
$$= P\left(\sup_{0\leqslant t\leqslant 1} |W(t)| > K\right).$$

It is well known that $\sup_{0 \leqslant t \leqslant 1} |W(t)|$ is finite with probability one, so that for any $t > 0$, there exist K and N so that for $n > N$,

$$P\left(\sup_{1 \leqslant k \leqslant n} \sigma^{-1} n^{-1/2} \left| \sum_{j=1}^{k} Y_j \right| > K \right) < \varepsilon. \tag{5.29}$$

Also, letting $Z_k = U^k Z_0$,

$$\begin{aligned} P\left(\sup_{0 \leqslant k \leqslant n} \sigma^{-1} n^{-1/2} |Z_k| \geqslant \delta \right) &\leqslant (n+1) P((\sigma^2 n)^{-1} Z_0^2 \geqslant \delta^2) \\ &\leqslant (n+1)(n\sigma^2\delta^2)^{-1} \int_{Z_0 \geqslant \sigma\delta n^{1/2}} Z_0^2 \, dP \to 0 \end{aligned} \tag{5.30}$$

as $n \to \infty$ since $Z_0 \in L^2$. If we can show that

$$\sup_{0 \leqslant t \leqslant 1} |\xi_n t) - \theta_n(t)| \xrightarrow{\text{p}} 0,$$

we shall have the desired result.

Now,

$$\begin{aligned} &\sup_{0 \leqslant t \leqslant 1} |\xi_n(t) - \theta_n(t)| \\ &\quad = \sup_{1 \leqslant k \leqslant n} \left| \sigma^{-1} n^{-1/2} \sum_{j=1}^{k} Y_j - \sigma_n^{-1} S_k \right| \\ &\quad \leqslant \sup_{1 \leqslant k \leqslant n} \sigma^{-1} n^{-1/2} \left| \sum_{j=1}^{k} (Y_j - X_j) \right| + \sup_{1 \leqslant k \leqslant n} \sigma^{-1} n^{-1/2} |S_k| \, |1 - \sigma n^{1/2} \sigma_n^{-1}| \\ &\quad \leqslant 2 \sup_{0 \leqslant k \leqslant n} \sigma^{-1} n^{-1/2} |Z_k| + \sup_{1 \leqslant k \leqslant n} \sigma^{-1} n^{-1/2} \left| \sum_{j=1}^{k} Y_j \right| |1 - \sigma n^{1/2} \sigma_n^{-1}| \\ &\qquad + 2 \sup_{0 \leqslant k \leqslant n} \sigma^{-1} n^{-1/2} |Z_k| \, |1 - \sigma n^{1/2} \sigma_n^{-1}|, \end{aligned}$$

using (5.27). The first term of this last expression goes to zero in probability because of (5.30), the second and third because of (5.29) and (5.30), respectively, together with $n^{-1}\sigma_n^2 \to \sigma^2$ as $n \to \infty$. This completes the proof of the CLT.

To prove the LIL we define a sequence of random functions $\zeta_n(\cdot)$ on $[0,1]$ by $\zeta_n(0) = 0$,

$$\zeta_n(t) = [\phi(n\sigma^2)]^{-1} \left(\sum_{j=1}^{k} Y_j + (nt - k) Y_{k+1} \right),$$

$$k \leqslant nt \leqslant k+1, \quad k = 0, 1, \ldots, n-1.$$

From Corollary 4.2 we know that $\{\zeta_n, n > e/\sigma^2\}$ is relatively compact and the set of its a.s. limit points coincides with K. Thus, the proof will be complete if we can show that

$$\sup_{0 \leqslant t \leqslant 1} |\eta_n(t) - \zeta_n(t)| \xrightarrow{\text{a.s.}} 0 \tag{5.31}$$

as $n \to \infty$.

To establish (5.31) we first consider

$$\begin{aligned} &\sup_{0 \leqslant t \leqslant 1} |\zeta_n(t) - \phi(n\sigma^2)[\phi(\sigma_n^2)]^{-1}\zeta_n(t)| \\ &\qquad \leqslant |1 - \phi(n\sigma^2)[\phi(\sigma_n^2)]^{-1}| \sup_{0 \leqslant t \leqslant 1} |\zeta_n(t)|. \end{aligned} \tag{5.32}$$

Now,

$$\begin{aligned} \limsup_{n \to \infty} \sup_{0 \leqslant t \leqslant 1} |\zeta_n(t)| &\leqslant \sup_{x \in K} \sup_{0 \leqslant t \leqslant 1} |x(t)| \quad \text{a.s.} \\ &\leqslant 1 \quad \text{a.s.}, \end{aligned}$$

so that from (5.32) and $n^{-1}\sigma_n^2 \to \sigma^2$ we conclude that

$$\sup_{0 \leqslant t \leqslant 1} |\zeta_n(t) - \phi(n\sigma^2)[\phi(\sigma_n^2)]^{-1}\zeta_n(t)| \xrightarrow{\text{a.s.}} 0. \tag{5.33}$$

Now, writing $Z_k = U^k Z_0$, consider

$$\begin{aligned} \sup_{0 \leqslant t \leqslant 1} |\eta_n(t) - \phi(n\sigma^2)[\phi(\sigma_n^2)]^{-1}\zeta_n(t)| &= \sup_{1 \leqslant k \leqslant n} [\phi(\sigma_n^2)]^{-1}|Z_k - Z_0| \quad \text{(by (5.27))} \\ &\leqslant 2 \sup_{0 \leqslant k \leqslant n} [\phi(\sigma_n^2)]^{-1}|Z_k|, \end{aligned}$$

which goes almost surely to zero if

$$\sup_{0 \leqslant k \leqslant n} [\phi(n\sigma^2)]^{-1}|Z_k| = (n\sigma^2 \log\log n\sigma^2)^{-1/2} \sup_{0 \leqslant k \leqslant n} |Z_k| \xrightarrow{\text{a.s.}} 0,$$

and thus if

$$(n \log\log n)^{-1/2} Z_n \xrightarrow{\text{a.s.}} 0.$$

But, this last result follows from the Borel–Cantelli lemma since $Z_0 \in L^2$. Thus

$$\sup_{0 \leqslant t \leqslant 1} |\eta_n(t) - \phi(n\sigma^2)[\phi(\sigma_n^2)]^{-1}\zeta_n(t)| \xrightarrow{\text{a.s.}} 0, \tag{5.34}$$

and (5.31) follows from (5.33) and (5.34), which completes the proof. ∎

The condition (5.24) does not appear to simplify in any really convenient way in general. However, it is just the condition $EZ_0^2 < \infty$ which can be viewed in an alternative way using (5.18). Also, we are often free to choose a convenient $\mathscr{M}_0$. For example, if $\mathscr{M}_0$ is the σ-field generated by X_k, $k \leqslant 0$, then the condition (5.24) becomes just

$$E\left(\sum_{k=0}^{\infty} E(X_k|\mathscr{M}_{-1})\right)^2 < \infty. \tag{5.35}$$

A useful sufficient condition for (5.24), which seems to be more widely applicable, can be obtained by the use of Schwarz's inequality. This result is embodied in the following corollary which is due to Heyde and Scott (1973).

Corollary 5.4. *The results of Theorem 5.5 hold if* (5.24) *and* (5.25) *are replaced by*

$$\sum_{m=1}^{\infty} \{(E[E(X_0|\mathscr{M}_{-m})]^2)^{1/2} + (E[X_0 - E(X_0|\mathscr{M}_m)]^2)^{1/2}\} < \infty. \tag{5.36}$$

Condition (5.36) is obvious from an upper bound on the expression for EZ_0^2 obtained from (5.18) using Schwarz's inequality. Note that (5.25) is automatically satisfied under (5.36).

We shall now give some applications of Theorem 5.5 and Corollary 5.4 to illustrate the use of these various conditions. First, we shall obtain the following result, which is due to Heyde and Scott (1973) and should be contrasted with that of Corollary 5.3.

Corollary 5.5. *If $\{X_k\}$ is a stationary uniform mixing process with $EX_0 = 0$, $E|X_0|^{2+\delta} < \infty$, some $\delta \geqslant 0$, and $\sum_1^\infty [\phi(n)]^{(1+\delta)/(2+\delta)} < \infty$, then the results of Theorem 5.5 apply.*

Proof. We first note that the remarks made in the first paragraph of the proof of Corollary 5.3 apply and our present situation fits within the framework of Theorem 5.5 with $\mathscr{F}_{-\infty}^0$ as the $\mathscr{M}_0$ therein.

Next we proceed to check (5.36). This is a simple matter using Theorem A.6 of Appendix III. We have

$$E(X_0|\mathscr{M}_m) = X_0, \qquad m \geqslant 0,$$

and

$$\begin{aligned} E[E(X_0|\mathscr{M}_{-m})]^2 &= E[X_0 E(X_0|\mathscr{M}_{-m})] \\ &\leqslant 2[\phi(m) E|E(X_0|\mathscr{M}_{-m})|^{(2+\delta)/(1+\delta)}]^{(1+\delta)/(2+\delta)} \\ &\quad \times [E|X_0|^{2+\delta}]^{1/(2+\delta)} \\ &\leqslant 2[\phi(m)]^{(1+\delta)/(2+\delta)}\{E|E(X_0|\mathscr{M}_{-m})|^2\}^{1/2}[E|X_0|^{2+\delta}]^{1/(2+\delta)} \end{aligned}$$

using Liapunov's inequality, so that

$$\{E[E(X_0|\mathscr{M}_{-m})]^2\}^{1/2} \leqslant 2[\phi(m)]^{(1+\delta)/(2+\delta)}[E|X_0|^{2+\delta}]^{1/(2+\delta)},$$

and (5.36) holds under the conditions of the corollary. This completes the proof. ∎

To illustrate the improvement of (5.24) over (5.36) we return to the stationary linear process $\{x(n)\}$ introduced in (5.10) and given by

$$x(n) - \mu = \sum_{j=-\infty}^{\infty} \beta(j)\varepsilon(n-j), \qquad \sum_{j=-\infty}^{\infty} \beta^2(j) < \infty,$$

where the $\varepsilon(n)$ are independent and identically distributed with mean zero and variance σ^2. One widely applicable model which gives rise to this process is the standard version of the mixed autoregression and moving average process [e.g., Hannan (1970, Chapter 1)]. Another is the stationary Gaussian process with absolutely continuous spectral density [Ibragimov and Linnik (1971, Theorem 16.7.1 and Chapter 17, Section 3)]. In the latter case the $\varepsilon(n)$ are themselves normally distributed.

Suppose that $x(1), x(2), \ldots, x(N)$ is a sample of N consecutive observations from the process $\{x(n)\}$ and $\bar{x}$ denotes the sample mean. It follows from the ergodic theorem that $\bar{x} \xrightarrow{\text{a.s.}} \mu$ as $N \to \infty$, and it is of special interest in time-series analysis to obtain central limit and iterated logarithm results which give information on the rate of this convergence. Here we have $X_k = x(k) - \mu$ and take $\mathscr{M}_k$ as the σ-field generated by $\varepsilon(j), j \leqslant k$. Then

$$x_l = E(X_{-l}|\mathscr{M}_0) - E(X_{-l}|\mathscr{M}_{-1}) = \beta(-l)\varepsilon(0)$$

and Theorem 5.5 applies if

$$\sum_{n=1}^{\infty}\left(\left(\sum_{l=n}^{\infty}\beta(l)\right)^2 + \left(\sum_{l=n}^{\infty}\beta(-l)\right)^2\right) < \infty. \tag{5.37}$$

On the other hand, the corresponding results based on the use of condition (5.36) hold if

$$\sum_{n=1}^{\infty}\left(\sum_{|l| \geqslant n}\beta^2(l)\right)^{1/2} < \infty. \tag{5.38}$$

The condition (5.37) represents a significant improvement over (5.38) in the case where the β continually vary in sign. Note, for example, the case $\beta(n) = 0, n < 0, \beta(n) = (-1)^n n^{-1}, n \geqslant 1$. If the β are ultimately all positive, an example where (5.37) holds but (5.38) does not is provided by $\beta(n) = 0, n < 0$, $\beta(n) \sim Cn^{-3/2}(\log n)^{-1}$ as $n \to \infty$ for some $C > 0$. These results should be compared with that of Corollary 5.2, where a weaker condition is used to obtain a 1-dimensional central limit theorem.

5.5. The Central Limit Theorem via Approximation to the Distribution of the Stationary Process

The emphasis so far given in this chapter has been on approximating the stationary process by an appropriately related martingale. For the central limit theorem, however, an alternative approach is to work with the distribution of the sum rather than the sum itself. This is a convenient approach for dealing with asymptotic independence conditions which are phrased in terms of mixing coefficients or in terms of the maximal correlation. We shall illustrate by working in terms of the latter.

As usual we suppose that $\{X_k, -\infty < k < \infty\}$ is a stationary sequence of random variables defined on a probability space $(\Omega, \mathscr{F}, P)$. We denote by $\mathscr{F}_a^b$ the σ-field generated by the random variables X_k, $-\infty \leqslant a \leqslant k \leqslant b \leqslant \infty$, and write $L^2(\mathscr{F}_a^b)$ for the set of all $\mathscr{F}_a^b$-measurable random variables with finite variance.

The *maximal correlation* coefficient $\rho(n)$ between the past $\{X_k, k \leqslant 0\}$ and the future $\{X_k, k \geqslant n\}$, $n > 0$, of the sequence $\{X_k\}$ is defined by the equation

$$\rho(n) = \sup \frac{|E(x - Ex)(y - Ey)|}{[E(x - Ex)^2 E(y - Ey)^2]^{1/2}},$$

where the supremum is taken over all random variables $x \in L^2(\mathscr{F}_{-\infty}^0)$ and $y \in L^2(\mathscr{F}_n^\infty)$. The condition $\rho(n) \to 0$ as $n \to \infty$, which was first studied by Hirschfeld (1935) and Gebelein (1941), is clearly a condition of asymptotic independence between the past and future of the sequence $\{X_k\}$. We shall proceed to relate it to the conditions of strong and uniform mixing which have been discussed earlier in the chapter (see Corollaries 5.1, 5.3, 5.5).

The process $\{X_k\}$ is said to satisfy the strong mixing condition if

$$\sup_{A \in \mathscr{F}_{-\infty}^0,\, B \in \mathscr{F}_n^\infty} |P(A \cap B) - P(A)P(B)| = \alpha(n) \downarrow 0$$

as $n \to \infty$. Then, if $I(A)$ denotes the indicator set of A,

$$\begin{aligned} \rho(n) &\geqslant \sup_{A \in \mathscr{F}_{-\infty}^0,\, B \in \mathscr{F}_n^\infty} \frac{|E(I(A) - EI(A))(I(B) - EI(B))|}{[E(I(A) - EI(A))^2 E(I(B) - EI(B))^2]^{1/2}} \\ &= \sup_{A \in \mathscr{F}_{-\infty}^0,\, B \in \mathscr{F}_n^\infty} \frac{|P(A \cap B) - P(A)P(B)|}{[P(A)(1 - P(A))P(B)(1 - P(B))]^{1/2}} \geqslant 4\alpha(n), \end{aligned} \tag{5.39}$$

since $x(1 - x) \leqslant 1/4$ for $0 \leqslant x \leqslant 1$. In the particular case where the X_k have a Gaussian distribution, $\alpha(n)$ and $\rho(n)$ are asymptotically equivalent. Indeed, it has been shown by Kolmogorov and Rozanov [e.g., Ibragimov and Rozanov (1978)] that

$$4\alpha(n) \leqslant \rho(n) \leqslant 2\pi\alpha(n).$$

On the other hand, the process $\{X_k\}$ is said to satisfy the more stringent *uniform mixing* condition if

$$\sup_{A \in \mathscr{F}_{-\infty}^{0},\, B \in \mathscr{F}_{n}^{\infty}} |P(B|A) - P(B)| = \phi(n) \downarrow 0$$

as $n \to \infty$. The Hölder-type inequality for uniform mixing processes which is provided in Theorem A.6 of Appendix III gives for random variables x, y which are measurable with respect to $\mathscr{F}_{-\infty}^{0}, \mathscr{F}_{n}^{\infty}$, respectively, with $Ex^2 < \infty$, $Ey^2 < \infty$,

$$|Exy - ExEy| \leqslant 2[\phi(n)]^{1/2}[Ex^2 Ey^2]^{1/2},$$

so that

$$\rho(n) \leqslant 2\phi^{1/2}(n). \tag{5.40}$$

The condition $\rho(n) \to 0$ is thus intermediate between uniform mixing and strong mixing.

We shall establish the following result which is due to Ibragimov (1975) but whose proof is incomplete.

Theorem 5.6. *Let $\{X_k\}$ denote a stationary and ergodic sequence with $EX_0 = 0$, $EX_0^2 < \infty$, and put $S_n = \sum_{k=1}^{n} X_k$, $\sigma_n^2 = ES_n^2$. Suppose that $\rho(n) \to 0$ as $n \to \infty$. Then, either $\sup_n \sigma_n^2 < \infty$ or $\sigma_n^2 = nh(n)$, where $\{h(n), n = 1, 2, \ldots\}$ is a slowly varying sequence. If $\sigma_n^2 \to \infty$ as $n \to \infty$ and $E|X_0|^{2+\delta} < \infty$ for some $\delta > 0$, then S_n/σ_n converges in distribution to the standard normal law.*

Proof. Suppose that $\sigma_n^2 \to \infty$. We shall first prove that for every integer $k = k(n) \leqslant A < \infty$,

$$\lim_{n \to \infty} \frac{\sigma_{nk(n)}^2}{k(n)\sigma_n^2} = 1. \tag{5.41}$$

Write

$$\xi_j = \sum_{s=1}^{n} X_{(j-1)n+(j-1)r+s}, \qquad 1 \leqslant j \leqslant k,$$

$$\eta_j = \sum_{s=1}^{r} X_{jn+(j-1)r+s}, \qquad 1 \leqslant j \leqslant k-1, \quad \eta_k = -\sum_{s=1}^{(k-1)r} X_{nk+s},$$

where $r = [\log \sigma_n^2]$, $[x]$ denoting the integer part of x. We have

$$\sigma_n^2 \leqslant \sum_{i=1}^{n}\sum_{j=1}^{n} E|X_i X_j| \leqslant n^2 EX_0^2, \tag{5.42}$$

so that $r = O(\log n)$.

Now, $S_{nk} = \sum_{j=1}^{k} \xi_j + \sum_{j=1}^{k} \eta_j$,
so that

$$\sigma_{nk}^2 = \sum_{j=1}^{k} E\xi_j^2 + 2 \sum_{1 \leqslant i < j \leqslant k} E\xi_i\xi_j + \sum_{i=1}^{k} \sum_{j=1}^{k} (E\xi_i\eta_k + E\eta_i\eta_j). \tag{5.43}$$

Using stationarity,

$$E\xi_j^2 = \sigma_n^2,$$

while, from the definition of maximal correlation,

$$|E\xi_i\xi_j| \leqslant \rho(|i-j|)(E\xi_i^2)^{1/2}(E\xi_j^2)^{1/2} \leqslant \rho(r)\sigma_n^2 \tag{5.44}$$

for $i \neq j$. Further, using Schwarz's inequality and (5.42),

$$|E\xi_j\eta_i| \leqslant (E\xi_j^2)^{1/2}(E\eta_i^2)^{1/2} \leqslant \sigma_n kr\sigma_1 = O(\sigma_n \log \sigma_n) \tag{5.45}$$

and

$$|E\eta_i\eta_j| \leqslant (E\eta_i^2)^{1/2}(E\eta_j^2)^{1/2} \leqslant k^2r^2\sigma_1^2 = O(\log \sigma_n)^2. \tag{5.46}$$

Since r increases with n, $\rho(r) = o(1)$ as $n \to \infty$ and hence the result (5.41) follows from (5.43)–(5.46).

For any fixed $\lambda > 0$ take $k(n) = n^{-1}[\lambda n](\leqslant \lambda)$. Then (5.41) gives

$$\lim_{n \to \infty} \frac{\sigma_{[\lambda n]}^2}{[\lambda n]\sigma_n^2} = 1,$$

so that $\sigma_n^2 = nh(n)$, where $\{h(n), n = 1, 2, \ldots\}$ is a slowly varying sequence. We note that $H(x) = h([x])$, $x \geqslant 1$, is a slowly varying function.

Now we represent the sum S_n in the form

$$S_n = \sum_{i=0}^{k-1} U_i + \sum_{i=0}^{k} V_i, \tag{5.47}$$

where

$$U_i = \sum_{ip+iq+1}^{(i+1)p+iq} X_j, \qquad 0 \leqslant i \leqslant k-1,$$

$$V_i = \sum_{(i+1)p+iq+1}^{(i+1)p+(i+1)q} X_j, \qquad 0 \leqslant i \leqslant k-1, \quad V_k = \sum_{kp+kq+1}^{n} X_j,$$

and, writing

$$\lambda(n) = \max((\rho([n^{1/4}]))^{1/3}, (\log n)^{-1}),$$

we define p, q, and k by

$$p = \max([n\rho([n^{1/4}])/\lambda(n)], [n^{7/8}/\lambda(n)]), \qquad q = [n^{1/4}]$$

and

$$k = [n/(p+q)] \sim n/p \qquad \text{as} \quad n \to \infty.$$

We begin our analysis of $\sigma_n^{-1}S_n$ by showing that $\sigma_n^{-1}\sum_{i=0}^k V_i \xrightarrow{\text{p}} 0$ and hence is asymptotically negligible. In fact,

$$\begin{aligned} E\left(\sigma_n^{-1}\sum_{i=0}^{k} V_i\right)^2 &= \sigma_n^{-2}\sum_{0\leqslant i,\,j\leqslant k-1} E(V_iV_j) + 2\sigma_n^{-2}\sum_{0\leqslant i\leqslant k-1} EV_iV_k + \sigma_n^{-2}EV_k^2 \\ &\leqslant \sigma_n^{-2}k^2EV_0^2 + 2\sigma_n^{-2}k(EV_0^2EV_k^2)^{1/2} + \sigma_n^{-2}EV_k^2 \\ &\leqslant \sigma_n^{-2}k^2\sigma_q^2 + 2\sigma_n^{-2}k\sigma_q\sigma_{q'} + \sigma_n^{-2}\sigma_{q'}^2, \end{aligned}$$

where $q' = n - (p+q)[n/(p+q)] \leqslant p+q$ is the number of terms in V_k. Since $\sigma_n^2 = nh(n)$ is regularly varying, we observe that for integral $l < n$,

$$\frac{lh(l)}{nh(n)} = O\left(\frac{l}{n}\right)^{2/3} \qquad \text{as} \quad n \to \infty,$$

and hence

$$\frac{k^2\sigma_q^2}{\sigma_n^2} \sim \frac{nqh(q)}{p^2h(n)} = O\left(\frac{n^2}{p^2}\left(\frac{q}{n}\right)^{2/3}\right) = O(n^{-1/4}\lambda^2(n)) \to 0,$$

$$\frac{k\sigma_q\sigma_{q'}}{\sigma_n^2} = k\left(\frac{qh(q)}{nh(n)}\,\frac{q'h(q')}{nh(n)}\right)^{1/2} = O\left(k\left(\frac{q'q}{n^2}\right)^{2/3}\right) = O\left(\left(\frac{q^2}{np}\right)^{1/3}\right) \to 0,$$

and

$$\frac{\sigma_{q'}^2}{\sigma_n^2} = \frac{qh(q)}{nh(n)} = O\left(\frac{q}{n}\right)^{2/3} \to 0,$$

so that $\sigma_n^{-1}\sum_{i=0}^k V_i \xrightarrow{\text{p}} 0$ via Chebyshev's inequality. Thus, in order to show that $\sigma_n^{-1}S_n \xrightarrow{\text{d}} N(0,1)$, it suffices to show that $\sigma_n^{-1}\sum_{i=0}^{k-1} U_i \xrightarrow{\text{d}} N(0,1)$ as $n \to \infty$.

As the first step toward establishing $\sigma_n^{-1}\sum_{i=0}^{k-1} U_i \xrightarrow{\text{d}} N(0,1)$, we shall show that the distribution of $\sigma_n^{-1}\sum_{i=0}^{k-1} U_i$ is asymptotically equivalent to that of the martingale composed of a sum of k independent and identically distributed random variables, each with the distribution $\sigma_n^{-1}U_0$. Indeed, for $2 \leqslant l \leqslant k-1$ we note that $\exp\{it\sigma_n^{-1}(U_0 + \cdots + U_{l-1})\}$ is measurable with respect to $\mathscr{F}_{-\infty}^{lp+(l-1)q}$, and $\exp\{it\sigma_n^{-1}U_l\}$ is measurable with respect to $\mathscr{F}_{lp+lq+1}^{\infty}$, so that we have

$$\left|E\exp\left\{it\sigma_n^{-1}\sum_{j=0}^{l} U_j\right\} - E\exp\left\{it\sigma_n^{-1}\sum_{j=0}^{l-1} U_j\right\}E\exp\{it\sigma_n^{-1}U_l\}\right| \leqslant 16\rho(q)$$

and hence, summing over $2 \leqslant l \leqslant k-1$,

$$\left|E\exp\left\{it\sigma_n^{-1}\sum_{j=0}^{k-1} U_j\right\} - (E\exp\{it\sigma_n^{-1}U_0\})^k\right| \leqslant 16k\rho(q) \to 0 \qquad (5.48)$$

as $n \to \infty$.

Now, using the Lindeberg condition for central limit convergence of triangular arrays of r.v. which are independent in each row, $\{E(\exp(it\sigma_n^{-1}U_0))\}^k \to$

$\exp(-\frac{1}{2}t^2)$ as $n \to \infty$ provided that

$$kEU_0^2/\sigma_n^2 = k\sigma_p^2/\sigma_n^2 \to 1 \tag{5.49}$$

as $n \to \infty$ and, for any $\varepsilon > 0$,

$$\lim_{n\to\infty} k \int_{|z|>\varepsilon} z^2 \, dP(\sigma_n^{-1} U_0 \leqslant z) = 0. \tag{5.50}$$

The checking of the requirement (5.49) was omitted by Ibragimov (1975) following the corresponding omission in the proof of Theorem 18.4.1 of Ibragimov and Linnik (1971). Thus, if (5.49) and (5.50) hold, we obtain the required $\sigma_n^{-1} \sum_{i=0}^{k-1} U_i \xrightarrow{d} N(0,1)$ as $n \to \infty$.

To check (5.49) we square (5.47) and take expectations to obtain

$$\begin{aligned}\sigma_n^2 = \sum_{i=0}^{k-1} EU_i^2 + 2 \sum_{0 \leqslant i < j \leqslant k-1} EU_iU_j + \sum_{i=0}^{k-1}\sum_{j=0}^{k-1} EU_iV_j \\ + 2\sum_{i=0}^{k-1} EU_iV_k + 2\sum_{i=0}^{k-1} EV_iV_k + \sum_{i=0}^{k-1} EV_i^2 + EV_k^2.\end{aligned} \tag{5.51}$$

Further, using stationarity,

$$EU_i^2 = \sigma_p^2, \qquad EV_i^2 = \sigma_q^2, \qquad 0 \leqslant i \leqslant k-1, \qquad EV_k^2 = \sigma_{q'}^2,$$

and corresponding to (5.44)–(5.46),

$$\begin{aligned}
&|EU_iU_j| \leqslant \rho(|i-j|)\sigma_p^2 \leqslant \rho(q)\sigma_p^2, \qquad i \neq j, \quad 0 \leqslant i,j \leqslant k-1,\\
&|EU_iV_j| \leqslant \sigma_p\sigma_q, \qquad 0 \leqslant i,j \leqslant k-1,\\
&|EV_iV_j| \leqslant \sigma_q^2, \qquad 0 \leqslant i,j \leqslant k-1,\\
&|EU_iV_k| \leqslant \rho(q)\sigma_p\sigma_{q'}, \qquad 0 \leqslant i \leqslant k-1,\\
&|EV_iV_k| \leqslant \sigma_q\sigma_{q'}, \qquad 0 \leqslant i \leqslant k-1.
\end{aligned}$$

Then, (5.51) gives

$$\begin{aligned}\sigma_n^2 = k\sigma_p^2 + O(k^2\rho(q)\sigma_p^2) + O(k^2\sigma_p\sigma_q) \\ + O(k\rho(q)\sigma_p\sigma_{q'}) + O(k\sigma_q\sigma_{q'}) + k\sigma_q^2 + \sigma_{q'}^2,\end{aligned}$$

and (5.49) holds since, as $n \to \infty$, $k\rho(q) \to 0$ by definition,

$$\frac{k\sigma_q}{\sigma_p} = k\left(\frac{qh(q)}{ph(p)}\right)^{1/2} = O\left(k\left(\frac{q}{p}\right)^{2/3}\right) \to 0,$$

$$\frac{\rho(q)\sigma_{q'}}{\sigma_p} = O(\rho(q)) \to 0,$$

$$\frac{\sigma_q\sigma_{q'}}{\sigma_p^2} = O\left(\left(\frac{q'q}{p^2}\right)^{2/3}\right) \to 0, \qquad \frac{\sigma_q^2}{\sigma_p^2} = O\left(\left(\frac{q}{p}\right)^{2/3}\right) \to 0.$$

Our final major step will be to show that there exists a constant $A > 0$ such that $E|S_n|^{2+\delta} \leqslant A\sigma_n^{2+\delta}$ (and hence $\{\sigma_n^{-2}S_n^2, n \geqslant 1\}$ is uniformly integrable). Then (5.50) follows since

$$\begin{aligned} k\int_{|z|\geqslant\varepsilon} z^2\,dP(\sigma_n^{-1}U_0 \leqslant z) &= k\sigma_n^{-2}E(S_p^2 I(|S_p| > \varepsilon\sigma_n)) \\ &\leqslant k\varepsilon^{-\delta}\sigma_n^{-(2+\delta)}E|S_p|^{2+\delta} \\ &\leqslant Ak\varepsilon^{-\delta}\sigma_n^{-(2+\delta)}\sigma_p^{2+\delta} \\ &\sim A\varepsilon^{-\delta}(pn^{-1})^{\delta}(h(p)(h(n))^{-1})^{1+\delta/2} \to 0, \end{aligned}$$

h being slowly varying. This will complete the proof.

In showing that $E|S_n|^{2+\delta} \leqslant A\sigma_n^{2+\delta}$ for some $0 < A < \infty$, we restrict consideration to $0 < \delta \leqslant 1$. Put $\tilde{S}_n = \sum_{n+r+1}^{2n+r} X_j$. We have

$$\begin{aligned} E|S_n + \tilde{S}_n|^{2+\delta} &\leqslant E\{(S_n + \tilde{S}_n)^2(|S_n|^{\delta} + |\tilde{S}_n|^{\delta})\} \\ &\leqslant 2a_n + E\{S_n^2|\tilde{S}_n|^{\delta}\} + E\{\tilde{S}_n^2|S_n|^{\delta}\} \\ &\quad + 2E\{|S_n|^{1+\delta}|\tilde{S}_n|\} + 2E\{|\tilde{S}_n|^{1+\delta}|S_n|\}, \end{aligned} \tag{5.52}$$

where $a_n = E|S_n|^{2+\delta}$. Then, using Hölder's inequality and the definition of ρ we obtain

$$\begin{aligned} E\{S_n^2|\tilde{S}_n|^{\delta}\} &\leqslant (E|S_n|^{2+\delta})^{(2-\delta)/(2+\delta)}(E|S_n|^{1+\delta/2}|\tilde{S}_n|^{1+\delta/2})^{2\delta/(2+\delta)} \\ &\leqslant \sigma_n^{2\delta}a_n^{(2-\delta)/(2+\delta)} + (\rho(r))^{2\delta/(2+\delta)}a_n \end{aligned}$$

and

$$\begin{aligned} E\{|S_n|^{1+\delta}|\tilde{S}_n|\} &\leqslant (E|S_n|^{2+\delta})^{\delta/(2+\delta)}(E|S_n|^{1+\delta/2}|\tilde{S}_n|^{1+\delta/2})^{2/(2+\delta)} \\ &\leqslant a_n^{\delta/(2+\delta)}\sigma_n^2 + (\rho(r))^{2/(2+\delta)}a_n \\ &\leqslant \sigma_n^{2\delta}a_n^{(2-\delta)/(2+\delta)} + (\rho(r))^{2/(2+\delta)}a_n, \end{aligned}$$

with analogous estimates for $E\{\tilde{S}_n^2|S_n|^{\delta}\}$ and $E\{|\tilde{S}_n|^{1+\delta}|S_n|\}$. The use of these estimates in (5.52) gives

$$E|S_n + \tilde{S}_n|^{2+\delta} \leqslant (2 + 6(\rho(r))^{2\delta/(2+\delta)})a_n + 6\sigma_n^{2\delta}a_n^{(2-\delta)/(2+\delta)},$$

and since $\rho(r) \to 0$ as $r \to \infty$, we can for any given $\varepsilon > 0$ choose r so large that

$$E|S_n + \tilde{S}_n|^{2+\delta} \leqslant (2 + \varepsilon)a_n + 6\sigma_n^{2\delta}a_n^{(2-\delta)/(2+\delta)}. \tag{5.53}$$

Next, using Minkowski's inequality,

$$\begin{aligned} a_{2n} &= E\left|S_n + \tilde{S}_n + \sum_{n+1}^{n+r} X_j - \sum_{2n+1}^{2n+r} X_j\right|^{2+\delta} \\ &\leqslant E|S_n + \tilde{S}_n|^{2+\delta}\left(1 + \frac{2a_r^{1/(2+\delta)}}{(E|S_n + \tilde{S}_n|^{2+\delta})^{1/(2+\delta)}}\right)^{2+\delta}, \end{aligned} \tag{5.54}$$

while, if $\rho(r) \leqslant \frac{1}{2}$,

$$(E|S_n + \tilde{S}_n|^{2+\delta})^{2/(2+\delta)} \geqslant E(S_n + \tilde{S}_n)^2 \geqslant 2\sigma_n^2(1 - \rho(r)) \geqslant \sigma_n^2 \to \infty.$$

Thus, from (5.53) and (5.54) we see that for any $\varepsilon > 0$ there exists an N such that for all $n > N$,

$$a_{2n} \leqslant (2 + \varepsilon)a_n + 7\sigma_n^{2\delta} a_n^{(2-\delta)/(2+\delta)}. \tag{5.55}$$

Now put $b_n = a_n/\sigma_n^{2+\delta}$. Since $\sigma_{2n}^2/\sigma_n^2 \to 2$ as $n \to \infty$, (5.55) ensures that for all sufficiently large n,

$$b_{2n} \leqslant (1 + \varepsilon)2^{-\delta/2} b_n + 7b_n^{(2-\delta)/(2+\delta)}.$$

Hence, there exist constants λ, $0 < \lambda < 1$, and $B > 0$ such that for all n,

$$b_{2n} \leqslant \lambda b_n + B,$$

and therefore

$$\sup_r b_{2^r} = A_1 < \infty. \tag{5.56}$$

Now let $2^l \leqslant n < 2^{l+1}$ and write n in binary form,

$$n = v_0 2^l + v_1 2^{l-1} + \cdots + v_l, \qquad v_0 = 1, \quad v_j = 0 \text{ or } 1.$$

We can then express S_n in the form

$$S_n = (X_1 + \cdots + X_{i_1}) + (X_{i_1+1} + \cdots + X_{i_2}) + \cdots + (X_{i_l+1} + \cdots + X_n),$$

where the number of terms in the jth parenthesis is $v_j 2^{l-j}$. Thus, using Minkowski's inequality and (5.56),

$$\begin{aligned} b_n &\leqslant \left| \sum_{j=0}^{l} \{\sigma_n^{-(2+\delta)} E|X_1 + \cdots + X_{v_j 2^{l-j}}|^{2+\delta}\}^{1/(2+\delta)} \right|^{2+\delta} \\ &\leqslant A_1 \left(\sigma_n^{-1} \sum_{j=0}^{l} \sigma_{2^{l-j}} \right)^{2+\delta}. \end{aligned} \tag{5.57}$$

But, as a simple consequence of the representation theorem for slowly varying functions,

$$\frac{h(2^{l-j})}{h(n)} = \frac{n\sigma_{2^{l-j}}^2}{2^{l-j}\sigma_n^2} = O\left(\left(\frac{n}{2^{l-j}} \right)^{1/3} \right)$$

as $n \to \infty$, and hence

$$\sigma_n^{-1} \sum_{j=0}^{l} \sigma_{2^{l-j}} = O(1) \qquad \text{as} \quad n \to \infty.$$

The required result then follows from (5.57) and this completes the proof. ∎

6
Estimation of Parameters from Stochastic Processes

6.1. Introduction

There is a substantial literature on the subject of estimation of the parameters of stochastic processes, but a close examination reveals that the subject is still in its comparative infancy. Books in the area, such as Billingsley (1961a), Mihoc and Craiu (1972), and Roussas (1972), deal principally with a very specialized context, that of stationary ergodic Markov chains. This represents, of course, the first step of generalization of the results of inference for the classical situation of random sampling (i.i.d. r.v.). The reason for this choice of context is not difficult to see. The classical i.i.d. results can be mimicked quite closely, and a more general context introduces a different order of complication for which the mathematical tools have not been available until very recently. Indeed, it is recent martingale limit results which provide the principal tools for a general discussion of large sample inference for stochastic processes based on the likelihood and its derivatives, although the recognition of their role in this dates back at least to Billingsley (1961a, p. 7 footnote) and Silvey (1961).

The recent work of Weiss and Wolfowitz on maximum probability (MP) estimators provides an alternative approach of considerable generality. However, for stochastic processes, the MP estimator is ordinarily even more difficult to obtain and to work with than the familiar maximum likelihood (ML) estimator. We shall merely refer the reader to a detailed discussion of the MP estimator in Weiss and Wolfowitz (1974). Martingale limit theory plays only an indirect role in this work.

In Section 6.2 we shall consider the ML estimator and show that it continues to enjoy many of the optimality properties of the i.i.d. case. For detailed discussion of the classical i.i.d. results see, for example, Rao (1973) and Zacks (1971). In situations where it can be calculated conveniently, the use of the ML estimator can ordinarily be recommended, although the scope for pathologies of the kind which emerge in the classical case is appreciably widened in the general context. A discussion of these pathologies is given, for example, in Weiss and Wolfowitz (1974). We shall confine our attention to the scalar case as most of the results presented do not (at present) admit complete extension to the multivariate case. Complications arise in the provision of subsidiary results such as a multivariate extension of the Kronecker lemma [see Anderson and Moore (1976)].

In cases where the ML estimator is impracticable it is usually necessary to resort to estimation procedures which do not necessarily have any particular optimality properties. In Section 6.3 we consider the method of conditional least squares (CLS). This procedure is one of choosing the estimator to minimize the sums of squares of the actual errors of best prediction. It has the advantage that, for the classical scaling, a discussion of the multivariate case poses no additional complications, but optimality questions are not treated. In various cases where the underlying process is Gaussian it can produce the ML estimators. Furthermore, under quite mild restrictions the CLS estimators are, as with the ML estimators, strongly consistent and asymptotically normally distributed.

One of the examples in Section 6.3 concerns the stationary autoregressive process and results relating to this process are given in generalized form in Section 6.4. This section is concerned with estimation based on the autocorrelations of a stationary linear time-series. The autocorrelations are, of course, generally quadratic functions of the parameters of the process, so the issue here is not a direct one of parameter estimation.

Finally, in Section 6.5 we consider the method of moments and particularly its role in the estimation of parameters of nonlinear time-series models. This important and newly developing area generates estimation problems which are not usually amenable to more sophisticated methods. Again it is frequently the case that martingale methods lead to strong consistency and asymptotic normality results for the estimators.

6.2. Asymptotic Behaviour of the Maximum Likelihood Estimator

In the case of estimating a parameter on the basis of observations which are i.i.d. or come from a stationary ergodic Markov chain, it is well known that the maximum likelihood (ML) estimator enjoys certain optimality properties. Among the most significant of these is a result of Schmetterer (1966), extending one of Rao (1963), that, subject to suitable regularity conditions, the ML estimator is the best consistent continuously asymptotically normal estimator in the sense of having minimum asymptotic variance. This was generalized by Weiss and Wolfowitz (1966), who showed that the ML estimator produces the best asymptotic probability of concentration in symmetric intervals. In this section we shall obtain analogous results for the more complex case where the sample is from a general stochastic process.

We consider a sample $X_1, X_2, \ldots, X_n$ of consecutive observations from some stochastic processes whose distribution depends on a single parameter θ, $\theta \in \Theta$, Θ being an open interval. Let $L_n(\theta)$ be the likelihood function associated with $X_1, X_2, \ldots, X_n$ and suppose that $L_n(\theta)$ is differentiable with respect to θ and $E_\theta(d \log L_n(\theta)/d\theta)^2 < \infty$ for each n. Suppose in addition that if $P_n(X_1, \ldots, X_n) = L_n(\theta)$ is the joint probability (density) function of $X_1, \ldots, X_n$, then $\sum_{x_n} P_n(x_1, \ldots, x_n)(\int P_n(x_1, \ldots, x_n)\, dx_n)$ can be differentiated twice with respect to θ under the summation (integration) sign. We shall write $\mathscr{F}_k$ for the σ-field generated by $X_1, \ldots, X_k$, $k \geqslant 1$, take $\mathscr{F}_0$ as the trivial σ-field, and set $L_0 = 1$. Then, writing

$$\frac{d \log L_n(\theta)}{d\theta} = \sum_{i=1}^{n} \frac{d}{d\theta}[\log L_i(\theta) - \log L_{i-1}(\theta)] = \sum_{i=1}^{n} u_i(\theta), \tag{6.1}$$

we have $E_\theta(u_i(\theta)|\mathscr{F}_{i-1}) = 0$ a.s., so that $\{d \log L_n(\theta)/d\theta, \mathscr{F}_n, n \geqslant 1\}$ is a square-integrable martingale. In addition, we set

$$I_n(\theta) = \sum_{i=1}^{n} E_\theta(u_i^2(\theta)|\mathscr{F}_{i-1}) \tag{6.2}$$

and note that, under the conditions imposed above, $v_i(\theta) = du_i(\theta)/d\theta$ satisfies

$$E_\theta(u_i^2(\theta)|\mathscr{F}_{i-1}) = -E_\theta(v_i(\theta)|\mathscr{F}_{i-1}) \quad \text{a.s.} \tag{6.3}$$

Also, writing

$$J_n(\theta) = \sum_{i=1}^{n} v_i(\theta),$$

we note that $\{J_n(\theta) + I_n(\theta), \mathscr{F}_n, n \geqslant 1\}$ is a martingale.

The quantity $I_n(\theta)$ is a form of conditional information which reduces to the standard Fisher information in the case where the X_i are independent random variables. The role of $I_n(\theta)$ in the stochastic process estimation context is a vital one and will emerge below. Aspects of this role are discussed in Heyde (1975b, 1977c) and Heyde and Feigin (1975). The importance of $I_n(\theta)$ is a partial consequence of the following expansion. We suppose that θ is the true parameter value. Then, we can use Taylor's expansion to write for $\theta' \in \Theta$,

$$\begin{aligned}\frac{d}{d\theta}\log L_n(\theta') &= \sum_{i=1}^{n} u_i(\theta') \\ &= \sum_{i=1}^{n} u_i(\theta) + (\theta' - \theta)\sum_{i=1}^{n} v_i(\theta_n^*) \\ &= \sum_{i=1}^{n} u_i(\theta) - (\theta' - \theta)I_n(\theta) + (\theta' - \theta)(J_n(\theta_n^*) + I_n(\theta)), \quad (6.4)\end{aligned}$$

while $\theta_n^* = \theta + \gamma(\theta' - \theta)$ with $\gamma = \gamma(X_1, \ldots, X_n, \theta)$ satisfying $|\gamma| < 1$. Various sufficient conditions can be provided to ensure the existence of a consistent root of the likelihood equation. For example, one such condition is obtained by supposing that $I_n(\theta) \xrightarrow{\text{a.s.}} \infty$ as $n \to \infty$ and that, for any $\delta > 0$ for which $N_0(\delta) = (\theta - \delta, \theta + \delta) \subseteq \Theta$, there exists $k(\delta) > 0$ and $h(\delta) \downarrow 0$ satisfying

$$\liminf_{n\to\infty} P_\theta\left\{\sup_{\theta' \in \Theta \backslash N_0(\delta)} [I_n(\theta)]^{-1}[\log L_n(\theta') - \log L_n(\theta)] < -k(\delta)\right\} \geqslant 1 - h(\delta).$$

This condition is closely related to the classical conditions of Wald (1948). For other related literature see M. M. Rao (1966), Weiss (1971, 1973), Bhat (1974), Caines (1975), Basawa, Feigin and Heyde (1976) and references cited therein.

Using Theorem 2.18 we see that

$$[I_n(\theta)]^{-1}\sum_{i=1}^{n} u_i(\theta) \xrightarrow{\text{a.s.}} 0 \qquad (6.5)$$

provided $I_n(\theta) \xrightarrow{\text{a.s.}} \infty$ as $n \to \infty$, since then

$$\sum_{1}^{\infty}[I_n(\theta)]^{-2}E_\theta(u_n^2(\theta)|\mathscr{F}_{n-1}) < \infty \quad \text{a.s.}$$

is automatically satisfied. This last result follows since, for $\{a_j\}$ any sequence of positive constants and $b_n = \sum_1^n a_j$,

$$\begin{aligned}\sum_{1}^{\infty}\left(\sum_{1}^{n} a_j\right)^{-2} a_n &= \sum_{1}^{\infty} b_n^{-2}(b_n - b_{n-1}) \qquad (b_0 = 0) \\ &= \sum_{1}^{\infty} b_n(b_n^{-2} - b_{n+1}^{-2}) \\ &= \sum_{1}^{\infty} b_n(b_n^{-1} - b_{n+1}^{-1})(b_n^{-1} + b_{n+1}^{-1}) \\ &\leqslant 2\sum_{1}^{\infty}(b_n^{-1} - b_{n+1}^{-1}) \leqslant 2b_1^{-1} < \infty.\end{aligned}$$

From (6.4) and (6.5) we see that the likelihood equation has a root $\hat{\theta}_n$ which is strongly consistent for θ (i.e., $\hat{\theta}_n \xrightarrow{\text{a.s.}} \theta$ as $n \to \infty$) if $I_n(\theta) \xrightarrow{\text{a.s.}} \infty$ and

$$\limsup_{n\to\infty}[I_n(\theta)]^{-1}|I_n(\theta) + J_n(\theta_n^*)| < 1 \quad \text{a.s.}$$

In order to utilize most effectively the ML estimator from large samples it is necessary for $\sum_{i=1}^{n} u_i(\theta)$ to converge in distribution, when appropriately normalized, to some proper limit law. Furthermore, it can be arranged in many cases that this limit law is normal, which is most convenient for confidence interval purposes. The most effective general norming seems to be provided by $I_n^{1/2}(\theta)$ and, indeed, $[I_n(\theta)]^{-1/2}\sum_{i=1}^{n} u_i(\theta)$ converges in distribution to normality under quite wide-ranging circumstances. Results of this kind appear in Chapter 3 and an extension of them, relevant to the present context, will be given below in Proposition 6.1.

If

$$[I_n(\theta)]^{-1/2}\sum_{i=1}^{n} u_i(\theta) \xrightarrow{\text{d}} N(0,1)$$

and, in addition to the abovementioned conditions, $J_n(\theta)/I_n(\theta) \to -1$ in probability as $n \to \infty$ uniformly on compacts of θ, we have from (6.4) that

$$I_n^{1/2}(\theta)(\hat{\theta}_n - \theta) \xrightarrow{\text{d}} N(0,1).$$

Then, if $T_n = T_n(X_1,\ldots,X_n)$ is any consistent estimator of θ for which $I_n^{1/2}(\theta)(T_n - \theta) \xrightarrow{\text{d}} N(0,\gamma^2(\theta))$, where $\gamma(\theta)$ is bounded and continuous in θ, we shall show, under suitable regularity conditions, that $\gamma^2(\theta) \geqslant 1$. That is, the ML estimator is optimal within this class since $\gamma(\theta) \equiv 1$ when T_n is the ML estimator.

Of course it is desirable to be able to make comparisons between the ML estimator and those other estimators T_n for which $I_n^{1/2}(\theta)(T_n - \theta)$ converges in distribution to a proper law (not necessarily normal) or perhaps does not even converge. Such comparisons have been made in the case where the $\{X_i\}$ form a stationary ergodic Markov chain, by using the concept of asymptotic efficiency in the Wolfowitz sense [see Roussas (1972, Chapter 5)]. Minor modifications of the standard theory as presented in the book of Roussas lead to comparisons of the kind

$$\begin{aligned}&\lim_{n\to\infty} P_\theta(-c < (E_\theta I_n(\theta))^{1/2}(\hat{\theta}_n - \theta) < b)\\ &\qquad \geqslant \limsup_{n\to\infty} P_\theta(-c + w(\theta) < (E_\theta I_n(\theta))^{1/2}(T_n - \theta) < W(\theta) + b)\end{aligned}$$

for arbitrary positive b and c and certain $W(\theta) \geqslant w(\theta)$. This inequality holds under conditions which are quite similar to those of this section but (at present) require that $I_n(\theta)(E_\theta I_n(\theta))^{-1} \xrightarrow{\text{p}} 1$ as $n \to \infty$. It is, however, the cases where this last condition is not satisfied that pose the real interest and challenge in the treatment of stochastic process estimation. Our arguments in this section are principally concerned with the commonly occurring circumstances under which $I_n(\theta)(E_\theta I_n(\theta))^{-1} \xrightarrow{\text{p}} \eta^2(\theta)$ as $n \to \infty$, η being a random variable in general.

The abovementioned difficulties can be avoided if we restrict consideration to symmetric concentration intervals. Then, using ideas of Weiss and Wolfowitz (1966), we shall show that

$$\lim_{n\to\infty} P_\theta(-c<(E_\theta I_n(\theta))^{1/2}(\hat{\theta}_n-\theta)<c) \geqslant \limsup_{n\to\infty} P_\theta(-c<(E_\theta I_n(\theta))^{1/2}(T_n-\theta)<c)$$

for all $c > 0$.

We now proceed to formalize our conditions. Suppose that $(X_1, \ldots, X_n)$ possesses a density $P_n(X_1, \ldots, X_n)$ $(=L_n(\theta))$, which is continuous in θ, with respect to a σ-finite measure μ_n, and that the following two assumptions are satisfied.

Assumption 1. *$I_n(\theta) \xrightarrow{\text{a.s.}} \infty$, $I_n(\theta)/E_\theta I_n(\theta) \xrightarrow{\text{p}} \eta^2(\theta)$ $(>0$ a.s.) for some r.v. η, and $J_n(\theta)/I_n(\theta) \xrightarrow{\text{p}} -1$ as $n \to \infty$, the convergences in probability being uniform on compacts of θ.*

Assumption 2. *For $\delta > 0$, suppose $|\theta_n - \theta| \leqslant \delta/(E_\theta I_n(\theta))^{1/2}$. Then,*

(i) *$E_{\theta_n} I_n(\theta_n) = E_\theta I_n(\theta)(1 + o(1))$ as $n \to \infty$,*
(ii) *$I_n(\theta_n) = I_n(\theta)(1 + o(1))$ a.s. as $n \to \infty$,*
(iii) *$J_n(\theta_n) = J_n(\theta) + o(I_n(\theta))$ a.s. as $n \to \infty$.*

These assumptions are not severe and conditions may be imposed directly on the stochastic process to ensure that they hold. Interestingly, the assumptions are enough to provide a central limit theorem for the normed martingale $(I_n(\theta))^{-1/2}\sum_{i=1}^n u_i(\theta)$ without the intervention of any further conditions (such as uniform asymptotic negligibility) on the differences $u_i(\theta)$.

Proposition 6.1. *Under Assumptions 1 and 2,*

$$[(E_\theta I_n(\theta))^{-1/2}\, d\log L_n(\theta)/d\theta,\ I_n(\theta)/E_\theta I_n(\theta)] \xrightarrow{\text{d}} (\eta(\theta)N(0,1), \eta^2(\theta)),$$

where η and N are independent.

Proof. Fix $c > 0$ and let $\theta_n = \theta + c(E_\theta I_n(\theta))^{-1/2}$. We set

$$\Lambda_n = \log[L_n(\theta_n)/L_n(\theta)]$$

and use Taylor's expansion to obtain

$$\Lambda_n = (\theta_n - \theta) \sum_{i=1}^n u_i(\theta) + \tfrac{1}{2}(\theta_n - \theta)^2 J_n(\theta_n^{**})$$

for $\theta_n^{**} \in [\theta,\theta_n]$. Then, writing

$$W_n(\theta) = (E_\theta I_n(\theta))^{-1/2} \sum_{i=1}^{n} u_i(\theta),$$

$$V_n(\theta) = -(E_\theta I_n(\theta))^{-1} J_n(\theta_n^{**}),$$

taking exponentials and rearranging, we find that

$$e^{cW_n(\theta)} L_n(\theta) = e^{c^2 V^2(\theta)/2} L_n(\theta_n). \tag{6.6}$$

Now let $A = A(\theta)$ be a continuity point of $\eta^2(\theta)$; that is, $P_\theta(\eta^2(\theta) = A) = 0$. Then, under Assumptions 1 and 2, $V_n(\theta) \xrightarrow{\text{p}} \eta^2(\theta)$ under P_{θ_n} and

$$P_{\theta_n}(|V_n(\theta)| \leqslant A) \to P_\theta(\eta^2(\theta) \leqslant A)$$

as $n \to \infty$.

Next let f be a bounded continuous function on $(-\infty,\infty)$ with $f(x) = 0$ *for* $|x| > A$. Then, using (6.6) together with Theorem 5.2 (iii) of Billingsley (1968),

$$\begin{aligned} E_\theta[f(V_n(\theta))e^{cW_n(\theta)} \,|\, |V_n(\theta)| \leqslant A] &= E_{\theta_n}[f(V_n(\theta))e^{c^2 V_n(\theta)/2} \,|\, |V_n(\theta)| \leqslant A] \\ &\to E_\theta[f(\eta^2(\theta))e^{c^2\eta^2(\theta)/2} | \eta^2(\theta) \leqslant A] \\ &= E_\theta[f(\eta_A^2(\theta))e^{c^2\eta_A^2(\theta)/2}], \end{aligned}$$

where $\eta_A(\theta)$ has the distribution of $\eta(\theta)$ conditional on $\eta^2(\theta) \leqslant A$. But

$$E_\theta[f(\eta_A^2(\theta))e^{c^2\eta_A^2(\theta)/2}] = E_\theta[f(\eta_A^2(\theta))e^{c\eta_A(\theta)N(0,1)}],$$

where η_A is independent of N, and hence the joint distribution of $W_n(\theta)$, $V_n(\theta)$, conditional on $|V_n(\theta)| \leqslant A$, converges to that of $(\eta_A(\theta)N(0,1), \eta_A^2(\theta))$. The required result then follows upon letting $A \to \infty$. ∎

The result of Proposition 6.1 together with the continuous mapping theorem (Theorem A.3 of Appendix II) ensures that for $g(x,y)$ any continuous function of two variables,

$$g\left[(E_\theta I_n(\theta))^{-1/2} \frac{d \log L_n(\theta)}{d\theta}, \frac{I_n(\theta)}{E_\theta I_n(\theta)}\right] \xrightarrow{\text{d}} g[\eta(\theta)N(0,1), \eta^2(\theta)],$$

where η and N are independent. This result will be used many times in the discussion below.

It should also be noted, in the light of Proposition 6.1 and the earlier discussion in this section, that Assumptions 1 and 2 ensure that there exists an ML estimator $\hat{\theta}_n$ of θ for which $\hat{\theta}_n \xrightarrow{\text{a.s.}} \theta$ and $I_n^{1/2}(\theta)(\hat{\theta}_n - \theta) \xrightarrow{\text{d}} N(0,1)$ as $n \to \infty$.

We are now in a position to establish the following result, which extends that of Heyde (1978b).

Theorem 6.1. *Fix* $c > 0$ *and let* $\theta_n = \theta + 2c(E_\theta I_n(\theta))^{-1/2}$. *Suppose that* T_n *is any estimator of* θ *which satisfies the condition that for any* $\theta \in \Theta$,

$$\lim_{n\to\infty} \{P_\theta((E_\theta I_n(\theta))^{1/2}(T_n - \theta) > -c) - P_{\theta_n}((E_\theta I_n(\theta))^{1/2}(T_n - \theta_n) > -c)\} = 0.$$

Under Assumptions 1 *and* 2 *there is a maximum likelihood estimator* $\hat{\theta}_n$ *which is consistent for* $\theta \in \Theta$ *and for which*

$$\lim_{n\to\infty} P_\theta(-c < (E_\theta I_n(\theta))^{1/2}(\hat{\theta}_n - \theta) < c)$$
$$\geqslant \limsup_{n\to\infty} P_\theta(-c < (E_\theta I_n(\theta))^{1/2}(T_n - \theta) < c).$$

Furthermore, if

$$[(E_\theta I_n(\theta))^{1/2}(T_n - \theta), I_n(\theta)/E_\theta I_n(\theta)] \xrightarrow{d} [\eta(\theta)N(0,\gamma^2(\theta)), \eta^2(\theta)],$$

where η *and* N *are independent and* $\gamma(\theta)$ *is bounded, then* $\gamma^2(\theta) \geqslant 1$.

Proof. Write $P_{\theta_n}^{(n)}$ and $P_\theta^{(n)}$ for the probability measures corresponding to $L_n(\theta_n)$ and $L_n(\theta)$, respectively. We first need to establish the contiguity of the sequences $\{P_{\theta_n}^{(n)}, n \geqslant 1\}$ and $\{P_\theta^{(n)}, n \geqslant 1\}$, and this is done via the checking of condition (S_3) of Roussas (1972, p. 11). Later, and indeed wherever possible without confusion, we shall drop the superscript (n).

As in the proof of Proposition 6.1, we define

$$\Lambda_n = \log[L_n(\theta_n)/L_n(\theta)]$$

and use Taylor's expansion to obtain

$$\Lambda_n = (\theta_n - \theta)\sum_{i=1}^{n} u_i(\theta) + \tfrac{1}{2}(\theta_n - \theta)^2 J_n(\theta_n^{**}) \tag{6.7}$$

for $\theta^{**} \in [\theta,\theta_n]$. Then, in view of Assumptions 1 and 2 and Proposition 6.1,

$$\Lambda_n \xrightarrow{d} 2cN(0,1)\eta(\theta) - 2c^2\eta^2(\theta) \tag{6.8}$$

as $n \to \infty$, η and N being independent. Furthermore,

$$E_\theta\{\exp(2cN(0,1)\eta(\theta) - 2c^2\eta^2(\theta))\}$$
$$= E_\theta[E_\theta\{\exp(2cN(0,1)\eta(\theta))|\eta(\theta)\}\exp(-2c^2\eta^2(\theta))] = 1$$

since η and N are independent. This completes the establishment of the desired contiguity.

To prove the main result of the theorem we begin by considering the Bayesian problem of deciding whether $L_n(\theta)$ or $L_n(\theta_n)$ is the true probability (density) function of $\mathbf{X} = (X_1, \ldots, X_n)$. The *a priori* probability of each probability (density) function is taken as $\frac{1}{2}$. Let $d(\mathbf{X})$ be the decision function and suppose that the loss is -1 when a correct decision is made and 0 in the case of an incorrect decision. Then, the Bayes solution corresponds to the d which minimizes the Bayes risk

$$R(d) = -\tfrac{1}{2}P_{\theta_n}(d(\mathbf{X}) = \theta_n) - \tfrac{1}{2}P_\theta(d(\mathbf{X}) = \theta). \tag{6.9}$$

However, if E_* denotes the expectation with respect to the posterior distribution of the parameter,

$$\begin{aligned} &P_{\theta_n}(d(\mathbf{X}) = \theta_n) + P_\theta(d(\mathbf{X}) = \theta) \\ &\quad = E_*[P(\theta_n|\mathbf{X})I(d(\mathbf{X}) = \theta_n) + P(\theta|\mathbf{X})I(d(\mathbf{X}) = \theta)] \\ &\quad = E_*[P(\theta_n|\mathbf{X})I(d(\mathbf{X}) = \theta_n) + P(\theta|\mathbf{X})(1 - I(d(\mathbf{X}) = \theta_n))] \\ &\quad = E_*[P(\theta|\mathbf{X}) + \{P(\theta_n|\mathbf{X}) - P(\theta|\mathbf{X})\}I(d(\mathbf{X}) = \theta_n)], \end{aligned}$$

which is maximized if $d(\mathbf{X}) = \theta_n$ when

$$P(\theta_n|\mathbf{X}) > P(\theta|\mathbf{X}).$$

But

$$P(\theta|\mathbf{X}) = \frac{L_n(\theta)}{L_n(\theta) + L_n(\theta_n)}, \qquad P(\theta_n|\mathbf{X}) = \frac{L_n(\theta_n)}{L_n(\theta) + L_n(\theta_n)},$$

and hence the Bayes solution corresponds to deciding for θ_n when

$$L_n(\theta_n) > L_n(\theta). \tag{6.10}$$

We now need to translate the condition (6.9) into one on an ML estimator $\hat{\theta}_n$ whose existence follows from the comments which preface this proof. We have, using (6.4) and (6.7),

$$\Lambda_n = -2c(E_\theta I_n(\theta))^{-1/2}(\hat{\theta}_n - \theta)J_n(\theta_n^*) + 2c^2(E_\theta I_n(\theta))^{-1}J_n(\theta_n^{**}),$$

where $\theta_n^* = \theta + \gamma_1(X_1,X_2,\ldots,X_n,\theta)(\hat{\theta}_n - \theta)$ with $|\gamma_1| < 1$. Thus, employing Assumptions 1 and 2, we have

$$\{\Lambda_n > 0\} = \{\hat{\theta}_n > \theta + c(E_\theta I_n(\theta))^{-1/2}(1 + o_p(1))\} \tag{6.11}$$

($o_p(1)$ denoting a term which tends in probability to zero as $n \to \infty$), while

$$\lim_{n\to\infty} P_\theta(\Lambda_n = 0) = \lim_{n\to\infty} P_{\theta_n}(\Lambda_n = 0) = 0. \tag{6.12}$$

Now, using Proposition 6.1 and an application of Theorem 7.1, Chapter 1, of Roussas (1972), which rests on the contiguity of $\{P_{\theta_n}\}$ and $\{P_\theta\}$, we obtain

$$\begin{aligned} \lim_{n\to\infty} P_\theta((E_\theta I_n(\theta))^{1/2}(\hat{\theta}_n - \theta) \leqslant y) &= \lim_{n\to\infty} P_{\theta_n}((E_\theta I_n(\theta))^{1/2}(\hat{\theta}_n - \theta_n) \leqslant y) \\ &= P((\eta(\theta))^{-1}N(0,1) \leqslant y) \end{aligned} \tag{6.13}$$

for any y, $-\infty < y < \infty$, where η and N are independent. The minimum Bayes risk in our decision problem is therefore

$$\begin{aligned}-\tfrac{1}{2}[P_{\theta_n}(\Lambda_n > 0) + P_\theta(\Lambda_n \leqslant 0)] &= -\tfrac{1}{2}[P_{\theta_n}(\hat{\theta}_n > \theta_n - c(E_\theta I_n(\theta))^{-1/2}(1 + o_p(1)) \\ &\qquad + P_\theta(\hat{\theta}_n \leqslant \theta + c(E_\theta I_n(\theta))^{-1/2}(1 + o_p(1))] \\ &\to -\tfrac{1}{2}[1 + P_\theta(|(\eta(\theta))^{-1}N(0,1)| < c)] \\ &= -\tfrac{1}{2}[1 + \lim_{n\to\infty} P_\theta(|(E_\theta I_n(\theta))^{1/2}(\hat{\theta}_n - \theta)| < c)]\end{aligned}$$

as $n \to \infty$ using (6.9) and (6.11)–(6.13).

Now suppose that $T_n = T_n(X_1, \ldots, X_n)$ is an estimator of θ satisfying the conditions of the theorem. We consider the decision function which accepts $L_n(\theta)$ when $T_n \leqslant \theta + c(E_\theta I_n(\theta))^{-1/2}$ and accepts $L_n(\theta_n)$ otherwise. Then, the Bayes risk of this decision function is

$$\begin{aligned}&-\tfrac{1}{2}[P_{\theta_n}(T_n > \theta + c(E_\theta I_n(\theta))^{-1/2}) + P_\theta(T_n \leqslant \theta + c(E_\theta I_n(\theta))^{-1/2})] \\ &\quad = -\tfrac{1}{2}[P_{\theta_n}(T_n > \theta_n - c(E_\theta I_n(\theta))^{-1/2}) + P_\theta(T_n \leqslant \theta + c(E_\theta I_n(\theta))^{-1/2}] \\ &\quad = -\tfrac{1}{2}[1 + P_\theta(|(E_\theta I_n(\theta))^{1/2}(T_n - \theta)| < c) \\ &\qquad + P_{\theta_n}(T_n > \theta_n - c(E_\theta I_n(\theta))^{-1/2}) - P_\theta(T_n > \theta - c(EI_n(\theta))^{-1/2})].\end{aligned}$$

Thus, under the conditions of the theorem,

$$\lim_{n\to\infty} P_\theta(|(E_\theta I_n(\theta))^{1/2}(\hat{\theta}_n - \theta)| > c) \geqslant \limsup_{n\to\infty} P_\theta(|(E_\theta I_n(\theta))^{1/2}(T_n - \theta)| < c)$$

as required.

To establish the second part of the theorem we note that the form of convergence result prescribed for T_n ensures that

$$\lim_{n\to\infty} P_\theta(|(E_\theta I_n(\theta))^{1/2}(T_n - \theta)| < c) = P(|\gamma(\theta)(\eta(\theta))^{-1}N(0,1)| < c),$$

where η and N are independent, and the required result follows in view of (6.13). ∎

It is interesting to compare Theorem 6.1 with a similar result for the maximum probability (MP) estimator [Theorem 3.1 of Weiss and Wolfowitz (1974)]. The ML and MP estimators are asymptotically equivalent under appropriate conditions.

Theorem 6.1 provides a convenient analog of results for the more classical context in which $I_n(\theta)/E_\theta I_n(\theta) \xrightarrow{p} 1$, but the norming by $E_\theta I_n(\theta)$ may be very inconvenient in the case where $\eta(\theta) \neq 1$ a.s. Indeed, the distribution of $\eta(\theta)$ may be intractable and, of course, the dependence on θ may make confidence interval arguments awkward. An example where $\eta(\theta)$ is difficult to cope with in general is in the estimation of the mean of the offspring distribution of a supercritical Bienaymé–Galton–Watson branching process. In this case, all that can be said about $\eta(\theta)$ is that the Laplace transform $\phi(s) = E\exp(-s\eta(\theta))$ satisfies

$$\phi(\theta s) = f(\phi(s)), \qquad \operatorname{Re} s \geqslant 0,$$

with $\phi'(0) = -1$ [see, e.g., Harris (1963)]. Here f is the probability generating function of the offspring distribution of the underlying branching process. The functional relationship uniquely determines ϕ but an explicit form can only be found in a very few special cases.

The possibility of suppressing $\eta(\theta)$ through the asymptotically normal form $I_n^{1/2}(\theta)(\hat{\theta}_n - \theta)$ is clearly attractive. Furthermore, asymptotic normality of $I_n^{1/2}(\theta)(\hat{\theta}_n - \theta)$ holds under rather more general circumstances than convergence in distribution of $(E_\theta I_n(\theta))^{1/2}(\hat{\theta}_n - \theta)$, and indeed in a variety of cases in which $I_n(\theta)/E_\theta I_n(\theta)$ does not even converge in distribution. Nevertheless, in cases where $I_n(\theta)/E_\theta I_n(\theta) \xrightarrow{\text{p}} \eta^2(\theta)$ and $\eta(\theta)$ has a known distribution, other normalizations may sometimes lead to better tests of hypotheses about θ, at least for parameter values in certain intervals. The problem of producing optimal procedures for testing hypotheses about θ is complex, and categorical answers do not seem possible [e.g., Basawa and Scott (1976, 1977), Feigin (1978), Sweeting (1978)]. We shall confine our attention to the problem of estimation.

In view of the role of $I_n(\theta)$ in producing asymptotic normality for $(I_n(\theta))^{-1/2}\, d\log L_n(\theta)/d\theta$, and the desirability of asymptotic normality of estimators, we shall introduce the following definition [Heyde (1975b); Heyde and Feigin (1975)].

Definition. *An estimator T_n of θ is said to be asymptotically first-order efficient if*

$$I_n^{1/2}(\theta)\left[T_n - \theta - \gamma(\theta) I_n^{-1}(\theta)\frac{d\log L_n(\theta)}{d\theta}\right] \xrightarrow{\text{p}} 0$$

as $n \to \infty$ for some $\gamma(\theta)$ not depending on n or the observations.

In the classical case of i.i.d. observations this definition reduces to that of Rao (1973, pp. 348–349). Furthermore, the motivation behind the general definition is the same as for the independence case. Under the conditions we have imposed, $\{d\log L_n(\theta)/d\theta\}$ is a martingale and (6.5) ensures the consistency of an asymptotically first-order efficient estimator T_n for θ if $I_n(\theta) \xrightarrow{\text{a.s.}} \infty$ as $n \to \infty$. The definition is particularly aimed at circumstances under which $[I_n(\theta)]^{-1/2}\, d\log L_n(\theta)/d\theta \xrightarrow{\text{d}} N(0,1)$, for then $I_n^{1/2}(\theta)(T_n - \theta) \xrightarrow{\text{d}} N(0,\gamma^2(\theta))$ and, under the conditions of Theorem 6.1, $\gamma^2(\theta) \geqslant 1$. Of course the

ML estimator is asymptotically first-order efficient and indeed is optimal in the sense of possessing minimum $\gamma^2(\theta)$.

The choice of definition of asymptotic first-order efficiency does require some elaboration, since there would necessarily be a clear preference for an estimator with the smallest possible $\gamma(\theta)$. However, under relatively mild conditions, asymptotic first-order efficiency itself implies $\gamma(\theta) = 1$. Under these circumstances we obtain

$$I_n^{1/2}(\theta)(T_n - \hat{\theta}_n) \xrightarrow{\text{p}} 0,$$

so that first-order efficient estimators are asymptotically equivalent to the ML estimator in the sense of leading to the same confidence statements. Conditions under which asymptotic first-order efficiency is sufficient to ensure $\gamma(\theta) = 1$ are given in Theorem 6.2 below. This result is analogous to, and extends that of, Rao (1963, Lemma 3) for the case where the X_i are i.i.d. r.v. In it we make use of the concept of continuous convergence. We shall say that $\{f_n\}$ converges *continuously* in θ to f if $f_n(\theta_n) \to f(\theta)$ whenever $\theta_n \to \theta$.

Theorem 6.2. *Suppose that Assumptions* 1 *and* 2 *are satisfied and that the estimator* T_n *is asymptotically first-order efficient with scale constant* $\gamma(\theta)$ *which is continuous in* θ. *Then,*

$$[I_n^{1/2}(\theta)(T_n - \theta), I_n(\theta)/E_\theta I_n(\theta)] \xrightarrow{\text{d}} [N(0,\gamma^2(\theta)), \eta^2(\theta)]. \tag{6.14}$$

If the convergence in (6.14) *holds continuously in* θ *then* $\gamma(\theta) = 1$ *and hence*

$$I_n^{1/2}(\theta)(T_n - \hat{\theta}_n) \xrightarrow{\text{p}} 0$$

as $n \to \infty$, $\hat{\theta}_n$ *denoting the ML estimator.*

Proof. The result (6.14) follows from Proposition 6.1 and the definition of asymptotic first-order efficiency.

Now suppose that (6.14) holds continuously in θ. Let θ_0 be an arbitrary interior point of Θ and let $\theta_n = \theta_0 + \delta/(E_{\theta_0} I_n(\theta_0))^{1/2}$, where $\delta > 0$ is chosen so small that θ_1 also belongs to Θ. From (6.14) we have

$$I_n^{1/2}(\theta)(T_n - \theta)(\gamma(\theta))^{-1} \xrightarrow{\text{d}} N(0,1)$$

continuously in θ. We consider the size α test of the hypothesis $\theta = \theta_0$ whose critical region is

$$I_n^{1/2}(\theta_0)(T_n - \theta_0)(\gamma(\theta_0))^{-1} \geqslant \lambda_n.$$

Then, as $n \to \infty$, $\lambda_n \to z_\alpha$, the upper $\alpha\%$ point of the distribution of $N(0,1)$.

The power of this test at θ_n is

$$\begin{aligned}\beta_n(\theta_n) &= P_{\theta_n}(I_n^{1/2}(\theta_0)(T_n - \theta_0)(\gamma(\theta_0))^{-1} \geqslant \lambda_n) \\ &= P_{\theta_n}\Big(I_n^{1/2}(\theta_n)(T_n - \theta_n)(\gamma(\theta_0))^{-1} \geqslant \\ &\qquad I_n^{1/2}(\theta_n)(\theta_0 - \theta_n)(\gamma(\theta_0))^{-1} + \left(\frac{I_n^{1/2}(\theta_n)}{I_n^{1/2}(\theta_0)}\lambda_n\right)\Big) \\ &\to P(N(0,1) \geqslant z_\alpha - \delta\eta(\theta_0)(\gamma(\theta_0))^{-1}) \qquad (6.15)\end{aligned}$$

as $n \to \infty$.

On the other hand, we can also find the asymptotic behavior of $\beta_n(\theta_n)$ using the contiguity of $\{P_{\theta_n}\}$ and $\{P_\theta\}$. Indeed, using Theorem 7.1, Chapter 1, of Roussas (1972) together with the appropriate version of (6.8), we have

$$\begin{aligned}\beta_n(\theta_n) &= P_{\theta_n}(I_n^{1/2}(\theta_0)(T_n - \theta_0)(\gamma(\theta_0))^{-1} \geqslant \lambda_n) \\ &\to P_{\theta_0}(Z(\theta_0) \geqslant z_\alpha),\end{aligned}$$

where $Z(\theta_0)$ has its characteristic function given by

$$E_{\theta_0}\{\exp(itZ(\theta_0))\} = E_{\theta_0}\{\exp(itN(0,1) + \delta\eta(\theta_0)N(0,1) - \tfrac{1}{2}\delta^2\eta^2(\theta_0)\},$$

η and N being independent. But

$$\begin{aligned}&E_{\theta_0}[E_{\theta_0}\{\exp(it + \delta\eta(\theta_0))N(0,1)|\eta(\theta_0)\}\exp(-\tfrac{1}{2}\delta^2\eta^2(\theta_0))] \\ &\quad = E_{\theta_0}[\exp(\tfrac{1}{2}(it + \delta\eta(\theta_0))^2 - \tfrac{1}{2}\delta^2\eta^2(\theta_0)] \\ &\quad = E_{\theta_0}[\exp(-\tfrac{1}{2}t^2 + it\delta\eta(\theta_0))],\end{aligned}$$

and hence

$$\beta_n(\theta_n) \to P(N(0,1) \geqslant z_\alpha - \delta\eta(\theta_0)). \qquad (6.16)$$

The result $\gamma(\theta_0) = 1$ follows from (6.15) and (6.16) and the proof is complete since θ_0 is an arbitrary interior point of Θ. ∎

For the explicit resolution of optimality questions concerning estimators of θ, the vital piece of information is a suitably tractable expression for $d\log L_n(\theta)/d\theta$. This will of course be in its simplest form when

$$\frac{d\log L_n(\theta)}{d\theta} = I_n(\theta)(\hat{\theta}_n - \theta), \qquad \theta \in \Theta, \quad n \geqslant 1, \qquad (6.17)$$

and then the statistic $\hat{\theta}_n$ is clearly the ML estimator. We shall focus attention on this functional form, recalling that in the case of independent X_i the form is obtained if and only if the probability (density) functions belong to the exponential family and θ is the so-called natural parameter for the problem. We begin by showing that (6.17) in turn specifies a factorization for $I_n(\theta)$. This result is due to Heyde and Feigin (1975).

Theorem 6.3. *If the factorization* (6.17) *holds, then*

$$I_n(\theta) = \phi(\theta)H_n(X_1, \ldots, X_{n-1})$$

for some functions ϕ *and* H_n, *where* ϕ *does not involve the* X_i *and* H_n *does not involve* θ. *Conversely, if*

$$\frac{d \log L_n(\theta)}{d\theta} = \phi(\theta)H_n(X_1, \ldots, X_{n-1})(\hat{\theta}_n - \theta), \qquad n \geqslant 1 \qquad (H_1 = \text{const.}) \tag{6.18}$$

for sequences $\{H_n\}$ *and* $\{\hat{\theta}_n\}$ *of statistics, then* $I_n(\theta) = \phi(\theta)H_n(X_1, \ldots, X_{n-1})$.

Proof. Suppose firstly that the factorization (6.17) holds for $n \geqslant 1$. Then, differentiation and recombination gives

$$\frac{I_n'(\theta)}{I_n(\theta)} = \frac{(d^2 \log L_n(\theta)/d\theta^2) + I_n(\theta)}{(d \log L_n(\theta)/d\theta)}. \tag{6.19}$$

Also,

$$\begin{aligned} E_\theta\left(\frac{d^2 \log L_n(\theta)}{d\theta^2}\middle| \mathscr{F}_{n-1}\right) &= I_n'(\theta)E(\hat{\theta}_n - \theta | \mathscr{F}_{n-1}) - I_n(\theta) \\ &= I_n'(\theta)E\left((I_n(\theta))^{-1} \frac{d \log L_n(\theta)}{d\theta}\middle| \mathscr{F}_{n-1}\right) - I_n(\theta) \\ &= I_n'(\theta)(I_n(\theta))^{-1} \frac{d \log L_{n-1}(\theta)}{d\theta} - I_n(\theta), \end{aligned}$$

while (6.3) gives

$$E_\theta\left(\frac{d^2 \log L_n(\theta)}{d\theta^2}\middle| \mathscr{F}_{n-1}\right) = -(I_n(\theta) - I_{n-1}(\theta)) + \frac{d^2 \log L_{n-1}(\theta)}{d\theta^2},$$

and hence

$$\frac{I_n'(\theta)}{I_n(\theta)} = \frac{(d^2 \log L_{n-1}(\theta)/d\theta^2) + I_{n-1}(\theta)}{(d \log L_{n-1}(\theta)/d\theta)}. \tag{6.20}$$

From (6.19) and (6.20) we have

$$\frac{d}{d\theta} \log I_n(\theta) = \frac{d}{d\theta} \log I_{n-1}(\theta) = \cdots = \frac{d}{d\theta} \log I_1(\theta) = C(\theta),$$

say, and the general solution of this system is of the form

$$I_n(\theta) = \phi(\theta)H_n(X_1, \ldots, X_{n-1}),$$

$I_n(\theta)$ being $\mathscr{F}_{n-1}$-measurable.

To establish the converse result we take differences in (6.18) to obtain

$$u_n(\theta) = \phi(\theta)[H_n(X_1,\ldots,X_{n-1})(\hat{\theta}_n - \theta) - H_{n-1}(X_1,\ldots,X_{n-2})(\hat{\theta}_{n-1} - \theta)],$$

for $n \geqslant 2$, and

$$u_1(\theta) = \phi(\theta)H_1(\hat{\theta}_1 - \theta),$$

from which we readily deduce that

$$\begin{aligned} E_\theta(u_n^2(\theta)|\mathscr{F}_{n-1}) &= -E_\theta(v_n(\theta)|\mathscr{F}_{n-1}) \\ &= \phi(\theta)[H_n(X_1,\ldots,X_{n-1}) - H_{n-1}(X_1,\ldots,X_{n-2})], \qquad n \geqslant 2, \\ E_\theta(u_1^2(\theta)) &= -E_\theta(v_1(\theta)) = \phi(\theta)H_1, \end{aligned}$$

and

$$I_n(\theta) = \phi(\theta)H_n(X_1,\ldots,X_{n-1})$$

follows. ∎

At this point it seems appropriate to discuss the concept of *sufficiency* in this context. Sufficiency can be characterized by the factorization theorem for the joint probability (density) function f of a sample of size n. That is, a statistic T_n is sufficient for θ if and only if

$$f(X_1,\ldots,X_n|\theta) = g(T_n,\theta)h_n(X_1,\ldots,X_n), \tag{6.21}$$

where h_n does not involve θ. In the case where the factorization (6.18) obtains, the pair $(H_n(X_1,\ldots,X_{n-1}),\hat{\theta}_n)$ is sufficient. In fact,

$$\frac{d\log L_n(\theta)}{d\theta} = I_n(\theta)(\hat{\theta}_n - \theta) = \phi(\theta)H_n(X_1,\ldots,X_{n-1})(\hat{\theta}_n - \theta)$$

implies

$$\log L_n(\theta) = H_n(X_1,\ldots,X_{n-1})(\hat{\theta}_n\Phi(\theta) - \Psi(\theta)) + K_n(X_1,\ldots,X_n), \tag{6.22}$$

where

$$\frac{d}{d\theta}\Phi(\theta) = \phi(\theta), \qquad \frac{d}{d\theta}\Psi(\theta) = \theta\phi(\theta),$$

and the function K_n does not involve θ. Clearly (6.22) corresponds to the form (6.21) with $T_n = (H_n(X_1,\ldots,X_{n-1}),\hat{\theta}_n)$. This last statistic is minimal-sufficient in the usual sense that any other sufficient statistic must be a function of H_n and $\hat{\theta}_n$.

The factorization (6.17) is amenable to detailed investigation in the case where $\{X_n, n \geqslant 0\}$ is a Markov process. We shall consider the time-homogeneous Markov process whose conditional probability (density) function of

X_n given X_{n-1} is $f(X_n|X_{n-1},\theta)$. Here we find it convenient to change the formulation a little and take

$$L_n(\theta) = \prod_{i=1}^{n} f(X_i|X_{i-1},\theta),$$

so that

$$\frac{d\log L_n(\theta)}{d\theta} = \sum_{i=1}^{n} \frac{d}{d\theta} \log f(X_i|X_{i-1},\theta) = \sum_{i=1}^{n} u_i(\theta)$$

and

$$u_i(\theta) = \frac{d}{d\theta} \log f(X_i|X_{i-1},\theta).$$

Suppose that

$$\frac{d\log L_n(\theta)}{d\theta} = I_n(\theta)(\hat{\theta}_n - \theta), \qquad n = 1, 2, \ldots.$$

Then, taking $n = 1$, we have

$$\frac{d}{d\theta} \log f(X_1|X_0,\theta) = I_1(\theta)(\hat{\theta}_1 - \theta) = \phi(\theta)H(X_0)(\hat{\theta}_1 - \theta).$$

Clearly there must be a single root $\tilde{\theta} = m(x,y)$ of the equation $(d/d\theta)f(x|y,\theta) = 0$ and then

$$\frac{d}{d\theta} \log f(x|y,\theta) = \phi(\theta)H(y)[m(x,y) - \theta]. \tag{6.23}$$

The equation (6.23) defines what we shall call a *conditional exponential family* for the problem under consideration.

Conversely, if (6.23) is satisfied, we have

$$\frac{d\log L_n(\theta)}{d\theta} = \phi(\theta) \sum_{i=1}^{n} H(X_{i-1})[m(X_i,X_{i-1}) - \theta] \tag{6.24}$$

and

$$u_i(\theta) = \phi(\theta)H(X_{i-1})[m(X_i,X_{i-1}) - \theta],$$

so that

$$E_\theta(u_i(\theta)|\mathscr{F}_{i-1}) = \phi(\theta)H(X_{i-1})[E_\theta(m(X_i,X_{i-1})|\mathscr{F}_{i-1}) - \theta] = 0 \quad \text{a.s.},$$

and hence

$$E_\theta(m(X_i,X_{i-1})|\mathscr{F}_{i-1}) = \theta \quad \text{a.s.}$$

Also, from (6.3),

$$E_\theta(u_i^2(\theta)|\mathscr{F}_{i-1}) = -E_\theta\left(\frac{d^2}{d\theta^2}\log f(X_i|X_{i-1},\theta)|\mathscr{F}_{i-1}\right) = \phi(\theta)H(X_{i-1}) \quad \text{a.s.,}$$

so that (6.24) can be rewritten as

$$\frac{d\log L_n(\theta)}{d\theta} = I_n(\theta)(\hat{\theta}_n - \theta),$$

where

$$\hat{\theta}_n = \left[\sum_{i=1}^{n} H(X_{i-1})\right]^{-1} \sum_{i=1}^{n} H(X_{i-1})m(X_i,X_{i-1}).$$

The conditional exponential form is thus necessary and sufficient for the derivative of the logarithm of the likelihood to be expressible in the form (6.17). Notice also that $\hat{\theta}_n$ is strongly consistent for θ provided $\sum_{i=1}^{\infty} H(X_{i-1})$ diverges a.s.

We shall complete this section by mentioning the conditional exponential families for two estimation problems. These examples have been deliberately chosen as ones in which the norming using random variables $I_n(\theta)$ is much more convenient than the constant norming using $E_\theta I_n(\theta)$. Further details are given in Heyde and Feigin (1975).

The first problem we shall mention is that of the estimation of the mean θ of the offspring distribution of a supercritical Bienaymé–Galton–Watson branching process on the basis of a sample $\{X_0,X_1,\ldots,X_n\}$ of consecutive generation sizes. Here we have $1 < \theta = E(X_1|X_0 = 1) < \infty$ and we shall suppose that $\sigma^2 = \operatorname{var}(X_1|X_0 = 1) < \infty$. It is not difficult to show in this case that the conditional exponential family is the family of power-series distributions. That is, the distributions for which

$$p_j = P(X_1 = j|X_0 = 1) = a_j\lambda^j[f(\lambda)]^{-1}, \qquad j = 0, 1, \ldots, \quad \lambda > 0,$$

where $a_j \geqslant 0$ and $f(\lambda) = \sum_{j=0}^{\infty} a_j\lambda^j$. For this family, $\theta = \theta(\lambda)$ is known to be a nonnegative monotone increasing function of λ (Patil, 1962), so that the parameterization can be equally well expressed in terms of θ. We have

$$\theta = \lambda f'(\lambda)[f(\lambda)]^{-1}, \qquad \sigma^2 = [(d/d\theta)\log\lambda]^{-1}$$

and

$$\frac{d}{d\theta}\log f(x|y,\theta) = \sigma^{-2}(x - \theta y).$$

A detailed discussion of the estimation problem in this case has been given by Heyde (1975b). It should be remarked that

$$I_n(\theta) = \sigma^{-2}\sum_{i=0}^{n-1} X_i$$

and that $I_n(\theta)/E_\theta I_n(\theta)$ converges almost surely to a nondegenerate random variable whose distribution is not known explicitly in most cases.

The second problem which we shall mention is that of estimating the parameter θ in a first-order autoregression

$$X_i = \theta X_{i-1} + \varepsilon_i,$$

where the ε_i are i.i.d. r.v. with mean zero and variance σ^2 and ε_i is independent of X_{i-1}. We wish to estimate θ on the basis of a sample $X_0, X_1, \ldots, X_n$, and the process is not necessarily subjected to the stability condition $|\theta| < 1$. In this case it is easily checked that the conditional exponential family contains only a single member, and this corresponds to the case where ε_1 is normally distributed.

A detailed discussion of this estimation problem has been given by Anderson (1959). It is interesting to note that the expression $\sum_{i=1}^n X_i X_{i-1}/\sum_{i=1}^n X_{i-1}^2$, which is the ML estimator when ε_1 is normally distributed, continues to be a strongly consistent estimator of θ whatever the distribution of ε_1 [as can easily be checked using the same reasoning as with (6.5)]. Furthermore,

$$\left[\sum_{i=1}^n X_{i-1}^2\right]^{1/2}\left[\left(\sum_{i=1}^n X_{i-1}^2\right)^{-1} \sum_{i=1}^n X_i X_{i-1} - \theta\right] \tag{6.25}$$

converges in distribution to a proper limit law whatever the distribution of ε_1. The limit law is normal if $|\theta| < 1$, but if $|\theta| > 1$ it is normal only if ε_1 is normally distributed. Curiously, if ε_1 is not normally distributed, the limit law is not even infinitely divisible. Note that the expression (6.25) is easily put into standard form as a martingale normalized by the square root of the sum of the conditional variances of its increments. Of course, if ε_1 is not normally distributed, the estimator $(\sum_{i=1}^n X_{i-1}^2)^{-1}\sum_{i=1}^n X_i X_{i-1}$ will not in general bear any special relation to the ML estimator. The above-mentioned results all extend from the case of the first-order autoregressive process to that of the autoregression of general order (M. M. Rao, 1961).

6.3. Conditional Least Squares

Now let $\{X_n, n = 1, 2, \ldots\}$ be a stochastic process defined on a probability space $(\Omega, \mathscr{F}, P_\theta)$, whose distribution depends on a (column) vector $\theta = (\theta_1, \ldots, \theta_p)'$ of unknown parameters with θ lying in some open set Θ of Euclidean p-space. Let $E_\theta(\cdot)$ and $E_\theta(\cdot|\cdot)$ denote expectation and conditional expectation under P_θ. We shall write, at least for the purposes of the general theory, $\theta^0 = (\theta_1^0, \ldots, \theta_p^0)'$ for the true value of θ and use $\mathscr{F}_n$ to denote the σ-field generated by $\{X_k, 1 \leqslant k \leqslant n\}$, while $\mathscr{F}_0$ is the trivial σ-field. We shall also suppose that $X_n \in L^1$, $n = 1, 2, \ldots$.

Given a sequence of observations $(X_1, \ldots, X_n)$, we can estimate θ by trying to minimize the sum of squares

$$Q_n(\theta) = \sum_{k=1}^{n} [X_k - E_\theta(X_k|\mathscr{F}_{k-1})]^2 \tag{6.26}$$

with respect to θ. The estimates will be taken to be the solutions of the system

$$\frac{\partial Q_n(\theta)}{\partial \theta_i} = 0, \qquad i = 1, 2, \ldots, p, \tag{6.27}$$

so we need to impose conditions to ensure the existence of the partial derivatives involved.

This procedure is one of minimization of the sum of squares of the actual errors of best prediction. Indeed, if the true parameter value were θ, $E_\theta(X_k|\mathscr{F}_{k-1})$ would be the best predictor of X_k based on $X_{k-1}, \ldots, X_1$ in the least squares sense. This is easily seen since for any $\hat{X}_k$ which is $\mathscr{F}_{k-1}$-measurable,

$$\begin{aligned} &E_\theta[(X_k - \hat{X}_k)^2|\mathscr{F}_{k-1}] \\ &\quad = E_\theta(X_k^2|\mathscr{F}_{k-1}) - 2\hat{X}_k E_\theta(X_k|\mathscr{F}_{k-1}) + \hat{X}_k^2 \\ &\quad = E_\theta[(X_k - E_\theta(X_k|\mathscr{F}_{k-1}))^2|\mathscr{F}_{k-1}] + (\hat{X}_k - E_\theta(X_k|\mathscr{F}_{k-1}))^2 \end{aligned}$$

is minimized by choosing $\hat{X}_k = E_\theta(X_k|\mathscr{F}_{k-1})$. The conditional least squares approach has been discussed in some detail by Klimko and Nelson (1978) although the methodology is implicit in the work of earlier authors. The approach leads to simple estimators in a wide variety of cases. Furthermore, under quite mild conditions the estimators are strongly consistent and, when appropriately normed, asymptotically normally distributed. However, they do not necessarily have any particular optimality properties. In cases which are somewhat distant from stationarity, the use of weighted conditional least squares can substantially improve the performance of estimators, but we shall not enter into a discussion of this refinement.

It is assumed throughout the discussion below that $E_\theta(X_n|\mathscr{F}_{n-1})$ is a.s. twice continuously differentiable with respect to θ in some neighborhood S of θ^0. All neighborhoods defined below will be taken to be contained in S. Then, for $\delta > 0$, $\|\theta - \theta^0\| < \delta$, and using θ^* to denote an appropriate intermediate point (not necessarily the same at each occurrence), we obtain from Taylor series expansion that

$$\begin{aligned} Q_n(\theta) &= Q_n(\theta^0) + (\theta - \theta^0)'(\partial Q_n(\theta)/\partial\theta)_{\theta=\theta^0} + \tfrac{1}{2}(\theta - \theta^0)'(\partial^2 Q_n(\theta)/\partial\theta^2)_{\theta=\theta^*}(\theta - \theta^0) \\ &= Q_n(\theta^0) + (\theta - \theta^0)'(\partial Q_n(\theta)/\partial\theta)_{\theta=\theta^0} + \tfrac{1}{2}(\theta - \theta^0)' V_n(\theta - \theta^0) \\ &\quad + \tfrac{1}{2}(\theta - \theta^0)' T_n(\theta^*)(\theta - \theta^0) \end{aligned} \tag{6.28}$$

where V_n is the $p \times p$ matrix of second partial derivatives of $Q_n(\theta)$ evaluated at θ^0 and

$$T_n(\theta^*) = (\partial^2 Q(\theta)/\partial\theta^2)_{\theta=\theta^*} - V_n. \tag{6.29}$$

The result (6.28) can be used to establish strong consistency of the conditional least squares estimators if V_n and T_n can be handled suitably. Note that $\frac{1}{2}V_n$ is a $p \times p$ matrix with (i,j)th element

$$\sum_{k=1}^{n}\left[\left\{\frac{\partial}{\partial\theta_i} E_\theta(X_k|\mathscr{F}_{k-1})\right\}_{\theta=\theta^0}\left\{\frac{\partial}{\partial\theta_j} E_\theta(X_k|\mathscr{F}_{k-1})\right\}_{\theta=\theta^0} - \left\{\frac{\partial^2}{\partial\theta_i\,\partial\theta_j} E_\theta(X_k|\mathscr{F}_{k-1})\right\}_{\theta=\theta^0}\{X_k - E_{\theta^0}(X_k|\mathscr{F}_{k-1})\}\right].$$

The usual strategy for dealing with V_n is to use a martingale strong law to show that (in the case of the classical norming)

$$n^{-1}\sum_{k=1}^{n}\left\{\frac{\partial^2}{\partial\theta_i\,\partial\theta_j} E_\theta(X_k|\mathscr{F}_{k-1})\right\}_{\theta=\theta^0}\{X_k - E_{\theta^0}(X_k|\mathscr{F}_{k-1})\} \xrightarrow{\text{a.s.}} 0$$

for all $1 \leqslant i \leqslant p$, $1 \leqslant j \leqslant p$, and use other strong laws, such as the ergodic theorem (when available) to show that the $p \times p$ matrix

$$\left(n^{-1}\sum_{k=1}^{n}\left\{\frac{\partial}{\partial\theta_i} E_\theta(X_k|\mathscr{F}_{k-1})\right\}_{\theta=\theta^0}\left\{\frac{\partial}{\partial\theta_j} E_\theta(X_k|\mathscr{F}_{k-1})\right\}_{\theta=\theta^0}\right) \xrightarrow{\text{a.s.}} V$$

for some positive definite $p \times p$ matrix V. On the other hand, T_n is assumed to be small in an appropriate sense; this amounts to a continuity type condition on $\partial^2 Q_n(\theta)/\partial\theta^2$. These ideas are formalized in the following theorem of Klimko and Nelson (1978).

Theorem 6.4. *Suppose that*

$$\lim_{\substack{n\to\infty\\ \delta\downarrow 0}} \sup (n\delta)^{-1}|T_n(\theta^*)_{ij}| < \infty \quad \text{a.s.}, \qquad 1 \leqslant i \leqslant p, \quad 1 \leqslant j \leqslant p, \tag{6.30}$$

$$(2n)^{-1}V_n \xrightarrow{\text{a.s.}} V, \tag{6.31}$$

where V is a positive definite (symmetric) $p \times p$ matrix of constants, and

$$n^{-1}(\partial Q_n(\theta)/\partial\theta_i)_{\theta=\theta^0} \xrightarrow{\text{a.s.}} 0, \qquad 1 \leqslant i \leqslant p. \tag{6.32}$$

Then, there exists a sequence of estimators $\{\hat{\theta}_n\}$ such that $\hat{\theta}_n \xrightarrow{\text{a.s.}} \theta^0$, and for any $\varepsilon > 0$ there is an event E with $P(E) > 1 - \varepsilon$ and an n_0 such that on E, for $n > n_0$, $\hat{\theta}_n$ satisfies the least squares equations (6.27) and Q_n attains a relative minimum at $\hat{\theta}_n$.

Proof. Given $\varepsilon > 0$ we can use (6.30)–(6.32) and Egorov's theorem [see, e.g., Loève (1977, p. 141)] to find an event E with $P(E) > 1 - \varepsilon$, a positive $\delta^* < \delta$, an $M > 0$, and an n_0 such that on E, for any $n > n_0$ and $\theta \in N_{\delta^*}$ (the open sphere of radius δ^* centered at θ^0), the following three conditions hold: (a) $|(\theta - \theta^0)'\{\partial Q_n(\theta)/\partial\theta\}_{\theta=\theta^0}| < n\delta^3$; (b) the minimum eigenvalue of $(2n)^{-1}V_n$ is greater than some $\Delta > 0$; and (c) $\frac{1}{2}(\theta - \theta^0)'T_n(\theta^*)(\theta - \theta^0) < nM\delta^3$. Hence, using (6.28) for θ on the boundary of N_{δ^*}, we have

$$\begin{aligned} Q_n(\theta) &\geqslant Q_n(\theta^0) + n(-\delta^3 + \delta^2\Delta - M\delta^3) \\ &= Q_n(\theta^0) + n\delta^2(\Delta - \delta - M\delta). \end{aligned}$$

Since $\Delta - \delta - M\delta$ can be made positive by initially choosing δ sufficiently small, $Q_n(\theta)$ must attain a minimum at some $\hat{\theta}_n = (\hat{\theta}_{n1}, \ldots, \hat{\theta}_{np})'$ in N_{δ^*}, at which point the least squares equations (6.27) must be satisfied on E for any $n > n_0$.

Now replace ε by $\varepsilon_k = 2^{-k}$ and δ by $\delta_k = k^{-1}$, $k = 1, 2, \ldots$, to determine a sequence of events $\{E_k\}$ and an increasing sequence $\{n_k\}$ such that the equations (6.27) have a solution on E_k for any $n > n_k$. For $n_k < n \leqslant n_{k+1}$, define $\hat{\theta}_n$ on E_k to be a solution of (6.27) within δ_k of θ^0 and at which Q_n attains a relative minimum, and define $\hat{\theta}_n$ to be zero off E_k. Then $\hat{\theta}_n \to \theta^0$ on $\liminf_{k\to\infty} E_k$ while $P(\liminf_{k\to\infty} E_k) = 1$ since

$$\begin{aligned} 1 - P\left(\liminf_{k\to\infty} E_k\right) &= P\left(\limsup_{k\to\infty} E_k^c\right) \\ &= \lim_{k\to\infty} P\left(\bigcup_{j=k}^{\infty} E_j^c\right) \\ &\leqslant \lim_{k\to\infty} \sum_{j=k}^{\infty} P(E_j^c) = \lim_{k\to\infty} \sum_{j=k}^{\infty} 2^{-j} = 0. \end{aligned}$$

This completes the proof. ∎

Joint asymptotic normality of the estimators obtained in Theorem 6.4 holds if the linear term in the Taylor expansion (6.28) has asymptotically a joint normal distribution. This property may often be verified by using the Cramér–Wold technique [see, e.g., Billingsley (1968, p. 48)] and an appropriate martingale central limit theorem applied to

$$\begin{aligned} &n^{-1/2}c'\{\partial Q_n(\theta)/\partial\theta\}_{\theta=\theta^0} \\ &\quad = -2n^{-1/2}\sum_{k=1}^{n}\left[\sum_{i=1}^{p} c_i\{\partial E_\theta(X_k|\mathscr{F}_{k-1})/\partial\theta_i\}_{\theta=\theta^0}\right][X_k - E_{\theta^0}(X_k|\mathscr{F}_{k-1})], \end{aligned} \tag{6.33}$$

where $c' = (c_1, c_2, \ldots, c_p)$ is an arbitrary nonzero vector of constants. Of course the term

$$\sum_{i=1}^{p} c_i \{\partial E_\theta(X_k | \mathscr{F}_{k-1})/\partial \theta_i\}_{\theta=\theta^0}$$

is $\mathscr{F}_{k-1}$-measurable, and hence from (6.33) we see that

$$\{n^{-1/2} c' \{\partial Q_n(\theta)/\partial \theta\}_{\theta=\theta^0}, \mathscr{F}_n, n \geqslant 1\}$$

is a martingale. If

$$\tfrac{1}{2} n^{-1/2} c' \{\partial Q_n(\theta)/\partial \theta\}_{\theta=\theta^0} \xrightarrow{\mathrm{d}} N(0, c'Wc)$$

for any nonzero vector c, W being a $p \times p$ covariance matrix, then

$$\tfrac{1}{2} n^{-1/2} \{\partial Q_n(\theta)/\partial \theta\}_{\theta=\theta^0} \xrightarrow{\mathrm{d}} N(0, W).$$

We shall establish the following theorem which is also from Klimko and Nelson (1978).

Theorem 6.5. *Suppose that the conditions of Theorem* 6.4 *hold and, in addition,*

$$\tfrac{1}{2} n^{-1/2} \{\partial Q_n(\theta)/\partial \theta\}_{\theta=\theta^0} \xrightarrow{\mathrm{d}} N(0, W),$$

$N(0,W)$ denoting a multivariate normal distribution with zero-mean vector and covariance matrix W, a $p \times p$ positive definite matrix. Then

$$n^{1/2}(\hat{\theta}_n - \theta^0) \xrightarrow{\mathrm{d}} N(0, V^{-1} W V^{-1})$$

as $n \to \infty$, where V is defined by (6.31).

Proof. Since we are dealing with an asymptotic result, we may assume that $\{\hat{\theta}_n\}$ satisfies the equations (6.27). Then, expanding the vector $n^{-1/2}\{\partial Q_n(\theta)/\partial \theta\}_{\theta=\hat{\theta}_n}$ in a Taylor series about θ^0, we obtain, using (6.29),

$$\begin{aligned} 0 &= n^{-1/2}\{\partial Q_n(\theta)/\partial \theta\}_{\theta=\hat{\theta}_n} \\ &= n^{-1/2}\{\partial Q_n(\theta)/\partial \theta\}_{\theta=\theta^0} + n^{-1}(V_n + T_n(\theta^*)) n^{1/2}(\hat{\theta}_n - \theta^0). \end{aligned} \tag{6.34}$$

Since $n^{-1}(V_n + T_n(\theta^*)) \xrightarrow{\text{a.s.}} 2V$, the limiting distribution of $n^{1/2}(\hat{\theta}_n - \theta^0)$ is the same as that of $-(2V)^{-1} n^{-1/2}\{\partial Q_n(\theta)/\partial \theta\}_{\theta=\theta^0}$. This yields the desired result. ∎

It is worth remarking that iterated logarithm results for $n^{1/2}(\hat{\theta}_n - \theta^0)$ can also be obtained via the equation (6.34).

Theorems 6.4 and 6.5 should be regarded as being indicative of a general approach to the problem rather than as providing specific conditions to be checked in practice. The results of Theorems 6.4 and 6.5 are clearly amenable to extension beyond the classical normings and to the case of weighted conditional least squares. Furthermore, concrete applications are best approached directly without explicit recourse to the Taylor expansion.

As an application of the methodology of conditional least squares, we shall consider the stationary autoregressive process $\{x(n)\}$ given by

$$\sum_{k=0}^{q} \beta(k)[x(n-k)-\mu] = \varepsilon(n),$$
$$\beta(0) = 1, \qquad E\varepsilon(n) = 0, \qquad E[\varepsilon(m)\varepsilon(n)] = \sigma^2\delta_{mn}, \tag{6.35}$$

where δ_{mn} is the Kronecker delta and we assume for technical reasons that

$$\sum_{k=0}^{q} \beta(k)z^k \neq 0, \qquad |z| \leqslant 1$$

[see, e.g., Hannan (1970, Chapter I)]. The model (6.35) is based on the empirical assumption that $x(n)$ depends on a finite past together with a stochastic disturbance $\varepsilon(n)$. A great deal of time-series analysis is concerned with the analysis of such models. In fact, any stationary purely nondeterministic (i.e., containing no deterministic component) process can be reasonably approximated by an autoregression of suitably high order (i.e., suitably high q).

In our present discussion of the model (6.35) we shall adopt the classical assumption that the $\varepsilon(n)$ are i.i.d. This will later be relaxed in Section 6.4. We have

$$\theta = (\mu, \beta(1), \ldots, \beta(q))'$$

and

$$E_\theta(x(n)|\mathscr{F}_{n-1}) = \mu - \sum_{k=1}^{q} \beta(k)[x(n-k)-\mu],$$

so that if our sample is $x(1), \ldots, x(n)$, we should take

$$Q_n(\theta) = \sum_{j=q+1}^{n} \left\{x(j) - \mu + \sum_{k=1}^{q} \beta(k)[x(j-k)-\mu]\right\}^2,$$

the starting value of the summation being dictated by the sample available. The system of equations (6.27) in this case produces a minor, but asymptotically equivalent, variant on the classical estimation equations, which are

$$\hat{\mu} = \bar{x} = n^{-1} \sum_{k=1}^{n} x(k),$$

$$c(0) + \sum_{j=1}^{q} \hat{\beta}(j)c(k-j) = 0, \qquad k = 1, 2, \ldots, q,$$

where

$$c(r) = c(-r) = n^{-1} \sum_{j=1}^{n-r} [x(j) - \bar{x}][x(j+r) - \bar{x}], \qquad r \geqslant 0,$$

[e.g., Hannan (1970, Chapter VI)]. The conditions of Theorem 6.4 are relatively straightforward to check, but the details are omitted as more general results are provided in Section 6.4.

As a further application we shall consider an estimation problem for a subcritical Bienaymé–Galton–Watson branching process $\{X_n, n = 0,1, \ldots\}$ with immigration. Suppose that the process has any distribution for X_0 subject to $EX_0^2 < \infty$, and that the means of the offspring and immigration distributions are m and λ, respectively, while the variances of these distributions are also finite. The problem is to estimate $\theta' = (m,\lambda)$. This model has been used widely in practice and an account of various applications is given in Heyde and Seneta (1972). The conditional least squares approach to this problem has been taken by Klimko and Nelson (1978).

For the model in question the $(n + 1)$st generation is obtained from the independent reproduction of each of the individuals in the nth generation, each with the basic offspring distribution, plus an independent immigration input with the immigration distribution. Thus,

$$E_\theta(X_{n+1} | \mathscr{F}_n) = mX_n + \lambda,$$

and solving the conditional least squares equations (6.27) leads to the estimators

$$\hat{m}_n = \frac{n \sum_{i=1}^{n} X_{i-1} X_i - \left(\sum_{i=1}^{n} X_{i-1} \right) \left(\sum_{i=1}^{n} X_i \right)}{n \sum_{i=1}^{n} X_{i-1}^2 - \left(\sum_{i=1}^{n} X_{i-1} \right)^2} \tag{6.36}$$

and

$$\hat{\lambda}_n = n^{-1} \left(\sum_{i=1}^{n} X_i - \hat{m}_n \sum_{i=1}^{n} X_{i-1} \right). \tag{6.37}$$

These estimators are essentially the same as those studied by Heyde and Seneta (1972, 1974) and later by Quine (1976, 1977), the only difference being whether or not an initial or final term is included in certain sums. Accordingly, both pairs of estimators have the same asymptotic behavior.

The conditions of Theorem 6.4 are not difficult to check, but it is simpler to obtain strong consistency of the estimators directly. This can be achieved via applications of Theorem 2.19, but since $\{X_n\}$ is a Markov process, the shortest route is via the ergodic theorem.

First we need to observe the existence of a stationary distribution. Write $P_n(s) = E(s^{X_n})$, $0 \leqslant s \leqslant 1$, for the probability generating function (p.g.f.) of X_n, and let $A(s)$ and $B(s)$ be the p.g.f.s corresponding to the offspring and immigration distributions, respectively. Write $A_k(s) = A_{k-1}(A(s))$ for the kth functional iterate of A. Then

$$P_n(s) = P_{n-1}(A(s))B(s),$$

and iteration yields

$$P_n(s) = P_0(A(s))B(s) \prod_{j=1}^{n-1} B(A_j(s)). \tag{6.38}$$

From (6.38) it is easy to see that the process $\{X_n\}$ has a unique stationary limiting distribution with p.g.f. $B(s)\prod_{j=1}^{\infty} B(A_j(s))$, provided this is proper. However, the convergence of $\prod_{j=1}^{\infty} B(A_j(s))$, or equivalently, $\sum_{j=1}^{\infty}[1 - B(A_j(s))]$, follows from

$$\frac{1 - B(A_j(s))}{1 - A_j(s)} \leqslant B'(1) = \lambda$$

and

$$1 - A_j(s) \leqslant 1 - A_j(0),$$

while

$$\frac{1 - A_j(0)}{1 - A_{j-1}(0)} = \frac{1 - A(A_{j-1}(0))}{1 - A_{j-1}(0)} \leqslant A'(1) = m < 1.$$

We shall write σ^2 and b^2 for the variances of the offspring and immigration distributions and set $c^2 = b^2 + \sigma^2\lambda(1 - m)^{-1}$. We shall also use r_i to denote the ith ordinary moment of the stationary distribution (with p.g.f. $B(s) \times \prod_{j=1}^{\infty} B(A_j(s))$). It can easily be shown that

$$r_1 = \lambda(1 - m)^{-1}, \qquad r_2 = c^2(1 - m^2)^{-1} + r_1^2.$$

If X_0 has the stationary distribution, then the process $\{X_n\}$ is stationary and ergodic. Ergodicity follows from the uniqueness of the stationary distribution [see, e.g., Billingsley (1961a, p. 52)]. Under these circumstances, we obtain from the ergodic theorem that

$$\begin{aligned} n^{-1} \sum_{i=1}^{n} X_i &\xrightarrow{\text{a.s.}} r_1 = \lambda(1 - m)^{-1}, \\ n^{-1} \sum_{i=1}^{n} X_i^2 &\xrightarrow{\text{a.s.}} r_2 = c^2(1 - m^2)^{-1} + r_1^2, \\ n^{-1} \sum_{i=1}^{n} X_i X_{i-1} &\xrightarrow{\text{a.s.}} mc^2(1 - m^2)^{-1} + r_1^2. \end{aligned} \tag{6.39}$$

Furthermore, the conclusions of the ergodic theorem continue to hold for any initial distribution [see, e.g., Billingsley (1961a, Theorem 1.1) and Révész (1968, Theorem 7.1.1)], and the results $\hat{m}_n \xrightarrow{\text{a.s.}} m$ and $\hat{\lambda}_n \xrightarrow{\text{a.s.}} \lambda$ follow immediately from (6.36), (6.37), and (6.39).

If the offspring and immigration distributions have finite third moments, then r_3 is also finite and a central limit result holds for $n^{1/2}(\hat{m}_n - m, \hat{\lambda}_n - \lambda)'$. To see this, observe first that

$$n^{-1/2}\begin{bmatrix} \dfrac{\partial Q}{\partial m} \\ \dfrac{\partial Q}{\partial \lambda} \end{bmatrix} = -2n^{-1/2}\begin{bmatrix} \sum_{i=1}^{n} X_{i-1}X_i - m\sum_{i=1}^{n} X_{i-1}^2 - \lambda \sum_{i=1}^{n} X_{i-1} \\ \sum_{i=1}^{n} X_i - m\sum_{i=1}^{n} X_{i-1} - \lambda n \end{bmatrix}, \tag{6.40}$$

and, after some algebra,

$$n^{1/2}\begin{bmatrix} \hat{m}_n - m \\ \hat{\lambda}_n - \lambda \end{bmatrix} = -\frac{1}{2}n^{-1/2}\left[n^{-1}\sum_{i=1}^{n} X_{i-1}^2 - n^{-2}\left(\sum_{i=1}^{n} X_{i-1}\right)^2\right]^{-1}$$
$$\times \begin{bmatrix} 1 & -n^{-1}\sum_{i=1}^{n} X_{i-1} \\ -n^{-1}\sum_{i=1}^{n} X_{i-1} & -n^{-1}\sum_{i=1}^{n} X_{i-1}^2 \end{bmatrix}\begin{bmatrix} \dfrac{\partial Q}{\partial m} \\ \dfrac{\partial Q}{\partial \lambda} \end{bmatrix}. \tag{6.41}$$

In view of the results (6.39) we have

$$\left[n^{-1}\sum_{i=1}^{n} X_{i-1}^2 - n^{-2}\left(\sum_{i=1}^{n} X_{i-1}\right)^2\right]^{-1}\begin{bmatrix} 1 & -n^{-1}\sum_{i=1}^{n} X_{i-1} \\ -n^{-1}\sum_{i=1}^{n} X_{i-1} & -n^{-1}\sum_{i=1}^{n} X_{i-1}^2 \end{bmatrix}$$
$$\xrightarrow{\text{a.s.}} c^{-2}(1-m^2)\begin{pmatrix} 1 & -r_1 \\ -r_1 & r_2 \end{pmatrix} = V^{-1}, \tag{6.42}$$

say. Finally, we shall establish that

$$n^{-1/2}\begin{bmatrix} \dfrac{\partial Q}{\partial m} \\ \dfrac{\partial Q}{\partial \lambda} \end{bmatrix} \xrightarrow{\text{d}} N(0,W), \tag{6.43}$$

where

$$W = \begin{pmatrix} \sigma_1^2 r_3 + \sigma_2^2 r_2 & \sigma_1^2 r_2 + \sigma_2^2 r_1 \\ \sigma_1^2 r_2 + \sigma_2^2 r_1 & \sigma_1^2 r_1 + \sigma_2^2 \end{pmatrix}$$

and hence, from (6.41)–(6.43),

$$n^{1/2}\begin{pmatrix} \hat{m}_n - m \\ \hat{\lambda}_n - \lambda \end{pmatrix} \xrightarrow{\text{d}} N(0, V^{-1}WV^{-1}). \tag{6.44}$$

To establish the central limit result (6.43) we may again suppose without loss of generality that the distribution of X_0 is the stationary one, so that $\{X_n\}$ is stationary and ergodic. The martingale central limit result which we shall employ (Corollary 3.2) provides mixing convergence, and hence the probability measure based on the stationary initial distribution may be replaced by any probability measure that is absolutely continuous with respect to it without perturbing the limit distribution. Furthermore, the state space of the process $\{X_n\}$ in general consists of an irreducible positive recurrent set J^* of states, including the state 0, plus (possibly) "ephemeral" states from which transition into J^* occurs in one step with probability one (Heyde and Seneta, 1972). Standard Markov chain theory ensures that for $j \in J^*$, the probability measure of the process based on starting at j is absolutely continuous with respect to the stationary probability measure [see, e.g., Billingsley (1961a, Theorem 1.3)]. The mixing property and summation with respect to the probabilities of a general initial distribution then yield the general convergence result.

We have

$$\frac{\partial Q}{\partial m} = -2 \sum_{i=1}^{n} X_{i-1}(X_i - E(X_i|\mathscr{F}_{i-1})),$$

and

$$\frac{\partial Q}{\partial \lambda} = -2 \sum_{i=1}^{n} (X_i - E(X_i|\mathscr{F}_{i-1})),$$

so that, assuming the process $\{X_k\}$ to be stationary and applying the Cramér–Wold device,

$$n^{-1/2}c' \begin{bmatrix} \dfrac{\partial Q}{\partial m} \\ \dfrac{\partial Q}{\partial \lambda} \end{bmatrix} = -2n^{-1/2} \sum_{i=1}^{n} (c_1 X_{i-1} + c_2)(X_i - E(X_i|\mathscr{F}_{i-1}))$$

$$\xrightarrow{\text{d}} N(0, 4E(c_1 X_0 + c_2)^2 (X_1 - mX_0 - \lambda)^2),$$

using Corollary 3.2 and (6.39). Furthermore,

$$\begin{aligned} E((c_1X_0 + c_2)^2(X_1 - mX_0 - \lambda)^2) &= E[E((X_1 - mX_0 - \lambda)^2|X_0)(c_1X_0 + c_2)^2] \\ &= E((X_0\sigma_1^2 + \sigma_2^2)(c_1X_0 + c_2)^2) \\ &= c'Wc, \end{aligned}$$

and hence (6.43) holds. This completes the proof of (6.44).

6.4. Quadratic Functions of Discrete Time Series

A great deal of time-series analysis is based upon quadratic functions of the data. In particular, many inferential results relate to theorems concerning the autocorrelations

$$r(j) = \frac{\sum_{n=1}^{N-j}\{x(n) - \bar{x}\}\{x(n+j) - \bar{x}\}}{\sum_{n=1}^{N}\{x(n) - \bar{x}\}^2}, \qquad 0 \leqslant j \leqslant N-1, \quad (6.45)$$

$$r(-j) = r(j),$$

$x(1), x(2), \ldots, x(N)$ being a sample of N consecutive observations on some process $\{x(n)\}$ and $\bar{x}$ the sample mean. It is well known that, under certain conditions on the process $\{x(n)\}$, a strong law of large numbers and a central limit theorem hold for $r(j)$ [see, e.g., Hannan (1970, Chapters IV and VI)]. In this section it is our object to show, using limit theorems for martingales, that the scope of the classical inferential theory can be appreciably widened in a natural way.

We shall be concerned here with a process of the form

$$\begin{aligned} &x(n) - \mu = \sum_{j=0}^{\infty} \alpha(j)\varepsilon(n-j), \qquad \sum_{j=0}^{\infty} \alpha^2(j) < \infty, \qquad \alpha(0) = 1; \\ &E\varepsilon(n) = 0, \qquad E\{\varepsilon(m)\varepsilon(n)\} = 0, \qquad m \neq n. \end{aligned} \quad (6.46)$$

If $\{x(n) - \mu\}$ is a stationary, purely nondeterministic (i.e., containing no deterministic component) process (Hannan, 1970, Chapter III), then it may be represented in this form with the $\varepsilon(n)$ as the linear prediction errors, having variance $\sigma^2 > 0$. As is well known, there will be many representations of such a stationary process in the form (6.46), but for only one of these will the $\varepsilon(n)$ be the prediction errors. However, our results extend beyond the stationary case so that we shall begin by assuming only that (6.46) holds for $n \geqslant 0$, together with other conditions to be discussed shortly. A process of the kind (6.46) arises in a wide variety of contexts, for example, from a mixed autoregressive and moving average process [Hannan (1970, Chapter I)], and as the response of a physically realizable filter to an uncorrelated sequence [Gikhman and Skorokhod (1969, Chapter 5)].

Now, the classical theory of inference for the process (6.46) usually requires that the $\varepsilon(n)$ be independent and identically distributed (i.i.d.) with zero mean and variance σ^2. The essential feature of our discussion here is that, subject to some reasonable additional conditions, the classical theory goes through if the independence assumption is replaced by the weaker condition

$$E(\varepsilon(n)|\mathscr{F}_{n-1}) = 0 \qquad \text{a.s.} \qquad \text{for all} \quad n, \tag{6.47}$$

where $\mathscr{F}_n$ is the σ-field generated by $\varepsilon(m)$, $m \leqslant n$. This requirement has a simple and natural interpretation in the case where $\{x(n)\}$ is stationary and the $\varepsilon(n)$ are the linear prediction errors, for then $\mathscr{F}_n$ is also the σ-field generated by the $x(m)$, $m \leqslant n$, so that, because of (6.47),

$$\varepsilon(n) = x(n) - E(x(n)|\mathscr{F}_{n-1}). \tag{6.48}$$

To see this, write $\mathscr{G}_n$ for the σ-field generated by $x(m)$, $m \leqslant n$. Clearly $\mathscr{F}_n \supseteq \mathscr{G}_n$ and, when the ε are the prediction errors,

$$\varepsilon(n) = x(n) - E(x(n)|\mathscr{G}_{n-1})$$

which is $\mathscr{G}_n$-measurable. Thus $\mathscr{G}_n = \mathscr{F}_n$ (or at least, $\mathscr{G}_n$ and $\mathscr{F}_n$ are contained in the completions of one another). Then $E(x(n)|\mathscr{F}_{n-1})$ is the best linear predictor and the best linear predictor is the best predictor (both in the least squares sense). Conversely, if this is so, (6.48) must hold, and hence (6.47) must hold also. Thus (6.47) is equivalent to the condition that the best predictor is the best linear predictor, both in the least squares sense. In the stationary case our additional conditions are, for example, the regularity condition (6.51) below, together with the requirement that $E(\varepsilon^2(n)|\mathscr{F}_{n-1}) = \sigma^2$ (const.) a.s. Our results imply that, subject to the mild regularity condition (6.51), the classical theory of inference for (6.46) goes through when the $\varepsilon(n)$ are the prediction errors, provided the best linear predictor is the best predictor and the prediction variance, given the past, is a constant.

To establish a strong law for autocorrelations we consider the process (6.46), where the $\varepsilon(n)$ satisfy the condition (6.47). We shall not require stationarity of $\{\varepsilon(n)\}$ but instead the condition

$$\lim_{N\to\infty} N^{-1} \sum_{n=1}^{N} E(\varepsilon^2(n)|\mathscr{F}_{n-1}) = \sigma^2 > 0 \quad \text{a.s.} \tag{6.49}$$

and the condition that there exists a random variable X with $EX^2 < \infty$ such that

$$P(|\varepsilon(n)| > u) \leqslant cP(|X| > u) \tag{6.50}$$

for some $0 < c < \infty$ and all $n \geqslant 1$, $u \geqslant 0$. If $\{x(n)\}$ is stationary, we modify (6.49) to

$$E(\varepsilon^2(n)|\mathscr{F}_{n-1}) = \sigma^2 > 0 \quad \text{a.s.,} \tag{6.49'}$$

and (6.50) is redundant. In connection with the central limit theorem we shall make use of the condition

$$\sum_{j=1}^{\infty} j^{1/2}[\alpha(j)]^2 < \infty. \tag{6.51}$$

Define

$$c(j) = c(-j) = N^{-1} \sum_{n=1}^{N-j} \{x(n) - \bar{x}\}\{x(n+j) - \bar{x}\}, \qquad j \geqslant 0,$$

where $\bar{x}$ is the sample mean of the $x(n)$, $n = 1, \ldots, N$. If $\{x(n)\}$ is stationary, its spectral density is

$$f(\lambda) = \frac{\sigma^2}{2\pi}\left|\sum_{0}^{\infty} \alpha(j)e^{ij\lambda}\right|^2,$$

and the autocovariances $\gamma(j)$ satisfy

$$\gamma(j) = \int_{-\pi}^{\pi} e^{ij\lambda} f(\lambda)\, d\lambda = \sigma^2 \sum_{u=0}^{\infty} \alpha(u)\alpha(u+j).$$

However, we may define $f(\lambda)$ and $\gamma(j)$ by these formulas whether or not $\{x(n)\}$ is stationary. We now have the following theorem [Hannan and Heyde, (1972)].

Theorem 6.6. *If* (6.47) *and* (6.50) *hold and* $\sum|\alpha(k)| < \infty$, $\bar{x}$ *converges a.s. to* μ. *If* (6.49) *is also true, then* $c(j)$ *converges in probability to* $\gamma(j)$. *If* $x(n)$ *is stationary,* $\bar{x}$ *converges a.s. to* μ, *and if* (6.47) *and* (6.49) *hold,* $c(j)$ *converges a.s. to* $\gamma(j)$.

Proof. If $\varepsilon(n)$ satisfies (6.47), (6.50), and $\Sigma|\alpha(k)| < \infty$, it is easily seen that $\bar{x}(N) = \bar{x}$ has a variance which is $O(N^{-1})$ as $N \to \infty$. Thus, if $N \geqslant m^2$, we have for some $K > 0$ that

$$E(\bar{x}(N) - \mu)^2 \leqslant Km^{-2}.$$

If $\varepsilon > 0$ and $N_m = m^2$, Chebyshev's inequality gives

$$\sum_{m=1}^{\infty} P(|\bar{x}(N_m) - \mu| > \varepsilon) \leqslant K\varepsilon^{-2} \sum_{m=1}^{\infty} m^{-2} < \infty,$$

and hence, from the Borel–Cantelli lemma,

$$\lim_{m\to\infty} \bar{x}(N_m) = \mu \quad \text{a.s.} \tag{6.52}$$

Moreover,

$$E\left[\max_{N_m \leqslant N < N_{m+1}} \left\{N^{-1}\sum_1^N (x(n)-\mu) - N^{-1}\sum_1^{N_m}(x(n)-\mu)\right\}^2\right]$$
$$\leqslant N_m^{-2} E\left[\left\{\sum_{N_m+1}^{N_{m+1}} |x(n)-\mu|\right\}^2\right]$$
$$= N_m^{-2} \sum_{j=N_m+1}^{N_{m+1}} \sum_{k=N_m+1}^{N_{m+1}} E|(x(j)-\mu)(x(k)-\mu)|$$
$$\leqslant \gamma(0) N_m^{-2}(N_{m+1}-N_m)^2 \leqslant 9\gamma(0)m^{-2},$$

using the fact that $2|xy| \leqslant x^2 + y^2$ to obtain the inequality second from the last. It then follows, as in the argument just used, that

$$\lim_{m\to\infty} \sup_{N_m \leqslant N < N_{m+1}} \left| N^{-1}\sum_1^N (x(n)-\mu) - N^{-1}\sum_1^{N_m}(x(n)-\mu)\right| = 0 \quad \text{a.s.} \tag{6.53}$$

Furthermore, by (6.52),

$$N^{-1}\sum_1^{N_m} x(n) = N^{-1}N_m\bar{x}(N_m) \xrightarrow{\text{a.s.}} \mu$$

uniformly in $N_m \leqslant N < N_{m+1}$, and the result $\bar{x}(N) \xrightarrow{\text{a.s.}} \mu$ follows via (6.53).
Since $\lim_{N\to\infty} \bar{x} = \mu$ a.s., it is clear that $c(j)$ has the same a.s. behavior as

$$c^*(j) = N^{-1}\sum_{n=1}^{N-j}(x(n)-\mu)(x(n+j)-\mu)$$
$$= \sum_{u=0}^{\infty}\sum_{v=0}^{\infty} \alpha(u)\alpha(v)N^{-1}\sum_{n=1}^{N-j}\varepsilon(n-u)\varepsilon(n+j-v).$$

Now

$$E\left|N^{-1}\sum_{n=1}^{N-j}\varepsilon(n-u)\varepsilon(n+j-v)\right| \leqslant \sup_{u,v} E|\varepsilon(u)\varepsilon(v)| < K < \infty$$

by virtue of (6.50), and if $\sum|\alpha(k)| < \infty$, then

$$\lim_{p\to\infty} E\left|\sum_{u=p+1}^{\infty}\sum_{v=0}^{\infty}\alpha(u)\alpha(v)N^{-1}\sum_{n=1}^{N-j}\varepsilon(n-u)\varepsilon(n+j-v)\right| = 0. \tag{6.54}$$

The same is true if, in the left term in (6.54), the first two sums are over $0 \leqslant u \leqslant p$, $p < v < \infty$. On the other hand,

$$N^{-1}\sum_{n=1}^{N} E\{\varepsilon(n)\varepsilon(n-k)|\mathscr{F}_{n-1}\} = 0 \quad \text{a.s.,} \qquad k > 0,$$

and it follows from (6.50) and Theorem 2.19 that

$$N^{-1} \sum_{n=1}^{N} \varepsilon(n)\varepsilon(n-k), \qquad k > 0,$$

and hence

$$N^{-1} \sum_{n=1}^{N-j} \varepsilon(n-u)\varepsilon(n+j-v), \qquad u \neq v-j,$$

converges in probability to zero. (In order to obtain the uniform bound on the distribution of $\varepsilon(n)\varepsilon(n-k)$ required to justify the application of Theorem 2.19, we note that

$$\begin{aligned} P(|\varepsilon(n)\varepsilon(n-k)| > u) &\leqslant P(\varepsilon^2(n) + \varepsilon^2(n-k) > 2u) \\ &\leqslant P(\varepsilon^2(n) > u) + P(\varepsilon^2(n-k) > u) \\ &\leqslant 2cP(X^2 > u) \end{aligned} \tag{6.55}$$

using (6.50).) Furthermore, by the same theorem together with (6.49), $N^{-1} \sum_{n=1}^{N-j} \varepsilon^2(n)$ converges in probability to σ^2. Thus

$$\sum_{u=0}^{p} \sum_{v=0}^{p} \alpha(u)\alpha(v) N^{-1} \sum_{n=1}^{N-j} \varepsilon(n-u)\varepsilon(n+j-v) \to \sigma^2 \sum_{u=0}^{p-j} \alpha(u)\alpha(u+j)$$

in probability as $N \to \infty$. It follows from this together with (6.54) and Markov's inequality that $c^*(j)$, and hence $c(j)$, converges in probability to $\gamma(j)$.

If $\{x(n)\}$ is stationary with an absolutely continuous spectrum, then $\bar{x} \xrightarrow{\text{a.s.}} \mu$ as $N \to \infty$. That $\bar{x}$ converges a.s. follows from the ergodic theorem, while the limit random variable can be identified as μ a.s. since $\bar{x} - \mu$ converges in the mean of order 2 to zero. This last result follows from

$$\begin{aligned} E(\bar{x} - \mu)^2 = N^{-2} \sum_{|j|<N} (N - |j|)\gamma(j) &= N^{-2} \int_{-\pi}^{\pi} \sum_{|j|<N} (N - |j|) e^{ij\lambda} f(\lambda)\, d\lambda \\ &= N^{-2} \int_{-\pi}^{\pi} \frac{\sin^2(\frac{1}{2}N\lambda)}{\sin^2(\frac{1}{2}\lambda)} f(\lambda)\, d\lambda \to 0 \end{aligned}$$

as $N \to \infty$ via the dominated convergence theorem.

In place of $c^*(j)$ consider

$$\bar{c}(j) = c^*(j) - \sum_{u=0}^{\infty} \alpha(u)\alpha(u+j) N^{-1} \sum_{n=1}^{N-j} \varepsilon^2(n-u).$$

Assuming $\{x(n)\}$ stationary and (6.47) and (6.49′), we may show that the mean of $\bar{c}(j)$ is zero and its variance converges to zero. The proof of the first is obvious and we prove the second, for simplicity, in the case $j = 0$. The variance is

$$\sum_{p,q}{}' \alpha(p)\alpha(q)\sum_{r,s}{}' \alpha(r)\alpha(s)\left[N^{-2}\sum_{m=1}^{N}\sum_{n=1}^{N} E\{\varepsilon(m-p)\varepsilon(m-q)\varepsilon(n-r)\varepsilon(n-s)\}\right],$$

where $\sum'_{u,v}$ stands for summation over values $0 \leqslant u,\ v < \infty$, with $u \neq v$. We evaluate the expectation using (6.47) and (6.49′). The only contribution comes when $m - p = n - r$ and $s = q - p + r$, or $m - p = n - s$ and $s = p - q + r$. Both sets of identifications give the same result and we take the first. The variance is

$$N^{-1}\sigma^4 \sum\sum\sum_{0 \leqslant p,\, q,\, r < \infty,\, |p-r| < N,\, q \neq p,\, q-p+r \geqslant 0} \alpha(p)\alpha(q)\alpha(r)\alpha(q-p+r)(1-N^{-1}|p-r|)$$

$$= N^{-1}\sigma^4\Bigg\{\sum\sum\sum_{0 \leqslant p,\, q,\, r < \infty,\, |p-r| < N,\, q-p+r \geqslant 0} \alpha(p)\alpha(q)\alpha(r)\alpha(q-p+r)(1-N^{-1}|p-r|)$$

$$- \sum\sum_{0 \leqslant p,\, r < \infty,\, |p-r| < N} (\alpha(p)\alpha(r))^2(1-N^{-1}|p-r|)\Bigg\}$$

$$\leqslant N^{-1}\sigma^4 \sum\sum\sum_{0 \leqslant p,\, q,\, r < \infty,\, |p-r| < N,\, q-p+r \geqslant 0} \alpha(p)\alpha(q)\alpha(r)\alpha(q-p+r)(1-N^{-1}|p-r|)$$

$$= N^{-1}\sigma^4 \sum_{j=-N+1}^{N-1} (1-N^{-1}|j|)\left\{\sum_{j=0}^{\infty} \alpha(p)\alpha(p+|j|)\right\}^2.$$

This is N^{-1} times the Cesàro sum of the Fourier series, evaluated at the origin, of the convolution of $f(\lambda)$ with itself and thus converges to zero [see, e.g., Hannan (1970, pp. 506–507)]. Thus $\bar{c}(0)$, and in the same way $\bar{c}(j)$, converges in probability to zero. However,

$$\sum_{u=0}^{\infty} \alpha(u)\alpha(u+j)N^{-1}\sum_{n=1}^{N-j} \varepsilon^2(n-u)$$

converges in probability to $\gamma(j)$ by the same kind of argument as was used earlier in the proof (the convergence of $\sum|\alpha(u)|$ not now being needed). Thus $c^*(j)$ and hence $c(j)$ converges in probability to $\gamma(j)$ and since, by the ergodic theorem, $c(j)$ converges almost-surely, it must converge almost surely to $\gamma(j)$. This completes the proof. ∎

Next we shall establish a central limit theorem for autocorrelations. From Theorem 6.6 we note that, depending on the conditions, $r(j) = c(j)/c(0)$ converges either in probability or almost surely to $\rho(j) = \gamma(j)/\gamma(0)$. From now on we shall confine ourselves to the stationary case, although results are available outside this context [see Hannan and Heyde (1972)] under less attractive assumptions. We shall obtain the following result.

Theorem 6.7. *Let $\{x(n)\}$ be stationary and satisfy* (6.47), (6.49′), *and* (6.51). *Then the joint distribution of $N^{1/2}\{r(j) - \rho(j)\}$, $1 \leqslant j \leqslant s$, converges to the s-variate multivariate normal distribution with zero mean vector and covariance matrix $W = (w_{ij})$, where*

$$\begin{aligned} w_{ij} &= \sum_{r=1}^{\infty} \{\rho(r+i) + \rho(r-i) - 2\rho(r)\rho(i)\} \\ &\quad \times \{\rho(r+j) + \rho(r-j) - 2\rho(r)\rho(j)\} \\ &= \sum_{r=-\infty}^{\infty} \{\rho(r)\rho(r+i-j) + \rho(r)\rho(r+i+j) + 2\rho^2(r)\rho(i)\rho(j) \\ &\quad - 2\rho(r)\rho(i)\rho(r+j) - 2\rho(r)\rho(j)\rho(r+i)\}. \end{aligned} \tag{6.56}$$

Proof. Let us take the case where μ is known to be zero since mean correction makes no difference to the truth of the theorem. We first note that the spectral density of $\{x(n)\}, f(\lambda)$, is square integrable over $[-\pi,\pi]$ under (6.51), since this is equivalent to

$$\sum_{k=0}^{\infty} \left\{ \sum_{j=1}^{\infty} \alpha(j)\alpha(j+k) \right\}^2 < \infty,$$

and

$$\begin{aligned} \sum_{k=0}^{\infty} \left\{ \sum_{j=1}^{\infty} j^{1/4}\alpha(j) j^{-1/4}\alpha(j+k) \right\}^2 &\leqslant \sum_{k=0}^{\infty} \left\{ \sum_{j=1}^{\infty} j^{1/2}\alpha^2(j) \right\} \left\{ \sum_{j=1}^{\infty} j^{-1/2}\alpha^2(j+k) \right\} \\ &\leqslant \left\{ \sum_{j=1}^{\infty} j^{1/2}\alpha^2(j) \right\} \sum_{k=0}^{\infty} \sum_{j=1}^{\infty} j^{-1/2}\alpha^2(j+k) \\ &\leqslant \left\{ \sum_{j=1}^{\infty} j^{1/2}\alpha^2(j) \right\} \sum_{k=1}^{\infty} \alpha^2(k) \sum_{j=1}^{k} j^{-1/2} < \infty \end{aligned}$$

under (6.51).

Now, as in the proof of Theorem 6.6,

$$N^2 E\bar{x}^2 = \int_{-\pi}^{\pi} \frac{\sin^2(\frac{1}{2}N\lambda)}{\sin^2(\frac{1}{2}\lambda)} f(\lambda)\, d\lambda.$$

But for $0 < \delta < \pi$ and using Schwarz's inequality,

$$\begin{aligned} &N^{-1/2} \int_{|\lambda| \leqslant \delta} \frac{\sin^2(\frac{1}{2}N\lambda)}{N \sin^2(\frac{1}{2}\lambda)} f(\lambda)\, d\lambda \\ &\leqslant N^{-1/2} \left\{ \int_{|\lambda| \leqslant \delta} \frac{\sin^2(\frac{1}{2}N\lambda)}{N \sin^2(\frac{1}{2}\lambda)} f^2(\lambda)\, d\lambda \int_{|\lambda| \leqslant \delta} \frac{\sin^2(\frac{1}{2}N\lambda)}{N \sin^2(\frac{1}{2}\lambda)}\, d\lambda \right\}^{1/2} \\ &\leqslant \left\{ 2\pi \int_{|\lambda| \leqslant \delta} f^2(\lambda)\, d\lambda \right\}^{1/2} \end{aligned}$$

(since $\int_{-\pi}^{\pi}[\sin^2(\frac{1}{2}N\lambda)/\sin^2(\frac{1}{2}\lambda)]\,d\lambda = 2\pi N$ and $\sin^2(\frac{1}{2}N\lambda) \leqslant N^2\sin^2(\frac{1}{2}\lambda)$), which can be made arbitrarily small by taking δ small since $f(\lambda) \in L^2(-\pi,\pi)$. Also,

$$\int_{\pi \geqslant |\lambda| > \delta} \frac{\sin^2(\frac{1}{2}N\lambda)}{N\sin^2(\frac{1}{2}\lambda)} f(\lambda)\,d\lambda = o(1)$$

as $N \to \infty$ since $\sin^2(\frac{1}{2}N\lambda)/N\sin^2(\frac{1}{2}\lambda)$ converges uniformly to zero outside any interval $(-\delta,\delta)$, $\delta > 0$. Hence, $N^{1/2}E\bar{x}^2 = o(1)$ as $N \to \infty$ and, using Markov's inequality, $N^{1/2}\bar{x}^2 \xrightarrow{\text{p}} 0$ as $N \to \infty$. Then, for any k, and using $O_p(N^{-1/2})$ for a random variable which is of order $N^{-1/2}$ in probability,

$$N^{1/2}\{c^*(k) - c(k)\} = N^{1/2}\bar{x}^2 + O_p(N^{-1/2}) \xrightarrow{\text{p}} 0$$

as $N \to \infty$, from which we deduce that

$$\begin{aligned} N^{1/2}\{r(j) - c^*(j)/c^*(0)\} &= N^{1/2}\{c(j) - c^*(j)\}/c(0) + N^{1/2}c^*(j)\{c^*(0) \\ &\quad - c(0)\}/c(0)c^*(0) \\ &\xrightarrow{\text{p}} 0. \end{aligned}$$

Thus, instead of $N^{1/2}\{r(j) - \rho(j)\}$, we may consider the limit distribution of $N^{1/2}\{c^*(j) - \rho(j)c^*(0)\}/c^*(0)$, and since $c^*(0)$ converges a.s. to $\gamma(0)$ by Theorem 6.6, we consider

$$N^{1/2}\{c^*(j) - \rho(j)c^*(0)\}. \tag{6.57}$$

We first show that we may omit all terms involving $\varepsilon^2(n)$ in (6.57). These terms are

$$\begin{aligned} &N^{-1/2}\left[\sum_{n=1}^{N-j}\sum_{k=0}^{\infty}\alpha(k)\alpha(k+j)\varepsilon^2(n-k) - \rho(j)\sum_{n=1}^{N}\sum_{k=0}^{\infty}\alpha^2(k)\varepsilon^2(n-k)\right] \\ &= N^{-1/2}\left[\sum_{k=0}^{\infty}\alpha(k)\alpha(k+j)\sum_{n=1-k}^{N-j-k}\varepsilon^2(n) - \rho(j)\sum_{k=0}^{\infty}\alpha^2(k)\sum_{n=1-k}^{N-k}\varepsilon^2(n)\right] \\ &= N^{-1/2}\left[\sum_{k=0}^{\infty}\alpha(k)\alpha(k+j)\left\{\sum_{n=1}^{N}\varepsilon^2(n) + T_{N,k}^{(j)}\right\}\right. \\ &\quad \left. - \rho(j)\sum_{k=0}^{\infty}\alpha^2(k)\left\{\sum_{n=1}^{N}\varepsilon^2(n) + T_{N,k}^{(0)}\right\}\right] \\ &= N^{-1/2}\left[\sum_{k=0}^{\infty}\alpha(k)\alpha(k+j)T_{N,k}^{(j)} - \rho(j)\sum_{k=0}^{\infty}\alpha^2(k)T_{N,k}^{(0)}\right], \end{aligned} \tag{6.58}$$

where

$$T_{N,k}^{(j)} = \sum_{n=1-k}^{N-j-k}\varepsilon^2(n) - \sum_{n=1}^{N}\varepsilon^2(n)$$

and

$$E|T_{N,k}^{(j)}| \leqslant \sigma^2 \min(2k+j, 2N-j).$$

Thus, taking the first term in (6.58), for example,

$$\begin{aligned} N^{-1/2} E\left\{\left|\sum_{k=0}^{\infty} \alpha(k)\alpha(k+j)T_{N,k}^{(j)}\right|\right\} &\leqslant N^{-1/2}\sigma^2 \sum_{k=0}^{\infty} |\alpha(k)\alpha(k+j)| \min(2k+j, 2N-j) \\ &< \sigma^2\left[\left\{N^{-1/2}\sum_{k=0}^{\infty} \alpha^2(k)\min(2k+j, 2N-j)\right\}^{1/2}\right. \\ &\quad \times \left.\left\{N^{-1/2}\sum_{k=0}^{\infty} \alpha^2(k+j)\min(2k+j, 2N-j)\right\}^{1/2}\right], \end{aligned}$$

using the Cauchy–Schwarz inequality. However,

$$\begin{aligned} N^{-1/2}\sum_{k=0}^{\infty} \alpha^2(k)\min(2k+j, 2N-j) &\leqslant 2\sum_{k=0}^{N-j} \alpha^2(k)k^{1/2}\{k/N\}^{1/2} \\ &\quad + jN^{-1/2}\sum_{k=0}^{N-j} \alpha^2(k) + 2\sum_{k=N-j}^{\infty} \alpha^2(k)k^{1/2}, \end{aligned}$$

which converges to zero. The same is true of the second factor and the second term in (6.58), and thus (6.58) converges in probability to zero.

Let us now fix $K > 0$ and set $x(n) = x_1(n) + x_2(n)$, where

$$x_2(n) = \sum_{j=K+1}^{\infty} \alpha(j)\varepsilon(n-j).$$

We also set

$$c_{uv}(k) = N^{-1}\sum_{n=1}^{N} x_u(n)x_v(n+k), \qquad u, v = 1, 2.$$

If $\gamma_{uv}(k) = Ec_{uv}(k)$, then

$$\gamma_{uv}(k) = \int_{-\pi}^{\pi} e^{ik\lambda} f_{uv}(\lambda)\, d\lambda$$

and

$$f_{11}(\lambda) = \sigma^2(2\pi)^{-1}\left|\sum_{0}^{K}\alpha(j)e^{ij\lambda}\right|^2, \qquad f_{22}(\lambda) = \sigma^2(2\pi)^{-1}\left|\sum_{K+1}^{\infty}\alpha(j)e^{ij\lambda}\right|^2,$$

$$f_{12}(\lambda) = \overline{f_{21}(\lambda)} = \sigma^2(2\pi)^{-1}\sum_{0}^{K}\alpha(j)e^{ij\lambda}\sum_{K+1}^{\infty}\alpha(j)e^{-ij\lambda}.$$

All of these functions are square integrable over $[-\pi,\pi]$ under condition (6.51), as can be seen from the argument on $f(\lambda)$ above.

Now (6.57) becomes

$$N^{1/2}[\{c_{11}(j) + c_{22}(j) + c_{12}(j) + c_{21}(j)\} - \rho(j)\{c_{11}(0) + c_{22}(0) + 2c_{12}(0)\}]. \tag{6.59}$$

Let $c'_{uv}(k)$ denote the expression $c_{uv}(k)$ with all terms involving $\varepsilon^2(n)$ omitted. We wish to show that the contribution to the primed form of (6.59) from the $c'_{uv}(k)$ for u, v not both equal to unity has a variance which, for all sufficiently large N, may be made arbitrarily small by taking K large. To this end we set

$$N^{1/2}\{c_{uv}(k) - Ec_{uv}(k)\} = N^{1/2}c'_{uv}(k) + N^{1/2}\{c''_{uv}(k) - Ec''_{uv}(k)\}, \tag{6.60}$$

where $c''_{uv}(k)$ contains all terms in $c_{uv}(k)$ involving $\varepsilon^2(n)$. The variance of (6.60) may be calculated as indicated in Hannan (1970, pp. 209–212), which shows that the correlation between the two terms on the right in (6.60) may be neglected for large K and N. Indeed, we may assume that the $\varepsilon(n)$ and $x(n)$ are Gaussian, since this does not effect the variance of $N^{1/2}c'_{uv}(k)$, which depends only on σ^2 and the $\alpha(j)$, and calculate the variance of (6.60) accordingly. In this way we obtain the variance of the left-hand term as an upper bound on the variance of $N^{1/2}c'_{uv}(k)$, since, on the Gaussian assumption, the two terms on the right-hand side of (6.60) are uncorrelated. The variance of the left-hand term in (6.60), on the Gaussian assumption, is found to be

$$\sum_{n=-N+1}^{N+1} (1 - |n|N^{-1})\{\gamma_{uu}(n)\gamma_{vv}(n) + \gamma_{uv}(n+k)\gamma_{vu}(n-k)\},$$

which converges to

$$2\pi \int_{-\pi}^{\pi} \{f_{uu}(\lambda)f_{vv}(\lambda) + |f_{uv}(\lambda)|^2 e^{2ik\lambda}\}\, d\lambda \tag{6.61}$$

as $N \to \infty$ because of the square integrability of the $f_{uv}(\lambda)$ and Parseval's theorem. However, as $K \to \infty$,

$$\int_{-\pi}^{\pi} f_{22}^2(\lambda)\, d\lambda \qquad \text{and} \qquad \int_{-\pi}^{\pi} f_{22}(\lambda)\, d\lambda$$

converge to zero because the Fourier series of a function in $L^p(-\pi,\pi)$, $1 < p < \infty$, converges in the L^p-norm to the function [e.g., Katznelson (1968, p. 50)]. In our case the function is $\sum_{K+1}^{\infty} \alpha(j)\exp(ij\lambda) \in L^4$. Thus, taking K sufficiently large, we may, if u, v are not both unity, make (6.61) arbitrarily small, and hence the variance of $N^{1/2}c'_{uv}(k)$, for all sufficiently large N, arbitrarily small.

By what is sometimes called Bernstein's lemma, the theorem will now follow if it is shown that $N^{1/2}[c'_{11}(j) - \rho(j)c'_{11}(0)]$, $1 \leqslant j \leqslant s$, are jointly asymptotically normal with a covariance matrix which converges, as K is increased, to $\gamma^2(0)W$. A statement of Bernstein's lemma follows. Let $\{x_N\}$ be a sequence of vector-valued random variables with zero mean such that for every $\varepsilon > 0$, $\zeta > 0$, $\eta > 0$ there exist sequences of random vectors $y_N(\varepsilon)$, $z_N(\varepsilon)$ such that $x_N = y_N(\varepsilon) + z_N(\varepsilon)$, where $y_N(\varepsilon)$ has a distribution converging to the multivariate normal distribution with zero mean vector and covariance matrix $A(\varepsilon)$, and

$$\lim_{\varepsilon\to 0} A(\varepsilon) = A, \qquad P(z_N'(\varepsilon)z_N(\varepsilon) > \zeta^2) < \eta.$$

Then the distribution of x_N tends to the multivariate normal with zero mean vector and covariance matrix A. The proof is straightforward [see, e.g., Hannan (1970, p. 242)].

Before proceeding to the final part of the proof, we first set

$$\rho'(j) = \sum_{k\leqslant K} \alpha(k)\alpha(k+j) \Big/ \sum_{k\leqslant K} \alpha^2(k).$$

We shall show that the joint distribution of the variables $N^{1/2}\{c_{11}'(j) - \rho'(j)c_{11}'(0)\}$, $1 \leqslant j \leqslant s$, converges to the multivariate normal distribution with zero mean vector and covariance matrix $\gamma^2(0)W^*$, where W^* is obtained from W by replacing $f(\lambda)$ by $f_{11}(\lambda)$. Since $N^{1/2}(\rho(j) - \rho'(j))c_{11}'(0)$ evidently converges in probability to zero and W^* converges to W as $K \to \infty$ because of the result, quoted above, on L^p-convergence, the theorem will then follow from Bernstein's lemma.

To establish the requisite convergence to normality we shall first show that $N^{1/2}\{c_{11}'(j) - \rho'(j)c_{11}'(0)\}$ converges in distribution to $N(0,\gamma^2(0)w_{jj}^*)$, where $W^* = (w_{ij}^*)$. Note that

$$N^{1/2}\{c_{11}'(j) - \rho'(j)c_{11}'(0)\} = N^{-1/2}\left[\sum_{n=1}^{N}\sum_{r=0}^{N}\sum_{s=-j}^{K-j}{}' \alpha(r)\alpha(s+j)\varepsilon(n-r)\varepsilon(n-s) - \rho'(j)\sum_{n=1}^{N}\sum_{r=0}^{K}\sum_{s=0}^{K}{}' \alpha(r)\alpha(s)\varepsilon(n-r)\varepsilon(n-s)\right], \tag{6.62}$$

where $\sum'$ indicates that the terms in the summation with $r = s$ are omitted. Then, changing the order of summation with respect to n and r, s in (6.2) and adding and subtracting a finite number of terms (the number not depending on N), we obtain

$$N^{1/2}\{c_{11}'(j) - \rho'(j)c_{11}'(0)\} = N^{-1/2}\sum_{r=1}^{2K+j} \delta_{r,K}^{(j)} \sum_{n=1}^{N} \varepsilon(n)\varepsilon(n-r) + O_p(N^{-1/2}),$$

where

$$\delta_{r,K}^{(j)} = \sum_{i=-K}^{K} [\alpha'(i)\alpha'(i+j+r) + \alpha'(i)\alpha'(i+j-r)$$
$$- \rho'(j)(\alpha'(i)\alpha'(i+r) + \alpha'(i)\alpha'(i-r))]$$

and $\alpha'(i) = \alpha(i)$ if $0 \leqslant i \leqslant K$, $\alpha'(i) = 0$ otherwise. Observe that

$$\sigma^4 \sum_{r=1}^{2K+j} [\delta_{r,K}^{(j)}]^2 = \gamma^2(0)w_{jj}^*.$$

We shall complete the proof of the 1-dimensional convergence result by showing that for any sequence of constants $c_1, \ldots, c_m$,

$$N^{-1/2} \sum_{r=1}^{m} c_r \sum_{n=1}^{N} \varepsilon(n)\varepsilon(n-r)$$

converges in distribution to $N(0, \sigma^4 \sum_{r=1}^{m} c_r^2)$. This is accomplished with the aid of Corollary 3.1.

Define $X_n = \sum_{r=1}^{m} c_r\varepsilon(n)\varepsilon(n-r)$, noting that $\{S_N = \sum_{n=1}^{N} X_n, \mathscr{F}_N, N \geqslant 1\}$ is a martingale. Let

$$V_N^2 = \sum_{n=1}^{N} E(X_n^2|\mathscr{F}_{n-1}) = \sum_{n=1}^{N} \sum_{r=1}^{m} \sum_{s=1}^{m} c_r c_s \varepsilon(n-r)\varepsilon(n-s)E(\varepsilon^2(n)|\mathscr{F}_{n-1})$$
$$= \sigma^2 \sum_{n=1}^{N} \left[\sum_{r=1}^{m} c_r\varepsilon(n-r)\right]^2$$

and

$$s_N^2 = EV_N^2 = NEX_0^2 = N\sigma^2 \sum_{r=1}^{m} c_r^2.$$

In order to apply Corollary 3.1 it suffices to show that

$$\text{(i)}\ s_N^{-2}V_N^2 \xrightarrow{\text{P}} 1 \qquad \text{and} \qquad \text{(ii)}\ s_N^{-2} \sum_{n=1}^{N} E(X_n^2 I(|X_n| \geqslant \varepsilon s_n)) \to 0$$

for any $\varepsilon > 0$. That (i) holds follows from an argument given in the proof of Theorem 6.6 based on the use of Theorem 2.19. The stationarity of $\{X_n\}$ ensures that (ii) holds. Corollary 3.1 then gives that $s_N^{-1}\sum_{n=1}^{N} X_n$ converges in distribution to $N(0,1)$ and hence $N^{-1/2}\sum_{r=1}^{m} c_r \sum_{n=1}^{N} \varepsilon(n)\varepsilon(n-r)$ converges in distribution to $N(0, \sigma^4 \sum_{r=1}^{m} c_r^2)$.

To complete the proof of Theorem 6.7 we have only now to observe that the argument used to show that $N^{1/2}\{c_{11}'(j) - \rho'(j)c_{11}'(0)\}$ converges in distribution to $N(0,\gamma^2(0)w_{jj}^*)$ can be carried through to show that the limiting distribution of $\sum_{j=1}^{s} c_j N^{1/2}\{c_{11}'(j) - \rho'(j)c_{11}'(0)\}$, $c' = (c_1, c_2, \ldots, c_s)$ being an arbitrary set of constants, is $N(0,\gamma^2(0)c'W^*c)$. This (Cramér–Wold device) assures that the joint limiting distribution of $N^{1/2}\{c'_{11}(j) - \rho'(j)c'_{11}(0)\}$, $1 \leqslant j \leqslant s$, is normal with mean-vector zero and covariance matrix $\gamma^2(0)W^*$. The proof is thus complete. ▮

The most obvious application of the strong law and central limit results given above is to the stationary autoregressive process $\{x(n)\}$ given by

$$\sum_{k=0}^{q} \beta(k)\{x(n-k) - \mu\} = \varepsilon(n),$$
$$\beta(0) = 1, \qquad E\varepsilon(n) = 0, \qquad E\{\varepsilon(m)\varepsilon(n)\} = \sigma^2\delta_{mn}, \tag{6.63}$$

where δ_{mn} is the Kronecker delta and we assume that

$$\sum_{k=0}^{q} \beta(k)z^k \neq 0, \qquad |z| \leqslant 1,$$

so that $x(n)$ can be represented in the form (6.44), and (6.51) is satisfied [see, e.g., Hannan (1970, Chapter I)]. This important model has already been discussed in Section 6.3.

The classical theory of inference for the model (6.63) has been developed on the basis of the assumption that the $\varepsilon(n)$ are i.i.d. r.v. This assumption is somewhat unrealistic but can sensibly be replaced by the martingale condition (6.47), which is equivalent to asking that the best linear predictor be the best predictor, both in the least squares sense. Since the autoregressive model is a linear model, such an approach seems eminently reasonable.

Under the classical assumption that the $\varepsilon(n)$ are i.i.d., the $\beta(j)$ and σ^2 are usually estimated through

$$\sum_{j=0}^{q} \hat{\beta}(j)c(k-j) = \delta_{0,k}\hat{\sigma}^2, \qquad k = 0, 1, \ldots, q$$

[see, e.g., Hannan (1970, Chapter VI)] and, remembering that $\beta(0) = 1$, we see that $\hat{\beta}(1), \ldots, \hat{\beta}(q)$ are linear functions only of the $r(j)$, $j = 1, 2, \ldots, q$, while $\hat{\sigma}^2$ is a linear function only of the $c(j)$, $j = 0, 1, \ldots, q$. It is clear from Theorem 6.6 that the estimation results continue to hold under the assumptions (6.47) and (6.49′) and that $\hat{\beta}(j) \xrightarrow{\text{a.s.}} \beta(j)$, $1 \leqslant j \leqslant q$, $\hat{\sigma}^2 \xrightarrow{\text{a.s.}} \sigma^2$, and $\bar{x} \xrightarrow{\text{a.s.}} \mu$. There has been much recent work related to these extensions and their multivariate generalizations. See, for example, Hannan (1973, 1976), Dunsmuir and Hannan (1976), Dunsmuir (1979), and Anderson and Taylor (1979).

The above estimation results can be further illuminated by providing detailed information on the rate of a.s. convergence via the use of iterated logarithm results for martingales and stationary processes. We shall, however, omit the details, which can be found in Heyde (1973, 1974d).

6.5. The Method of Moments

There are many situations in which estimation methods with known or presumed optimality properties are not tractable or even available. In such situations it is usually possible to resort to the classical *method of moments.* This technique, introduced by K. Pearson in the context of the Pearson system of distributions, consists of equating a convenient number of sample moments (or product moments) to their corresponding a.s. limits, which are functions of the unknown parameters. By considering as many such relationships as there are parameters to be estimated, and solving the resultant equations with respect to the parameters, estimates of the latter are obtained.

Estimation by this method most certainly cannot be expected to convey any optimality properties, but it is convenient and frequently leads to simple calculations.

As an illustration, let $Z_0 = 1, Z_1, Z_2, \ldots$ denote a supercritical Bienaymé–Galton–Watson branching process. Suppose that $1 < EZ_1 = m$ and $0 < \operatorname{var} Z_1 = \sigma^2 < \infty$, and that it is desired to estimate m and σ^2 on the basis of a single realization $\{Z_k, 0 \leqslant k \leqslant n+1\}$.

Since Z_{k+1} may be represented in the form

$$Z_{k+1} = Z_{1,k}^{(1)} + \cdots + Z_{1,k}^{(Z_k)}$$

where, conditional on Z_k, the $Z_{1,k}^{(i)}$ are i.i.d., each with the distribution of Z_1, we have

$$E(Z_{k+1}|Z_k) = mZ_k \quad \text{a.s.,}$$

$$E((Z_{k+1} - mZ_k)^2|Z_k) = \sigma^2 Z_k \quad \text{a.s.}$$

Now suppose that $P(Z_1 = 0) = 0$ for convenience. (If $P(Z_1 = 0) > 0$, then it is necessary to condition on nonextinction at each stage in the argument.) We note that $\{m^{-n}Z_n, n \geqslant 0\}$ is a nonnegative martingale and hence, from the martingale convergence theorem, $m^{-n}Z_n \xrightarrow{\text{a.s.}} W$ as $n \to \infty$ for some r.v. W. It is well known that W is nondegenerate and positive a.s. for the case in question [see, e.g., Harris (1963)] and hence $\hat{m} = Z_{n+1}Z_n^{-1} \xrightarrow{\text{a.s.}} m$. The method of moments suggests $\tilde{m}_n = n^{-1}\sum_{j=1}^{n} Z_{j+1}Z_j^{-1}$ as an estimator of m, and $\tilde{m}_n$ is indeed a strongly consistent estimator of m since $\hat{m}_n \xrightarrow{\text{a.s.}} m$. However, $\hat{m}_n$ is substantially superior. It can be shown, for example, that [Heyde and Leslie (1971, Theorem 2)]

$$\hat{m}_n - m = \sigma\zeta(n)(2Z_n^{-1}\log n)^{1/2}, \tag{6.64}$$

where $\zeta(n)$ has its set of a.s. limit points confined to $[-1,1]$ with $\limsup_{n\to\infty} \zeta(n) = +1$ a.s. and $\liminf_{n\to\infty} \zeta(n) = -1$ a.s. Then, summation in (6.64) shows that the rate of convergence of $\tilde{m}_n$ to m is vastly inferior to that of $\hat{m}_n$.

In the case of estimation of σ^2 there seems to be no alternative to the method of moments, which suggests the use of

$$\hat{\sigma}_n^2 = n^{-1} \sum_{k=1}^{n} (Z_{k+1} - \hat{m}_n Z_k)^2 Z_k^{-1}$$

as an estimator. This is, in fact, strongly consistent for σ^2 and, provided that $EZ_1^4 < \infty$, asymptotically normally distributed when appropriately normed.

To establish that $\hat{\sigma}_n^2 \xrightarrow{\text{a.s.}} \sigma^2$, one first observes that $\{(Z_{n+1} - mZ_n)^2 Z_n^{-1} - \sigma^2, n \geqslant 0\}$ is a martingale difference sequence, and truncation and application of Theorem 2.18 leads to

$$n^{-1} \sum_{k=1}^{n} (Z_{k+1} - mZ_k)^2 Z_k^{-1} \xrightarrow{\text{a.s.}} \sigma^2.$$

That m may be replaced by $\hat{m}_n = Z_{n+1} Z_n^{-1}$ without disturbing this a.s. convergence can be established using the result (6.64). Details are given in Heyde (1974c).

The method of moments provides the only approach of wide applicability for the estimation of parameters in the newly developing area of nonlinear time-series. Until recently the overwhelming majority of work on time-series analysis had been concerned with linear models. The theory developed has proved extremely useful, but nonlinear relationships are endemic in most scientific disciplines and need to be treated. Little success has so far been achieved in this enterprise. Part of the difficulty is, of course, the enormous range of possible models, but if sensible restrictions are imposed, many of these can be discarded. Indeed, because one of the main uses of time-series is for producing forecasts, it is reasonable to restrict consideration to non-linear models which are capable of producing forecasts. This approach has been adopted by Granger and Andersen (1978a,b) through the introduction of a property known as *invertibility*, which is described below.

We consider a model of the form

$$X_n = f(X_{n-1}, \ldots, X_{n-p}; \varepsilon_{n-1}, \ldots, \varepsilon_{n-q}) + \varepsilon_n, \tag{6.65}$$

where the ε_i are i.i.d. inputs which are unobserved and f is a known function. If this type of model is to be used for forecasting, then provided the function f itself does depend on some ε_{n-j}, $j \geqslant 1$, it is necessary to be able to estimate the unobserved values of ε from the observed values of X.

Suppose that q starting values ε_k, $k = -q+1, \ldots, 0$, for the $\{\varepsilon_n\}$ sequence are concocted and that the $\{X_n\}$ sequence is known. Let $\hat{\varepsilon}_n$, $n > 0$, be generated by

$$\hat{\varepsilon}_n = X_n - f(X_{n-1}, \ldots, X_{n-p}; \hat{\varepsilon}_{n-1}, \ldots, \hat{\varepsilon}_{n-q}),$$

and define the innovation estimated error series $\{e_n\}$ by

$$e_n = \varepsilon_n - \hat{\varepsilon}_n.$$

Granger and Andersen (1978a,b) describe the model (6.65) as *invertible* if

$$Ee_n^2 \to 0 \qquad \text{as} \quad n \to \infty.$$

Invertibility ensures the ability to forecast reasonably at least for large n. Indeed, writing $\mathscr{F}_j$ for the σ-field generated by $X_j, \ldots, X_1$, the optimal k step forecast, under the least squares criterion, of X_{n+k} given $X_n, \ldots, X_1$ is

$$E(X_{n+k}|\mathscr{F}_n) = E(\hat{\varepsilon}_{n+k}|\mathscr{F}_n) + f(X_{n+k-1}, \ldots, X_{n+k-p}; \hat{\varepsilon}_{n+k-1}, \ldots, \hat{\varepsilon}_{n+k-q})$$

and, in the case of an invertible model, $E(\hat{\varepsilon}_{n+k}|\mathscr{F}_n) \xrightarrow{\text{P}} 0$ as $n \to \infty$ since

$$\begin{aligned} E(E(\hat{\varepsilon}_{n+k}|\mathscr{F}_n))^2 &= E(E(\hat{\varepsilon}_{n+k} - \varepsilon_{n+k}|\mathscr{F}_n))^2 \\ &= E(E(e_{n+k}|\mathscr{F}_n))^2 \\ &\leqslant E(E(e_{n+k}^2|\mathscr{F}_n)) = Ee_{n+k}^2 \to 0 \end{aligned}$$

as $n \to \infty$.

Invertibility seems a sensible requirement of models and the application of the concept in the context of linear models produces no surprises. For nonlinear models the checking of invertibility is not always straightforward. An example of a noninvertible model is

$$X_n = \varepsilon_n + \alpha\varepsilon_{n-1}\varepsilon_{n-2}, \tag{6.66}$$

while the bilinear model

$$X_n = \varepsilon_n + \alpha X_{n-1}\varepsilon_{n-1} \tag{6.67}$$

is certainly invertible for $\alpha^2 EX_n^2 < 1$, which reduces, in the case where the ε_n are Gaussian, to $|\alpha(E\varepsilon_1^2)^{1/2}| < 0.605$ [Granger and Andersen (1978a, Chapter 8)].

The general techniques of estimation discussed hitherto in this chapter do not provide a convenient approach for models such as (6.66) and (6.67). If the distribution of ε_1 is known, in principle it is possible to use maximum likelihood, but the method seems intractable for the large-sample case. The method of moments, however, produces estimates without difficulty. The model (6.66) is treated via this route in Robinson (1977). We shall illustrate by treating the model (6.67).

Suppose that the ε_i are i.i.d. with $E\varepsilon_0 = 0$ and $\sigma^2 = E\varepsilon_0^2 < \infty$ while $\alpha^2\sigma^2 < 1$ to avoid exponential growth problems. If $E|\varepsilon_0|^3 < \infty$ and $E\varepsilon_0^3 = 0$, the estimation problem for α and σ^2 has a particularly simple resolution. We readily find that at least for k sufficiently large,

$$\begin{aligned} a &= EX_k = \alpha\sigma^2, \\ b &= EX_kX_{k-1} = \alpha[E\varepsilon_0^3 + 2\alpha\sigma^4], \end{aligned} \tag{6.68}$$

so that $b = 2\alpha^2\sigma^4$ if $E\varepsilon_0^3 = 0$. Then, given observations $X_1, X_2, \ldots, X_n$, we define

$$\hat{a}_n = n^{-1} \sum_{k=1}^{n} X_k, \qquad \hat{b}_n = n^{-1} \sum_{k=2}^{n} X_k X_{k-1}.$$

It is not difficult to verify that

$$\hat{a}_n \xrightarrow{\text{a.s.}} a, \qquad \hat{b}_n \xrightarrow{\text{a.s.}} b \tag{6.69}$$

as $n \to \infty$, and this provides a procedure for the estimation of α and σ^2 when $E\varepsilon_0^3 = 0$. Furthermore, if $E\varepsilon_0^4 < \infty$,

$$n^{1/2}(\hat{a}_n - a) \xrightarrow{\text{d}} N(0, \operatorname{var} X_0 + 2\operatorname{cov}(X_0, X_1)), \tag{6.70}$$

while if $E\varepsilon_0^6 < \infty$,

$$n^{1/2}(\hat{b}_n - b) \xrightarrow{\text{d}} N(0, \operatorname{var} X_0 X_1 + 2\operatorname{cov}(X_0 X_1, X_1 X_2) + 2\operatorname{cov}(X_0 X_1, X_2 X_3)). \tag{6.71}$$

The variances of the limit distributions in (6.70) and (6.71) can, of course, be expressed in terms of α and the higher moments of ε_0. For example, after some calculation (and still assuming that $E\varepsilon_0^3 = 0$), we find that

$$\operatorname{var} X_0 + 2\operatorname{cov}(X_0, X_1) = \sigma^2 + \alpha^2\sigma^4 + \alpha^2(1 - \alpha^2\sigma^2)^{-1} E\varepsilon_0^4.$$

Similar estimation procedures can be used on most simple nonlinear time-series models, although in most cases higher powers of X than the first are required. Martingale methods frequently can be used to establish strong consistency and asymptotic normality results for the estimators so obtained.

For simplicity we shall verify the results (6.69)–(6.71) in the particular case where the process $\{X_k, -\infty < k < \infty\}$ is (strictly) stationary. In any case the process approaches the stationary regime as $k \to \infty$, no matter what the starting mechanism. That the limit results continue to hold in general can be established via judicious use of the results of Chapters 2 and 3.

In the stationary case we have from repeated iteration of (6.67) that

$$X_n = \varepsilon_n + \alpha\varepsilon_{n-1}^2 + \sum_{k=2}^{\infty} \alpha^k \varepsilon_{n-k}^2 \prod_{j=0}^{k-1} \varepsilon_{n-j}, \tag{6.72}$$

which is well defined since $\sum_1^\infty |\alpha|^n (E|\varepsilon_0|)^n < \infty$ by virtue of $\alpha^2\sigma^2 < 1$ and $(E|\varepsilon_0|)^2 \leqslant E\varepsilon_0^2 = \sigma^2$. Denote by $(\Omega, \mathscr{F}, P)$ the basic probability space and let T be the ergodic measure-preserving transformation operating on points $\omega \in \Omega$ such that

$$\varepsilon_k(\omega) = \varepsilon_0(T^k\omega), \qquad -\infty < k < \infty.$$

Then it is clear from (6.72) that

$$X_k(\omega) = X_0(T^k\omega), \qquad -\infty < k < \infty,$$

and $\{X_k, -\infty < k < \infty\}$ is ergodic. The ergodic theorem, together with (6.68), then gives (6.69).

Next, we set $Y_k = X_k - EX_k$, $Z_k = X_k X_{k-1} - EX_k X_{k-1}$, $-\infty < k < \infty$, and let $\mathscr{M}_k$ denote the σ-field generated by ε_j, $j \leqslant k$. Then, it is easily seen that

$$\begin{aligned} E(Y_0 | \mathscr{M}_{-k}) &= 0, \qquad k \geqslant 2, \\ E(Z_0 | \mathscr{M}_{-k}) &= 0, \qquad k \geqslant 3, \end{aligned}$$

and the results (6.70) and (6.71) follow from Corollary 5.4. The variances of the limits are identified through

$$\begin{aligned} n^{-1} \operatorname{var}\left(\sum_1^n Y_k\right) &= n^{-1}\left[E\sum_1^n Y_k^2 + 2\sum_{j=1}^{n-1}\sum_{k=j+1}^{n} E(Y_j Y_k)\right] \\ &= n^{-1}[nEY_0^2 + 2(n-1)E(Y_0 Y_1)] \to EY_0 + 2E(Y_0 Y_1), \end{aligned}$$

and similarly

$$\begin{aligned} n^{-1} \operatorname{var}\left(\sum_1^n Z_k\right) &= n^{-1}\left[nEZ_0 + 2\sum_{j=1}^{n-1} E(Z_j Z_{j+1}) + 2\sum_{j=1}^{n-2} E(Z_j Z_{j+2})\right] \\ &\to EZ_0^2 + 2E(Z_0 Z_1) + 2E(Z_0 Z_2). \end{aligned}$$

7
Miscellaneous Applications

7.1. Exchangeable Sequences

The finite set $(X_1,X_2,\ldots,X_n)$ of random variables is said to be *exchangeable* if the joint distribution of $(X_{\pi 1},X_{\pi 2},\ldots,X_{\pi n})$ is the same as that of $(X_1,X_2,\ldots,X_n)$ for every permutation π. An infinite sequence $(X_1,X_2,\ldots)$ of random variables is said to be exchangeable if $(X_1,X_2,\ldots,X_n)$ is exchangeable for each $n \geqslant 2$. The set $(A_1,A_2,\ldots)$ of events is called exchangeable if the corresponding indicators $(I(A_1),I(A_2),\ldots)$ are exchangeable. The terminology *interchangeable* or *symmetrically dependent* is also sometimes used.

There have been many important uses of finite exchangeable sequences, particularly in conjunction with combinational arguments for random walk and ballot problems [e.g., Feller (1971)], but our emphasis here is on infinite sequences and properties which are consequent on limiting behavior.

It should be emphasized that finite exchangeable sets cannot necessarily be extended to infinite exchangeable sequences. For example [Galambos (1978, pp. 127–128)], let X_1, X_2, X_3 be r.v. which can take only the values 0 and 1. Let

$$P(X_j = 1) = \tfrac{1}{2}, \qquad j = 1,2,3,$$

and

$$P(X_1 = X_2 = 1) = P(X_1 = X_3 = 1) = P(X_2 = X_3 = 1) = 0.2.$$

It is easily checked that (X_1,X_2,X_3) is an exchangeable set and we suppose that it can be extended to a set $(X_1,X_2,\ldots,X_n)$ of exchangeable variables. Then,

$$\begin{aligned} 0 &\leqslant E\left(\sum_{i=1}^{n} I(X_i = 1)\right)^2 - \left(E\sum_{i=1}^{n} I(X_i = 1)\right)^2 \\ &= \sum_{i=1}^{n}\sum_{j=1}^{n} P(X_i = 1,\, X_j = 1) - \tfrac{1}{4}n^2 \\ &= \tfrac{1}{2}n + \sum\sum_{i\neq j} P(X_i = 1,\, X_j = 1) - \tfrac{1}{4}n^2 \\ &= \tfrac{1}{2}n + (0.2)n(n-1) - \tfrac{1}{4}n^2 = (0.3)n - (0.05)n^2, \end{aligned}$$

from which we deduce that $n \leqslant 6$. Note that in this example $\operatorname{corr}(X_1,X_2) < 0$, while the variables in an infinite exchangeable sequence are necessarily nonnegatively correlated. This last result can be seen from an examination of the terms in $\operatorname{var}(\sum_{i=1}^{n} X_i)$.

The question of extension is an important issue in, for example, the analysis of extreme values for dependent variables. If A_i, $i = 1, 2, \ldots, n$, are any events, then there is a corresponding set $(X_1,X_2,\ldots,X_n)$ of exchangeable variables, each taking only the values 0 or 1, such that if N_n is the number of A_i, $1 \leqslant i \leqslant n$, which occur,

$$P(N_n = k) = P\left(\sum_{i=1}^{n} X_i = k\right), \qquad 0 \leqslant k \leqslant n$$

[Kendall (1967); Galambos (1978, Chapter 3)].

An important review of exchangeability and its uses has been provided by Kingman (1978), and we follow his treatment of the central result, known as de Finetti's theorem, in the discussion below.

We shall say that a random variable is n-symmetric if it is a function of $(X_1,X_2,\ldots)$ which is unchanged under any permutation of the first n variables. For example,

$$X_1X_2X_3 + X_4X_6$$

is 3-symmetric but not 4-symmetric. Let $\mathscr{F}_n$ denote the σ-field generated by the n-symmetric random variables, and note that $\mathscr{F}_{n+1} \subseteq \mathscr{F}_n$.

Suppose that f is a measurable function for which

$$E|f(X_1)| < \infty.$$

If $A \in \mathscr{F}_n$, the exchangeability property gives

$$E\{f(X_j)I(A)\} = E\{f(X_1)I(A)\}$$

for $1 \leqslant j \leqslant n$, and hence

$$E\left\{n^{-1} \sum_{j=1}^{n} f(X_j)I(A)\right\} = E\{f(X_1)I(A)\}.$$

This last result holds for any $A \in \mathscr{F}_n$ and, since $n^{-1}\sum_{j=1}^{n} f(X_j)$ is $\mathscr{F}_n$-measurable,

$$n^{-1} \sum_{j=1}^{n} f(X_j) = E\{f(X_1)|\mathscr{F}_n\} \quad \text{a.s.}$$

Then applying the martingale convergence theorem (Theorem 2.5) to the martingale $\{E(f(X_1)|\mathscr{F}_n),\mathscr{F}_n, n \geqslant 1\}$, we have

$$\lim_{n\to\infty} n^{-1} \sum_{j=1}^{n} f(X_j) = E\{f(X_1)|\mathscr{F}_\infty\} \quad \text{a.s.,} \tag{7.1}$$

where $\mathscr{F}_\infty = \bigcap_{n=1}^{\infty} \mathscr{F}_n$.

The strong law of large numbers given by (7.1) is also an immediate consequence of the Birkhoff–von Neumann ergodic theorem (with $\mathscr{F}_\infty$ replaced by the invariant σ-field; see Appendix IV), but the argument just given has an important generalization which is utilized below. Note that if f is the indicator function of the interval $(-\infty,x]$, then (7.1) gives

$$\lim_{n\to\infty} n^{-1} \sum_{j=1}^{n} I(X_j \leqslant x) = F(x) \quad \text{a.s.,} \tag{7.2}$$

where

$$F(x) = P(X_1 \leqslant x | \mathscr{F}_\infty) \tag{7.3}$$

is a conditional distribution function.

Next we observe that the argument leading to (7.1) extends at once to give, for $n \geqslant k$,

$$[n(n-1)\cdots(n-k+1)]^{-1} \sum \prod_{i=1}^{k} I(X_{j_i} \leqslant x_i) = E\left\{\prod_{i=1}^{k} I(X_i \leqslant x_i) \middle| \mathscr{F}_n\right\},$$

where the summation extends over distinct $j_1, j_2, \ldots, j_n$ with $1 \leqslant j_i \leqslant n$, $i = 1, 2, \ldots, n$. Thus, again with the aid of the martingale convergence theorem, and observing that the contribution from terms with coincidences among the j_i is of order n^{k-1},

$$\begin{aligned} E\left\{\prod_{i=1}^{k} I(X_i \leqslant x_i) \middle| \mathscr{F}_\infty\right\} &= \lim_{n\to\infty} [n(n-1)\cdots(n-k+1)]^{-1} \sum \prod_{i=1}^{k} I(X_{j_i} \leqslant x_i) \\ &= \lim_{n\to\infty} n^{-k} \prod_{i=1}^{k} \sum_{j_i=1}^{n} I(X_{j_i} \leqslant x_i) \\ &= \prod_{i=1}^{k} F(x_i) \quad \text{a.s.,} \end{aligned} \tag{7.4}$$

using (7.2) and (7.3). This provides one form of a result which is known as de Finetti's theorem.

Theorem 7.1. *Let $\{X_1, X_2, \ldots\}$ form an exchangeable sequence. Then there is a σ-field conditional on which the X_j are independent and identically distributed.*

Theorem 7.1 can be stated in a number of different forms. Clearly $F(x)$ is for each x an $\mathscr{F}_\infty$-measurable random variable. Then, conditioning on the minimal sub-σ-field of $\mathscr{F}_\infty$ with respect to which each $F(x)$ is measurable, we have, in an obvious notation,

$$\begin{aligned} &P(X_1 \leqslant x_1, X_2 \leqslant x_2, \ldots, X_k \leqslant x_k | F) \\ &\quad = E\{P(X_1 \leqslant x_1, X_2 \leqslant x_2, \ldots, X_k \leqslant x_k | \mathscr{F}_\infty) | F\} = \prod_{i=1}^{k} F(x_i) \quad \text{a.s.} \end{aligned}$$

From this form we see that every infinite exchangeable sequence can be constructed by beginning with a sequence of i.i.d. r.v. having the same distribution function F, and then allowing F to vary randomly. The randomization in general destroys the independence while preserving exchangeability. For example, this is the framework which appears in the context of Bayesian statistics, where the parameters of a distribution are considered to be random variables. From this viewpoint, $X_1, X_2, \ldots$ are i.i.d. for a given value of the parameter λ of the common distribution function $F(\cdot,\lambda)$, say. Then, for each $n \geqslant 1$,

$$P\left(\bigcap_{i=1}^{n} \{X_i \leqslant x_i\}\right) = E \prod_{i=1}^{n} F(x_i,\lambda),$$

from which the exchangeability property follows.

There is a further variant on Theorem 7.1 which is intuitively appealing. The conditioning by the σ-field $\mathscr{F}_\infty$, or by the conditional distribution function F, can be replaced by conditioning on a single random variable. There is a random variable Y such that, conditional on Y, the X_j are independent, each with the distribution of X_1 conditional on Y. This variant was first established by Dynkin (1953), and the forms are all equivalent since the σ-fields generated by Y and by F are both identical with $\mathscr{F}_\infty$ up to events of probability zero or one [Olshen (1974)]. Olshen has also shown that Y is essentially unique in the sense that, if the X_j are conditionally independent given Z, then a version of Z is measurable with respect to the σ-field generated by Y. Furthermore, the σ-field generated by Y also corresponds to the invariant and tail σ-fields up to events of zero probability. Various aspects of the choice of conditioning are also discussed in Chow and Teicher (1978).

The random variable Y has a very simple interpretation in the case where the X_j can take only the values 0 or 1. The following theorem is essentially Kendall's (1967) modification of Rényi and Révész (1963).

Theorem 7.2. *If the variables in the exchangeable sequence* $(X_1,X_2,\ldots)$ *assume only the values* 0 *or* 1, *there is a random variable* Y *whose distribution is concentrated on* $[0,1]$ *such that*

$$P\left(\sum_{j=1}^{n} X_j = k \middle| Y\right) = \binom{n}{k} Y^k(1 - Y)^{n-k} \quad a.s. \tag{7.5}$$

and

$$Y = \lim_{n\to\infty} n^{-1} \sum_{j=1}^{n} X_j \quad a.s. \tag{7.6}$$

Proof. Using Theorem 7.1, there exists a σ-field $\mathscr{F}_\infty$ such that

$$P(X_1 = 1, \ldots, X_k = 1, X_{k+1} = 0, \ldots, X_n = 0 | \mathscr{F}_\infty) = (P(X_1 = 1|\mathscr{F}_\infty))^k (P(X_1 = 0|\mathscr{F}_\infty))^{n-k}.$$

Now set $Y = P(X_1 = 1|\mathscr{F}_\infty)$. Then, since Y is $\mathscr{F}_\infty$-measurable,

$$\begin{aligned} P(X_1 = 1, \ldots, X_k = 1, X_{k+1} = 0, \ldots, X_n = 0|Y) &= E(P(X_1 = 1, \ldots, X_k = 1, X_{k+1} = 0, \ldots, X_n = 0|\mathscr{F}_\infty)|Y) \\ &= E(Y^k(1-Y)^{n-k}|Y) = Y^k(1-Y)^{n-k} \quad \text{a.s.,} \end{aligned}$$

and the result (7.5) is simply deduced in view of the exchangeability property. The result (7.6) follows immediately from the stong law (7.1). ∎

Theorem 7.2 has a neat application in connection with the Polya urn scheme. For this scheme an urn originally contains b black and r red balls. Balls are drawn at random from the urn and after each drawing the ball is returned together with c balls of the colour drawn. Set $X_n = 1$ or 0 according as the nth drawing produces a ball which is black or red. It is easily checked that $(X_1, X_2, \ldots)$ is an exchangeable sequence. Now,

$$\begin{aligned} P(X_1 = 1, X_2 = 1, \ldots, X_n = 1) &= \frac{b(b+c)\cdots(b+(n-1)c)}{(b+r)(b+r+c)\cdots(b+r+(n-1)c)} \\ &= \frac{\Gamma(n+b/c)\Gamma((b+r)/c)}{\Gamma(n+(b+r)/c)\Gamma(b/c)} = EY^n, \end{aligned}$$

$n = 1, 2, \ldots,$ using Theorem 7.2. Furthermore, every distribution on a bounded interval is characterized by its moments [e.g., Feller (1971, pp. 225–227)]. A proof of this result, which is due to Hausdorff, can be based on exchangeability; see Rényi (1970, pp. 320–323). Inspection then shows that Y has a beta distribution with parameters b/c and r/c.

7.2. Limit Laws for Subsequences of Sequences of Random Variables

Motivated by a problem posed by Steinhaus and work of Austin, Rényi, and Révész, Komlós (1967) established the following remarkable theorem.

Theorem 7.3. *Let $\{X_n, n \geqslant 1\}$ be a sequence of random variables such that $\sup_{n\geqslant 1} E|X_n| < \infty$. Then there is a (nonrandom) subsequence $\{X_{n_k}, k \geqslant 1\}$, and a random variable X with $E|X| < \infty$, such that*

$$\lim_{n\to\infty} n^{-1} \sum_{k=1}^{n} X_{n_k} = X \quad a.s.$$

Theorem 7.3 was generalized by Chatterji (1970) to the case where $\sup_{n\geq 1} E|X_n|^p < \infty$ for fixed $p, 0 < p < 2$. Chatterji established the existence of a subsequence $\{X_{n_k}\}$ and a random variable X with $E|X|^p < \infty$ such that

$$\lim_{n\to\infty} n^{-1/p} \sum_{k=1}^{n} (X_{n_k} - X) = 0 \quad \text{a.s.}$$

Corresponding central limit and iterated logarithm laws for subsequences were subsequently proved. If $\sup_{n\geq 1} EX_n^2 < \infty$, then there exists a subsequence $\{X_{n_k}\}$ and random variables X with $EX^2 < \infty$ and $Y\,(\geq 0)$ with $E|Y| < \infty$ such that

$$\text{(i)} \qquad n^{-1/2} \sum_{k=1}^{n} (X_{n_k} - X) \xrightarrow{\text{d}} Y' N(0,1),$$

where Y' and $N(0,1)$ are independent, Y' having the distribution of Y [Chatterji (1974b)], and

$$\text{(ii)} \qquad \limsup_{n\to\infty} (2n \log\log n)^{-1/2} \sum_{k=1}^{n} (X_{n_k} - X) = Y \qquad \text{a.s.},$$

$$\liminf_{n\to\infty} (2n \log\log n)^{-1/2} \sum_{k=1}^{n} (X_{n_k} - X) = -Y \quad \text{a.s.}$$

[Berkes (1974)].

The same basic approach was used to prove all these results. Broadly, the idea was to first reduce from a general sequence $\{X_n\}$ to the case of a martingale difference sequence. Each of the theorems under consideration has, of course, a version for martingale differences. It was then shown that an arbitrary martingale difference sequence has a subsequence satisfying the appropriate conditions. A separate proof was required for each case. The results, however, suggested the following heuristic principle (Chatterji, 1974a): given a limit theorem for independent and identically distributed random variables (i.i.d. r.v.) under certain moment conditions, there exists an analogous theorem such that a tight but arbitrarily dependent sequence, which uniformly satisfies the same moment conditions, always contains a (nonrandom) subsequence satisfying this analogous theorem.

Any such general principle implicitly involves exchangeable sequences. Suppose that a limit theorem holds for some subsequence of every random sequence satisfying certain moment conditions. Then it holds for some subsequence of every exchangeable sequence which satisfies the moment conditions and, by the selection property, this subsequence has the same probabilistic structure as the original sequence. Thus, the limit theorem holds for every exchangeable sequence satisfying the appropriate uniform moment conditions. On the other hand, de Finetti's theorem (Theorem 7.1) shows that any limit theorem for i.i.d r.v. has an analog for exchangeable sequences.

The challenge of establishing a general result to embrace all the known special cases was partly met by Aldous (1977a,b). He made the heuristic principle precise in the case where the property is an "a.s. limit theorem" via the introduction of the concept of a "statute," defined below.

Let $\mathscr{P}(\mathbb{R})$ denote the space of probability measures on the real line $\mathbb{R}$. For $\lambda \in \mathscr{P}(\mathbb{R})$ write

$$|\lambda|_1 = \int x\lambda(dx)$$

with $|\lambda|_1 = \infty$ if $\int |x|\lambda(dx) = \infty$, and

$$|\lambda|_2 = \int (x - |\lambda|_1)^2 \lambda(dx)$$

with $|\lambda|_2 = \infty$ if $\int x^2\lambda(dx) = \infty$. Define a *statute* to be a measurable subset A of $\mathscr{P}(\mathbb{R}) \times \mathbb{R}^\infty$ such that, whenever $\lambda \in \mathscr{P}(\mathbb{R})$ and X_n, $n \geqslant 1$, are i.i.d. with distribution given by λ, then $(\lambda, X_1(\omega), X_2(\omega), \ldots) \in A$ a.s., the $\omega \in \Omega$ referring to a point in our probability space. For example, the statutes corresponding to the strong law of large numbers and the law of the iterated logarithm are

$$A_1 = \left\{ (\lambda, \mathbf{x}) : \lim_{n\to\infty} n^{-1} \sum_{j=1}^{n} x_i = |\lambda|_1 \text{ or } |\lambda|_1 = \infty \right\}$$

and

$$A_2 = \left\{ (\lambda, \mathbf{x}) : \limsup_{n\to\infty} (2n \log\log n)^{-1/2} \left(\sum_{i=1}^{n} x_i - n|\lambda|_1 \right) = |\lambda|_2^{1/2} \text{ or } |\lambda|_2 = \infty \right\},$$

respectively. Here $\mathbf{x} = (x_1, x_2, \ldots)$ denotes a generic point in $\mathbb{R}^\infty$. It seems clear that any "a.s. limit theorem" for i.i.d. r.v. may be represented by a statute.

The approach used by Aldous was to extract a subsequence which is close, in a certain sense, to an exchangeable sequence $\{Z_i\}$. Then for each such $\{Z_i\}$ there is a random measure $\mu: \Omega \to \mathscr{P}(\mathbb{R})$ such that for each statute A, $(\mu(\omega), Z_1(\omega), Z_2(\omega), \ldots) \in A$ a.s. The random measure μ is the one for which the regular conditional distribution of $(Z_1, Z_2, \ldots)$ given $\mathscr{F}\{\mu\}$, the σ-field generated by μ, is $\mu^* = \mu \times \mu \times \cdots$. Sometimes μ is called the *canonical random measure for* $\{Z_i\}$.

The principal result of Aldous (1977b) is given in the following theorem.

Theorem 7.4. *Let $\{X_n, n \geqslant 1\}$ be a tight sequence of random variables defined on a probability space $(\Omega, \mathscr{F}, P)$ and suppose that A is a statute for which $(\lambda, \mathbf{x}) \in A$ and $\sum |x_k' - x_k| < \infty$ implies $(\lambda, \mathbf{x}') \in A$. Then there exists a random* *measure* $\mu:\Omega \to \mathscr{P}(\mathbb{R})$ *and a subsequence* $\{X_{m_k}\}$ *all of whose subsequences* $\{X_{n_k}\}$ *satisfy* $(\mu(\omega), X_{n_1}(\omega), X_{n_2}(\omega), \ldots) \in A$ *a.s.*

Before this theorem can be applied it is necessary to note that $E|\mu(\omega)|^p \leqslant \limsup E|X_n|^p$, $0 < p < \infty$ [Lemma 2 of Aldous (1977b)]. Then, for example, application to the statutes A_1 and A_2 gives the results of Komlós (1967) and Berkes (1974), respectively. Aldous (1977b, Theorem 6) has also provided a result for weak limit theorems. It covers the subsequence form of the Donsker invariance principle for the central limit theorem, and hence the result of Chatterji (1974b), but its scope is not comparable with that of Theorem 7.4. There are, in addition, other convergence theorems for subsequences not covered by the above-mentioned results, for example, ones of Gaposhkin (1972) on L^p-convergence of series.

Although Aldous' methods rest on exchangeability, martingale methods are important in his proofs. These are, however, quite technical and somewhat beyond our scope. We shall merely provide a proof of Theorem 7.3 in which the use of martingale methods is central. We begin with a lemma.

Lemma 7.1. *If* $\sup_{n \geqslant 1} E|X_n| \leqslant K < \infty$ *and* $\{X_n I(|X_n| \leqslant k), n \geqslant 1\}$ *converges weakly in* L^1 *for each* $k \geqslant 1$ *to* η_k, *say, then there exists an integrable r.v.* η *such that* $\eta_k \xrightarrow{\text{a.s.}} \eta$ *and* $E|\eta_k - \eta| \to 0$ *as* $k \to \infty$.

Proof. Setting $\eta_0 = 0$ we have for $k \geqslant 1$,

$$X_n I(k-1 < |X_n| \leqslant k) \to \eta_k - \eta_{k-1} \quad (\text{weakly in } L^1).$$

(Chapter 3 contains a discussion of weak L^1-convergence.) Further, choosing $\theta = 1$ if $\eta_k - \eta_{k-1} \geqslant 0$, -1 otherwise, we have from weak L^1-convergence that

$$E[\theta X_n I(k-1 < |X_n| \leqslant k)] \to E[(\eta_k - \eta_{k-1})\theta] = E|\eta_k - \eta_{k-1}|.$$

But

$$|E[\theta X_n I(k-1 < |X_n| \leqslant k)]| \leqslant E[|X_n| I(k-1 < |X_k| \leqslant k)]$$

and hence, using Fatou's lemma,

$$E|\eta_k - \eta_{k-1}| \leqslant \liminf_{n\to\infty} E[|X_n| I(k-1 < |X_k| \leqslant k)]. \tag{7.7}$$

Now, for any integer N there exists an integer $m = m(N)$ such that

$$\begin{aligned}\sum_{k=1}^{n} \liminf_{n\to\infty} E[|X_n| I(k-1 < |X_n| \leqslant k)] &\leqslant \sum_{k=1}^{n} E[|X_m| I(k-1 < |X_m| \leqslant k)] + 1 \\ &\leqslant E|X_m| + 1,\end{aligned}$$

and hence

$$\sum_{k=1}^{N} \liminf_{n\to\infty} E[|X_n| I(k-1 < |X_n| \leqslant k)] \leqslant \sup_{n\geqslant 1} E|X_n| + 1.$$

This last result holds for each integer N, so that

$$\sum_{k=1}^{\infty} \liminf_{n\to\infty} E[|X_n| I(k-1 < |X_n| \leqslant k)] < \infty$$

and hence, in view of (7.7),

$$\sum_{k=1}^{\infty} E|\eta_k - \eta_{k-1}| < \infty. \tag{7.8}$$

The result (7.8) ensures the almost sure convergence of

$$\eta_n = \sum_{k=1}^{n} (\eta_k - \eta_{k-1})$$

to η, say, as $n \to \infty$. That

$$E|\eta_n - \eta| \to 0$$

as $n \to \infty$ also follows from (7.8) since

$$E|\eta_n - \eta| = E\left|\sum_{k=n+1}^{\infty} (\eta_k - \eta_{k-1})\right| \leqslant \sum_{k=n+1}^{\infty} E|\eta_k - \eta_{k-1}| \to 0. \quad \blacksquare$$

Proof of Theorem 7.3. Clearly $\{X_n I(|X_n| \leqslant k), n \geqslant 1\}$ is weakly compact in the space of square-integrable random variables (being bounded). Hence, using the Bolzano–Weierstrass theorem, and the usual diagonal choice procedure, we can choose a subsequence

$$\{X_{n_j} I(|X_{n_j}| \leqslant k), j = 1, 2, \ldots\}$$

such that for $k = 1, 2, \ldots,$

$$X_{n_j} I(|X_{n_j}| \leqslant k) \to \beta_k \quad \text{(weakly in } L^1) \tag{7.9}$$

as $j \to \infty$. Note that in view of Lemma 7.1, there exists an integrable random variable β such that $\beta_k \xrightarrow{\text{a.s.}} \beta$ as $k \to \infty$.

Next, we make further use of the Bolzano–Weierstrass theorem and the diagonal choice procedure to choose a subsequence $\{X_{r_j}\}$ of $\{X_{n_j}\}$ such that

$$\lim_{n\to\infty} P(k-1 < |X_{r_n}| \leqslant k) = a_k, \qquad k = 1, 2, \ldots, \tag{7.10}$$

and

$$P(k-1 < |X_{r_n}| \leqslant k) \leqslant a_k + 2^{-k} \qquad \text{for} \quad 0 \leqslant k \leqslant n^2. \tag{7.11}$$

Then, since

$$\infty > K = \sup_{n \geqslant 1} E|X_n| \geqslant E|X_{r_n}| \geqslant \sum_{k=1}^{\infty} (k-1)P(k-1 < |X_{r_n}| \leqslant k),$$

we have, using Fatou's lemma (for series),

$$\sum_{k=1}^{\infty} (k-1)a_k \leqslant \liminf_{n \to \infty} \sum_{k=1}^{\infty} (k-1)P(k-1 < |X_{r_n}| \leqslant k) \leqslant K,$$

so that

$$\sum_{1}^{\infty} k a_k < \infty. \tag{7.12}$$

Now,

$$\begin{aligned} P(|X_{r_k}| > k) &\leqslant \sum_{k<j<k^2} P(j-1 < |X_{r_k}| \leqslant j) + P(|X_{r_k}| \geqslant k^2) \\ &\leqslant \sum_{k<j<k^2} (a_j + 2^{-j}) + k^{-2}K \end{aligned}$$

by (7.11) and Markov's inequality. Thus,

$$\begin{aligned} \sum_{k=1}^{\infty} P(|X_{r_k}| > k) &\leqslant \sum_{k=1}^{\infty} \sum_{k<j<k^2} (a_j + 2^{-j}) + K \sum_{k=1}^{\infty} k^{-2} \\ &\leqslant \sum_{j=1}^{\infty} (a_j + 2^{-j}) \sum_{j^{1/2}<k<j} 1 + K \sum_{k=1}^{\infty} k^{-2} \\ &\leqslant \sum_{j=1}^{\infty} j(a_j + 2^{-j}) + K \sum_{k=1}^{\infty} k^{-2} < \infty \end{aligned} \tag{7.13}$$

in view of (7.12).

Further, as above,

$$\begin{aligned} E[X_{r_k}^2 I(|X_{r_k}| \leqslant k)] &\leqslant \sum_{0 \leqslant j \leqslant k} j^2 P(j-1 < |X_{r_k}| \leqslant j) \\ &\leqslant \sum_{0 \leqslant j \leqslant k} j^2 (a_j + 2^{-j}) \end{aligned}$$

by (7.11), so that

$$\begin{aligned} \sum_{k=1}^{\infty} k^{-2} E[X_{r_k}^2 I(|X_{r_k}| \leqslant k)] &\leqslant \sum_{k=1}^{\infty} k^{-2} \sum_{0 \leqslant j \leqslant k} j^2 (a_j + 2^{-j}) \\ &= \sum_{j=1}^{\infty} j^2 (a_j + 2^{-j}) \sum_{k \geqslant j} k^{-2} \\ &\leqslant 2 \sum_{j=1}^{\infty} j(a_j + 2^{-j}) < \infty, \end{aligned} \tag{7.14}$$

again in view of (7.12). We note also that (7.13) and (7.14) continue to hold for any subsequence $\{X_{t_k}\}$ of $\{X_{r_k}\}$, namely,

$$\sum_{k=1}^{\infty} P(|X_{t_k}| > k) < \infty, \qquad \sum_{k=1}^{\infty} k^{-2}E[X_{t_k}^2 I(|X_{t_k}| \leqslant k)] < \infty. \qquad (7.15)$$

Finally, we proceed to choose a further subsequence $\{X_{m_j}\}$ of $\{X_{r_j}\}$. Set $m_1 = r_1$ and

$$\gamma_{m_1} = X_{r_1} I(|X_{r_1}| \leqslant 1) - \beta_1.$$

Also, define the function $G_a(\cdot)$ by

$$G_a(x) = k/a \qquad \text{if} \quad k/a \leqslant x < (k+1)/a, \qquad k = 0, \pm 1, \pm 2, \ldots,$$

and set

$$\gamma'_{m_1} = G_{1/2}(\gamma_{m_1}).$$

Note that γ'_{m_1} takes only finitely many values, so that $\mathscr{F}(\gamma'_{m_1})$, the σ-field generated by γ'_{m_1}, contains only finitely many sets.

We set

$$\varepsilon_1 = \min\{P(A)\colon P(A) > 0,\, A \in \mathscr{F}(\gamma'_{m_1})\}$$

and let m_2 be the smallest $r_j > m_1$ for which

$$\max_{\{A: P(A) > 0,\, A \in \mathscr{F}(\gamma'_{m_1})\}} E[I(A)(X_{r_j} I(|X_{r_j}| \leqslant 2) - \beta_2)]| \leqslant \varepsilon_1/2.$$

In view of (7.9), m_2 is well defined. Then set

$$\gamma_{m_2} = X_{m_2} I(|X_{m_2}| \leqslant 2) - \beta_2, \qquad \gamma'_{m_2} = G_{\varepsilon_1/2^2}(\gamma_{m_2}).$$

Note that γ'_{m_1} and γ'_{m_2} take on only finitely many values, so that $\mathscr{F}(\gamma'_{m_1}, \gamma'_{m_2})$, the σ-field generated by γ'_{m_1}, and γ'_{m_2}, contains only finitely many sets. Put

$$\varepsilon_2 = \min\{P(A)\colon P(A) > 0,\, A \in \mathscr{F}(\gamma'_{m_1}, \gamma'_{m_2})\},$$

let m_3 be the smallest $r_j > m_2$ such that

$$\max_{\{A: P(A) > 0,\, A \in \mathscr{F}(\gamma'_{m_1}, \gamma'_{m_2})\}} |E[I(A)(X_{r_j} I(|X_{r_j}| \leqslant 3) - \beta_3)]| \leqslant \varepsilon_2/3,$$

and set

$$\gamma_{m_3} = X_{m_3} I(|X_{m_3}| \leqslant 3) - \beta_3, \qquad \gamma'_{m_3} = G_{\varepsilon_2/8}(\gamma_{m_3}).$$

Generally, we define

$$\varepsilon_n = \min\{P(A)\colon P(A) > 0,\, A \in \mathscr{F}(\gamma'_{m_1}, \ldots, \gamma'_{m_{n-1}})\},$$

let m_n be the smallest $r_j > m_{n-1}$ such that

$$\max_{\{A: P(A) > 0,\, A \in \mathscr{F}(\gamma'_{m_1}, \ldots, \gamma'_{m_{n-1}})\}} |E[I(A)(X_{r_j} I(|X_{r_j}| \leqslant n) - \beta_n)]| \leqslant \varepsilon_{n-1}/n,$$

and set

$$\gamma_{m_n} = X_{m_n} I(|X_{m_n}| \leqslant n) - \beta_n, \qquad \gamma'_{m_n} = G_{\varepsilon_{n-1}/2^n}(\gamma_{m_n}).$$

We observe that, writing $A = \{\gamma_{m_n} > 0\}$ and A^c for the complement of A,

$$\begin{aligned} |E[X_{m_n} I(|X_{m_n}| \leqslant n)\beta_n] - E\beta_n^2| &\leqslant E(|\beta_n \gamma_{m_n}|) \\ &\leqslant n[E(I(A)\gamma_{m_n}) - E(I(A^c)\gamma_{m_n})] \\ &\leqslant 2n\varepsilon_{n-1}/n \leqslant 2, \end{aligned}$$

so that

$$\begin{aligned} E\gamma_{m_n}^2 &\leqslant E[X_{m_n}^2 I(|X_{m_n}| \leqslant n)] + 2E\beta_n^2 - 2E[\beta_n X_{m_n} I(|X_{m_n}| \leqslant n)] \\ &\leqslant E[X_{m_n}^2 I(|X_{m_n}| \leqslant n)] + 4. \end{aligned} \tag{7.16}$$

Our construction gives

$$0 \leqslant \gamma_{m_k} - \gamma'_{m_k} \leqslant 2^{-k}\varepsilon_{k-1} \leqslant 2^{-k}, \tag{7.17}$$

so that

$$E(\gamma'_{m_k})^2 \leqslant E\gamma_{m_k}^2 \leqslant E[X_{m_k}^2 I(|X_{m_k}| \leqslant k)] + 4$$

in view of (7.16). Hence, using (7.15),

$$\sum_{k=1}^{\infty} k^{-2} E(\gamma'_{m_k})^2 < \infty,$$

and the martingale strong law of Theorem 2.18 gives

$$\lim_{n\to\infty} n^{-1} \sum_{k=1}^{n} [\gamma'_{m_k} - E(\gamma'_{m_k}|\gamma'_{m_1}, \ldots, \gamma'_{m_{k-1}})] = 0 \quad \text{a.s.} \tag{7.18}$$

Now, writing $\mathscr{F}_{k-1}$ for $\mathscr{F}(\gamma'_{m_1}, \ldots, \gamma'_{m_{k-1}})$, the σ-field generated by $\gamma'_{m_1}, \ldots, \gamma'_{m_{k-1}}$, we have a.s.

$$\begin{aligned} |E(\gamma'_{m_k}|\mathscr{F}_{k-1})| &\leqslant \max_{\{A\in\mathscr{F}_{k-1},\, P(A)>0\}} |E(\gamma'_{m_k}|A)| \\ &= \max_{\{A\in\mathscr{F}_{k-1},\, P(A)>0\}} [P(A)]^{-1} |E(I(A)\gamma'_{m_k})| \\ &\leqslant \varepsilon_{k-1}^{-1} \max_{\{A\in\mathscr{F}_{k-1},\, P(A)>0\}} |E(I(A)\gamma'_{m_k})| \\ &\leqslant \varepsilon_{k-1}^{-1} \max_{\{A\in\mathscr{F}_{k-1},\, P(A)>0\}} |E(I(A)\gamma_{m_k})| \\ &\quad + \varepsilon_{k-1}^{-1} \max_{\{A\in\mathscr{F}_{k-1},\, P(A)>0\}} |E[I(A)(\gamma_{m_k} - \gamma'_{m_k})]| \\ &\leqslant (k\varepsilon_{k-1})^{-1}\varepsilon_k + \varepsilon_{k-1}^{-1} E|\gamma_{m_k} - \gamma'_{m_k}| \\ &\leqslant 2k^{-1}, \end{aligned}$$

so that

$$E(\gamma'_{m_k}|\mathscr{F}_{k-1}) \xrightarrow{\text{a.s.}} 0$$

as $k \to \infty$, and hence

$$\lim_{n\to\infty} n^{-1} \sum_{k=1}^{n} E(\gamma'_{m_k}|\mathscr{F}_{k-1}) = 0 \quad \text{a.s.} \tag{7.19}$$

From (7.18) and (7.19) we obtain

$$\lim_{n\to\infty} n^{-1} \sum_{k=1}^{n} \gamma'_{m_k} = 0 \quad \text{a.s.}$$

and thus, in view of (7.17),

$$\lim_{n\to\infty} n^{-1} \sum_{k=1}^{n} \gamma_{m_k} = 0 \quad \text{a.s.} \tag{7.20}$$

But recalling the definition of $\{\gamma_{m_k}\}$ and that $\beta_n \xrightarrow{\text{a.s.}} \beta$ as $n \to \infty$, we have from (7.20) that

$$\lim_{n\to\infty} n^{-1} \sum_{k=1}^{n} X_{m_k} I(|X_{m_k}| \leqslant k) = \beta \quad \text{a.s.} \tag{7.21}$$

Furthermore, from (7.15),

$$P(|X_{m_k}| > k \text{ i.o.}) = 0,$$

and hence (7.21) gives

$$\lim_{n\to\infty} n^{-1} \sum_{k=1}^{n} X_{m_k} = \beta \quad \text{a.s.,} \tag{7.22}$$

are required. This completes the proof. ∎

7.3. Limit Laws for Subadditive Processes

The concept of a subadditive random process was introduced by Hammersley and Welsh (1965) as a tool for discussing problems of percolation in lattices and networks. Their paper contains a number of interesting applications and has provided motivation for much recent work; see, for example, Smythe and Wierman (1978) and references therein.

Hammersley and Welsh conjectured, and went some way toward proving, an ergodic theorem for subadditive processes, a result which was finally established by Kingman (1968). Kingman obtained a complete generalization to subadditive processes of the Birkhoff–von Neumann ergodic theorem for stationary random processes (see Appendix IV). He also provided surveys containing many interesting examples in Kingman (1973, 1976). Our development closely follows Kingman's work.

A *subadditive process* is a family of real valued random variables x_{st}, all defined on the same probability space $(\Omega,\mathscr{F},P)$, where the indices s and t $(s < t)$ run over the set I of nonnegative integers, which satisfy the following three axioms.

(S_1) Whenever $s < t < u$,

$$x_{su} \leqslant x_{st} + x_{tu}. \tag{7.23}$$

(S_2) The joint distributions of $\{x_{st}\}$ are the same as those of $\{x_{s+1,t+1}\}$.

(S_3) The expectation

$$g_t = Ex_{0t} \tag{7.24}$$

exists and satisfies

$$g_t \geqslant -At \tag{7.25}$$

for some constant A and all integers $t > 1$.

The subadditivity assumption (S_1) is the basic axiom. (S_2) is a condition of stationarity and (S_3) brings the random variables into L^1, as is certainly necessary for an ergodic theorem. The assumption (S_2) is sometimes inappropriate and needs to be weakened. A discussion of various interesting examples in this category has been provided by Hammersley (1974). We shall merely note that a convenient replacement for the ergodic theorem in this more general context has yet to be established.

From (S_2) we have

$$Ex_{st} = g_{t-s} \tag{7.26}$$

for all $s < t$, so that, taking expectations in (7.23),

$$g_{u-s} \leqslant g_{t-s} + g_{u-t}$$

or

$$g_{m+n} \leqslant g_m + g_n \qquad (m, n \geqslant 1). \tag{7.27}$$

We fix a positive integer k. Then for any $r \geqslant 1$ and $1 \leqslant s \leqslant k$, iteration using (7.27) yields

$$g_{rk+s} \leqslant rg_k + g_s.$$

Hence

$$\limsup_{r\to\infty} g_{rk+s}/(rk+s) \leqslant \limsup_{r\to\infty}(rg_k + g_s)/(rk+s) = g_k/k$$

and, since this holds for $s = 1, 2, \ldots, k$,

$$\limsup_{n\to\infty} g_n/n \leqslant g_k/k. \tag{7.28}$$

Now k is arbitrary, so taking lower limits as $k \to \infty$ in (7.28),

$$\limsup_{n\to\infty} g_n/n \leqslant \liminf_{k\to\infty} g_k/k.$$

Hence g_n/n converges to a limit which, in view of (7.28), cannot be $+\infty$. Condition (7.25) is just needed to ensure that the limit is not $-\infty$. We have thus established the existence of the finite limit

$$\gamma = \lim_{n\to\infty} g_n/n \tag{7.29}$$

and

$$\gamma = \inf_{n\geqslant 1} g_n/n. \tag{7.30}$$

The constant γ, which turns out to be of major importance, is called the *time constant* of the subadditive process.

The classical Birkhoff–von Neumann ergodic theorem deals with the particular case in which (7.23) is strengthened to

$$x_{su} = x_{st} + x_{tu},$$

in which case the process $\mathbf{x} = (x_{st}; s, t \in I, s < t)$ is called an *additive process.* Then the process $\{y_t, t \in I, t \geqslant 1\}$ defined by

$$y_t = x_{t-1,t} \tag{7.31}$$

is stationary and

$$x_{0t} = \sum_{r=1}^{t} y_r.$$

The classical ergodic theorem (see Appendix IV) shows that the limit

$$\lim_{t\to\infty} x_{0t}/t$$

exists with probability one. It has a complete generalization to the case of subadditive processes as given in the following theorem.

Theorem 7.5. *If* $\mathbf{x} = \{x_{st}; s, t \in I, s < t\}$ *is a subadditive process, then the finite limit*

$$\xi = \lim_{t\to\infty} x_{0t}/t \tag{7.32}$$

exists with probability one and in mean, and

$$E\xi = \gamma = \inf_{n \geq 1} g_n/n. \tag{7.33}$$

As with the classical ergodic theorem, the random variable ξ will in general be nondegenerate. If $\mathscr{I}$ denotes the σ-field of events defined in terms of $\mathbf{x}$ and invariant under the shift $\mathbf{x} \to \mathbf{x}^*$ where

$$x_{st}^* = x_{s+1,t+1},$$

then we shall show that ξ is an $\mathscr{I}$-measurable random variable with the explicit representation

$$\xi = \lim_{t \to \infty} t^{-1} E(x_{0t} | \mathscr{I}). \tag{7.34}$$

Thus, if $\mathscr{I}$ is trivial, as will be the case in many important examples, we have $\xi = \gamma$ a.s.

The key to the analysis of the asymptotic behavior of $\mathbf{x}$ and in particular to the proof of Theorem 7.5 is the decomposition which is given in the following theorem.

Theorem 7.6. *The subadditive process* $\mathbf{x} = (x_{st};\, s, t \in I,\, s < t)$ *can be decomposed into*

$$x_{st} = y_{st} + z_{st}, \tag{7.35}$$

where $\mathbf{y} = (y_{st};\, s, t \in I,\, s < t)$ *is an additive process satisfying*

$$\gamma(\mathbf{y}) = Ey_{01} = \gamma(\mathbf{x}) = \inf_{n \geq 1} n^{-1} Ex_{0n}$$

and $\mathbf{z} = (z_{st};\, s, t \in I,\, s < t)$ *is a nonnegative subadditive process with*

$$\gamma(\mathbf{z}) = \inf_{n \geq 1} n^{-1} Ez_{0n} = 0.$$

The decomposition (7.35) is certainly not uniquely determined by $\mathbf{x}$. For example, let $v_1, v_2, \ldots$ be independent $N(0,1)$ random variables and write

$$w_{st} = v_{s+1} + v_{s+2} + \cdots + v_t.$$

Then $\mathbf{w} = \{w_{st};\, s, t \in I,\, s < t\}$ is an additive process with time constant $\gamma(\mathbf{w}) = 0$. If

$$x_{st} = \max(w_{st}, 0),$$

then $\mathbf{x}$ is subadditive with

$$g_t = Ex_{0t} = (t/2\pi)^{1/2},$$

so that $\gamma(\mathbf{x}) = 0$. Two different decompositions satisfying the conditions of Theorem 7.6 are those with

$$y_{st} = 0, \qquad z_{st} = x_{st},$$

and

$$y_{st} = w_{st}, \qquad z_{st} = \max(-w_{st}, 0).$$

Proof of Theorem 7.6. The argument here was provided by Burkholder in the discussion to Kingman (1973). For an alternative proof based on the use of the martingale convergence theorem, see del Junco (1977). We shall show that there exists a stationary sequence of random variables $f_0, f_1, \ldots$ such that $Ef_0 = \gamma = \gamma(\mathbf{x})$ and

$$\sum_{k=s}^{t-1} f_k \leqslant x_{st}, \qquad s, t \in I, \quad 0 \leqslant s < t. \tag{7.36}$$

Then we can set $y_{st} = \sum_{k=s}^{t-1} f_k$ and $z_{st} = x_{st} - y_{st}$. Clearly $\mathbf{y} = (y_{st})$ is an additive process with $Ey_{01} = \gamma$. Further, $\mathbf{z} = (z_{st})$ is a nonnegative subadditive process and $\gamma(\mathbf{z}) = 0$ since

$$n^{-1}Ez_{0n} = n^{-1}E(x_{0n} - y_{0n}) = n^{-1}g_n - \gamma \to 0$$

as $n \to \infty$. This then establishes (7.35) and will complete the proof of the theorem.

To prove (7.36) we begin by introducing

$$f_{kn} = n^{-1} \sum_{r=1}^{n} (x_{k,\,k+r} - x_{k+1,\,k+r}). \tag{7.37}$$

Since $(x_{s+1,\,t+1})$ has the same distribution as (x_{st}), it is clear that $\{f_{kn}, k = 1,2, \ldots\}$ is a stationary sequence for each fixed n.

Let $s \leqslant k \leqslant t - 1$ and $n > t$. By subadditivity, $x_{kr} - x_{k+1,\,r} \leqslant x_{k,\,k+1}$, $r \geqslant k + 1$, so that

$$\begin{aligned} nf_{kn} &= \sum_{r=k+1}^{k+n} (x_{kr} - x_{k+1,\,r}) \\ &\leqslant \sum_{r=t+1}^{n} (x_{kr} - x_{k+1,\,r}) + tx_{k,\,k+1}. \end{aligned}$$

Therefore,

$$\begin{aligned} n\sum_{k=s}^{t-1} f_{kn} &\leqslant \sum_{r=t+1}^{n} (x_{sr} - x_{tr}) + t\sum_{k=s}^{t-1} x_{k,\,k+1} \\ &\leqslant \sum_{r=t+1}^{n} x_{st} + t\sum_{k=s}^{t-1} x_{k,\,k+1} \\ &= (n-t)x_{st} + t\sum_{k=s}^{t-1} x_{k,\,k+1}, \end{aligned}$$

and hence

$$\sum_{k=s}^{t-1} f_{kn} \leqslant x_{st} + n^{-1} w_{st}, \tag{7.38}$$

where

$$w_{st} = t\left\{\sum_{k=s}^{t-1} x_{k,\,k+1} - x_{st}\right\}.$$

Now $f_{0n} \leqslant x_{01}$ and, from (7.37),

$$Ef_{0n} = n^{-1} \sum_{r=1}^{n} (g_r - g_{r-1}) = n^{-1} g_n \geqslant \gamma.$$

Therefore,

$$\begin{aligned} E|f_{0n}| &\leqslant E|x_{01}| + E(x_{01} - f_{0n}) \\ &\leqslant E|x_{01}| + g_1 - \gamma, \end{aligned}$$

and using the SLLN for subsequences (Theorem 7.3), there is a sequence $n_1 < n_2 < \cdots$ of positive integers and an integrable random variable f_0 such that

$$A_{0j} = j^{-1} \sum_{i=1}^{j} f_{0n_i} \to f_0 \quad \text{a.s.}$$

as $j \to \infty$. Furthermore, in view of the stationarity of $\{f_{kn}, k = 1,2, \ldots\}$,

$$A_{kj} = j^{-1} \sum_{i=1}^{j} f_{kn_i}$$

also converges a.s., say to f_k, and $f_0, f_1, \ldots$ is a stationary sequence.

Next, let $\delta_j = j^{-1} \sum_{i=1}^{j} n_i^{-1}$. Then by (7.38),

$$\sum_{k=s}^{t-1} A_{kj} \leqslant x_{st} + \delta_j w_{st}, \tag{7.39}$$

and (7.36) follows upon letting $j \to \infty$.

It remains to show that $Ef_0 = \gamma$. By (7.36) with $s = 0$,

$$Ef_0 = t^{-1} \sum_{k=0}^{t-1} Ef_k \leqslant t^{-1} Ex_{0t} = t^{-1} g_t \to \gamma$$

as $t \to \infty$, so that $Ef_0 \leqslant \gamma$. To see that $Ef_0 \geqslant \gamma$, we recall that $Ef_{0n} = n^{-1} g_n \to \gamma$, which implies that $EA_{0j} \to \gamma$. Hence by (7.39), which implies that $x_{01} - A_{0j} \geqslant 0$, and Fatou's lemma,

$$g_1 - Ef_0 = E(x_{01} - f_0) \leqslant \liminf_{j\to\infty} E(x_{01} - A_{0j}) = g_1 - \gamma.$$

Therefore, $Ef_0 \geqslant \gamma$ and hence $Ef_0 = \gamma$, as required. ∎

Proof of Theorem 7.5. First we shall show that

$$\xi = \limsup_{t\to\infty} x_{0t}/t \tag{7.40}$$

is almost surely finite with

$$E\xi = \gamma, \tag{7.41}$$

and that

$$E|t^{-1}x_{0t} - \xi| \to 0 \tag{7.42}$$

as $t \to \infty$.

We fix $k \geqslant 1$ and write $N = N(t)$ for the integral part of t/k. The subadditive property gives

$$\begin{aligned} x_{st} &\leqslant \sum_{r=1}^{N} x_{(r-1)k,\, rk} + x_{Nk,\, t} \\ &\leqslant \sum_{r=1}^{N} y_r + w_N, \end{aligned}$$

where $y_r = x_{(r-1)k,\, rk}$ and

$$w_N = \sum_{j=1}^{k-1} |x_{Nk,\, Nk+j}|.$$

The sequence $\{y_r\}$ is stationary with $Ey_1 = g_k$, and hence the ergodic theorem (see Appendix IV) shows that

$$\xi_k = \lim_{N\to\infty} N^{-1} \sum_{r=1}^{N} y_r \tag{7.43}$$

exists with probability one and that $E\xi_k = g_k$. Furthermore, the distribution of w_N is the same for each N, and $Ew_1 < \infty$, so that for $\varepsilon > 0$,

$$\sum_{N=1}^{\infty} P(w_N \geqslant \varepsilon N) = \sum_{N=1}^{\infty} P(w_1 \geqslant \varepsilon N) \leqslant \varepsilon^{-1} E w_1 < \infty. \tag{7.44}$$

The Borel–Cantelli lemma now implies that $N^{-1}w_N \to 0$ a.s. as $N \to \infty$. Consequently, from (7.43) and (7.44),

$$\xi = \limsup_{t\to\infty} x_{0t}/t \leqslant \limsup_{N\to\infty} (Nk)^{-1} \left[\sum_{r=1}^{N} y_r + w_N\right] \leqslant \xi_k/k. \tag{7.45}$$

This holds for $k = 1,2,3,\ldots$, and since $E\xi_k = g_k$,

$$E\xi \leqslant \gamma = \lim_{k\to\infty} g_k/k. \tag{7.46}$$

Now consider the nonpositive subadditive process

$$u_{st} = x_{st} - \sum_{j=s+1}^{t} x_{j-1,j}$$

and set

$$\zeta_n = \sup_{t \geqslant n} u_{0t}/t.$$

We note that ζ_n decreases as n increases to the limit

$$\limsup_{t \to \infty} u_{0t}/t = \xi - \xi_1,$$

and in view of the monotone convergence,

$$\lim_{n \to \infty} E\zeta_n = E\left(\lim_{n \to \infty} \zeta_n\right) = E(\xi - \xi_1) = E\xi - g_1.$$

On the other hand,

$$\lim_{n \to \infty} E\zeta_n \geqslant \liminf_{n \to \infty} E(u_{0n}/n) = \liminf_{n \to \infty}(n^{-1}g_n - g_1) = \gamma - g_1,$$

so that $E\xi - g_1 \geqslant \gamma - g_1$, which, together with (7.46), gives (7.41).
Now we have $\zeta_n \geqslant u_{0n}/n$ and

$$\lim_{n \to \infty} E\zeta_n = \gamma - g_1 = \lim_{n \to \infty} Eu_{0n}/n,$$

so that

$$E|\zeta_n - u_{0n}/n| \to 0 \tag{7.47}$$

as $n \to \infty$. Further, by monotone convergence,

$$E|\zeta_n - \xi + \xi_1| = E(\zeta_n - \xi + \xi_1) \to 0 \tag{7.48}$$

as $n \to \infty$, while we obtain from the classical ergodic theorem that

$$E|n^{-1}(u_{0n} + x_{0n}) - \xi_1| \to 0 \tag{7.49}$$

as $n \to \infty$. The results (7.47)–(7.49) combine to give (7.42) as required.
Finally we turn to the representation of Theorem 7.6. Using (7.40) and (7.41) on the nonnegative subadditive process **z**, we have

$$\limsup_{t \to \infty} z_{0t}/t = \zeta \geqslant 0$$

with $E\zeta = 0$, and hence $\zeta = 0$ a.s. Thus,

$$\lim_{t \to \infty} z_{0t}/t = 0 \quad \text{a.s.},$$

while the classical ergodic theorem assures the existence of

$$\lim_{t\to\infty} y_{0t}/t,$$

so that the existence of

$$\lim_{t\to\infty} x_{0t}/t$$

is established. The limit must be ξ in view of (7.40), and (7.33) holds in view of (7.41). This completes the proof. ▌

To establish the representation (7.34) for the ξ we first note that the ergodic theorem enables us to identify

$$\xi = \lim_{t\to\infty} t^{-1}E(y_{0t}|\mathscr{I}) \quad \text{a.s.}$$

Hence, in order to establish (7.34) it suffices to show that

$$\lim_{t\to\infty} t^{-1}E(z_{0t}|\mathscr{I}) = 0 \quad \text{a.s.} \tag{7.50}$$

Let Φ_t be a version of the conditional expectation $E(z_{0t}|\mathscr{I})$. The subadditive property of z_{0t} ensures that

$$\Phi_{m+n} \leqslant \Phi_m + \Phi_n$$

for all $m, n \in I$, $m, n \geqslant 1$, and hence, following the argument leading up to (7.29),

$$\phi = \lim_{n\to\infty} \Phi_n/n$$

exists with probability one. Thus, the limit exists in (7.50). That it is zero a.s. follows from Fatou's lemma since the time constant of **z** is zero.

The representation of a subadditive process **x** given in Theorem 7.6, in terms of an additive process **y** plus a nonnegative subadditive process **z**, offers the possibility of proving various other limit results for x_{0t} via those for y_{0t} if z_{0t} may be neglected asymptotically, that is, if the intrinsic subadditivity of the process may be neglected asymptotically. Indeed, with a view to the CLT we note that

$$(x_{0t} - \gamma t)/t^{1/2} \qquad \text{and} \qquad (y_{0t} - \gamma t)/t^{1/2}$$

have the same limit distribution if

$$(g_t - \gamma t)/t^{1/2} \to 0$$

as $t \to \infty$. This follows from

$$\begin{aligned} E|t^{-1/2}(x_{0t} - \gamma t) - t^{-1/2}(y_{0t} - \gamma t)| &= t^{-1/2}E(x_{0t} - y_{0t}) \\ &= t^{-1/2}Ez_{0t} = t^{-1/2}(g_t - \gamma t). \end{aligned} \tag{7.51}$$

Similarly, with a view to the LIL we note that

$$(x_{0t} - \gamma t)/(2t \log \log t)^{1/2} \qquad \text{and} \qquad (y_{0t} - \gamma t)/(2t \log \log t)^{1/2}$$

have the same a.s. limit behavior if

$$\sum_{t \geqslant 3} E z_{0t}/(2t \log \log t)^{1/2} = \sum_{t \geqslant 3} (g_t - \gamma t)/(2t \log \log t)^{1/2} < \infty. \tag{7.52}$$

This result follows from (7.52) as an immediate consequence of the Borel–Cantelli lemma.

Requirements on the rate of convergence of g_t/t to γ such as those mentioned above are often satisfied in practice, for example, in many of the applications considered by Hammersley and Welsh (1965). The remaining problem is, of course, to say enough about the additive process $\mathbf{y}$. This is especially awkward since $\mathbf{y}$ is not uniquely specified, nor is there a particularly transparent construction for its increments. However, as we observed in the proof of Theorem 7.6, it is possible to find a sequence $n_1 < n_2 < \cdots$ of positive integers and an integrable random variable y_k such that

$$y_k = \lim_{N \to \infty} N^{-1} \sum_{j=1}^{N} n_j^{-1} \sum_{t=1}^{n_j} (x_{k,\,t+k} - x_{k+1,\,t+k}) \quad \text{a.s.} \tag{7.53}$$

for each $k \geqslant 0$, and $E y_k = \gamma$. The $\{y_k\}$ are stationary and we can choose

$$y_{st} = \sum_{k=s}^{t-1} y_k.$$

of course (7.53) does simplify drastically when $n^{-1} \sum_{t=1}^{n} (x_{0t} - x_{1,t})$ converges a.s., and the class of subadditive processes for which this holds may repay special consideration.

With the choice of $\mathbf{y}$ as given by (7.53), we can obtain a variety of CLT and LIL results with the aid of the theory of Chapter 5. We shall give one such result, together with an application, to illustrate the possibilities.

First we need some more notation. Let T be the measure preserving transformation such that

$$x_{s+1,\,t+1}(\omega) = x_{st}(T\omega) \qquad \text{for all} \quad s, t \in I, \quad s < t.$$

Write $\{\mathscr{M}_a^b, -\infty \leqslant a \leqslant b \leqslant \infty\}$ for the family of sub-σ-fields of $\mathscr{F}$ satisfying the conditions

(i) if $a \leqslant c \leqslant d \leqslant b$, then $\mathscr{M}_c^d \subseteq \mathscr{M}_a^b$,
(ii) for all $a \leqslant b$, $T^{-1} \mathscr{M}_a^b = \mathscr{M}_{a+1}^{b+1}$.

We define $\phi(n)$ by

$$\phi(n) = \sup\{|P(B|A) - P(B)|;\, A \in \mathscr{M}_{-\infty}^{k},\, B \in \mathscr{M}_{k+n}^{\infty},\, P(A) > 0,\, -\infty < k < \infty\},$$

and we shall suppose that $\phi(n) \to 0$ as $n \to \infty$. This is a condition of uniform mixing as discussed in Chapter 5. The following result adds a LIL to the CLT of Ishitani (1977).

Theorem 7.7. *Suppose that* $\{x_{st}, s, t \in I, s < t\}$ *is a subadditive process, with time constant* γ, *for which the following three conditions are satisfied*:

$$\sum_{n=1}^{\infty} [\phi(n)]^{1/2} < \infty; \tag{7.54}$$

there exists a random variable Ψ *with* $E\Psi^2 < \infty$ *such that*

$$|x_{0t} - x_{1t}| \leqslant \Psi \qquad \text{a.s.} \tag{7.55}$$

for all $t \geqslant 1$; *and*

$$\sum_{n=1}^{\infty} \sup_{t \geqslant 1} \{E[(x_{0t} - x_{1t}) - E(x_{0t} - x_{1t}|\mathscr{M}_{-\infty}^{n})]^2\}^{1/2} < \infty. \tag{7.56}$$

If

$$(g_t - t\gamma)/t^{1/2} \to 0 \qquad as \quad t \to \infty, \tag{7.57}$$

then $\lim_{n\to\infty} E|x_{0n} - n\gamma|/n^{1/2} = (2\sigma^2/\pi)^{1/2}$ *exists for* $0 \leqslant \sigma < \infty$, *and if* $\sigma > 0$,

$$(n\sigma^2)^{-1/2}(x_{0n} - n\gamma) \xrightarrow{d} N(0,1)$$

as $n \to \infty$. *Furthermore, if* (7.57) *is strengthened to*

$$\sum_{t \geqslant 3} (g_t - t\gamma)/(2t \log\log t)^{1/2} < \infty \tag{7.58}$$

and $\sigma > 0$, *then*

$$(2n\sigma^2 \log\log n)^{-1/2}(x_{0n} - n\gamma)$$

has for its set of a.s. limit points the closed interval $[-1,1]$.

Proof. In view of (7.57) and (7.58) and the considerations surrounding (7.51) and (7.52) we observe that it suffices to establish the CLT and LIL for the stationary process $\{y_k(\omega) = y_0(T^k\omega), -\infty < k < \infty\}$ prescribed in (7.53). This will be done by checking the condition (5.36) of Corollary 5.4 from which the result follows straightforwardly.

First we need to ensure that $Ey^2 < \infty$ and this follows easily from (7.55) and the definition of y since

$$Ey^2 = E\left(\lim_{N\to\infty} N^{-1} \sum_{j=1}^{N} n_j^{-1} \sum_{t=1}^{n_j} (x_{0t} - x_{1t})\right)^2 \leqslant E\Psi^2 < \infty.$$

Furthermore, the condition (7.56) guarantees that

$$E(y|\mathcal{M}^n_{-\infty}) = \lim_{N\to\infty} N^{-1} \sum_{j=1}^{N} n_j^{-1} \sum_{t=1}^{n_j} E(x_{0t} - x_{1t}|\mathcal{M}^n_{-\infty}) \quad \text{a.s.}$$

and, from (7.55),

$$\left| N^{-1} \sum_{j=1}^{N} n_j^{-1} \sum_{t=1}^{n_j} \{(x_{0t} - x_{1t}) - E(x_{0t} - x_{1t}|\mathcal{M}^n_{-\infty})\} \right| \leqslant \Psi + E(\Psi|\mathcal{M}^n_{-\infty}),$$

which is square integrable. Then, using Fatou's lemma and Minkowski's inequality,

$$\begin{aligned}
&[E(y - E(y|\mathcal{M}^n_{-\infty}))^2]^{1/2} \\
&\quad \leqslant \liminf_{N\to\infty} [E(N^{-1} \sum_{j=1}^{N} n_j^{-1} \sum_{t=1}^{n_j} \{(x_{0t} - x_{1t}) - E(x_{0t} - x_{1t}|\mathcal{M}^n_{-\infty})\})^2]^{1/2} \\
&\quad \leqslant \liminf_{N\to\infty} N^{-1} \sum_{j=1}^{N} n_j^{-1} \sum_{t=1}^{n_j} [E\{(x_{0t} - x_{1t}) - E(x_{0t} - x_{1t}|\mathcal{M}^n_{-\infty})\}^2]^{1/2} \\
&\quad \leqslant \sup_{t\geqslant 1} [E\{(x_{0t} - x_{1t}) - E(x_{0t} - x_{1t}|\mathcal{M}^n_{-\infty})\}^2]^{1/2},
\end{aligned}$$

so that, from (7.56),

$$\sum_{n=1}^{\infty} [E(y - E(y|\mathcal{M}^n_{-\infty}))^2]^{1/2} < \infty.$$

Furthermore,

$$E(E(y|\mathcal{M}^{-k}_{-\infty}))^2 = E(yE(y|\mathcal{M}^{-k}_{-\infty}))$$

and y is $\mathcal{M}_0^{\infty}$ – measurable, while $E(y|\mathcal{M}^{-k}_{-\infty})$ is $\mathcal{M}^{-k}_{-\infty}$ – measurable. The argument used in the proof of Corollary 5.5 can then be used to give

$$[E(E(y|\mathcal{M}^{-k}_{-\infty}))^2]^{1/2} \leqslant 2(\phi(k)Ey^2)^{1/2}$$

and hence

$$\sum_{k=1}^{\infty} [E(E(y|\mathcal{M}^{-k}_{-\infty}))^2]^{1/2} \leqslant 2(Ey^2)^{1/2} \sum_{k=1}^{\infty} [\phi(k)]^{1/2} < \infty$$

in view of (7.54). The conditions of Corollary 5.4 are therefore satisfied and the result of the theorem follows provided that we can identify Ey^2 with σ^2. However, if $Ey^2 > 0$, then the result of Corollary 5.4 ensures that

$$E|y_{0n} - n\gamma|/n^{1/2} \to (2Ey^2/\pi)^{1/2}$$

since

$$(y_{0n} - n\gamma)/n^{1/2} \to N(0, Ey^2)$$

and $E(y_{0n} - n\gamma)^2/n \to Ey^2$ gives the uniform integrability of $(y_{0n} - n\gamma)/n^{1/2}$ [using, e.g., Theorem 5.4 of Billingsley (1968)]. Thus, in view of (7.57),

$$E|x_{0n} - n\gamma|/n^{1/2} \to (2Ey^2/\pi)^{1/2}$$

as required. ∎

Theorem 7.7 and other similar results which can be established via the representation (7.53) seem to have a disappointingly narrow spectrum of applicability. Furthermore, the form of the conditions is certainly not tailor made to most of the subadditive processes that have been studied to date. This is despite the fact that (7.57) or (7.58) is often satisfied. One very important class which is not conveniently handled is that of the so-called *independent subadditive processes*, namely, those which have the property that for $t_1, t_2, \ldots, t_n \in I, 0 = t_0 < t_1 < \cdots < t_n$, the random variables $x_{t_{r-1}t_r}$, $r = 1,2, \ldots, n$, are independent.

We conclude this section with the sketch of an application of Theorem 7.7 to a problem concerning products of random matrices. Full details can be found in Ishitani (1977) together with Furstenberg and Kesten (1960).

Let $Z_1, Z_2, \ldots$ be a stationary random sequence of $k \times k$ matrices with strictly positive elements. Write T for the measure-preserving transformation such that

$$Z_{i+1}(\omega) = Z_i(T\omega), \qquad i = 1, 2, \ldots,$$

and define $\mathscr{M}_a^b$ as the σ-field generated by $Z_{a+1}, \ldots, Z_{b+1}$. We shall use $(A)_{ij}$ to denote the (i,j)th element of a matrix A and shall suppose that the matrices $\{Z_k\}$ satisfy

$$E[\log(Z_1)_{11}]^2 < \infty \tag{7.59}$$

and

$$1 \leqslant \max_{i,j}(Z_1)_{ij}/\min_{i,j}(Z_1)_{ij} \leqslant C \quad \text{a.s.} \tag{7.60}$$

for some constant C, $1 < C < \infty$.

For $s, t \in I$, $s < t$, set

$$_tY_s = Z_t Z_{t-1} \cdots Z_s. \tag{7.61}$$

Then it is easily checked that

$$\{x_{st} = -\log(_tY_{s+1})_{11},\ s, t \in I,\ s < t\} \tag{7.62}$$

is a subadditive process.

Now it has been shown by Furstenberg and Kesten (1960) that

$$g_{t+1} - g_t = \gamma + O[(1 - C^{-3})^t],$$

from which we readily find that the condition (7.59) is satisfied.

Next, we have

$$\frac{({}_tY_1)_{11}}{({}_tY_2)_{11}} = \frac{\sum_{i=1}^k \sum_{j=1}^k ({}_tY_3)_{1i}(Z_2)_{ij}(Z_1)_{j1}}{\sum_{i=1}^k ({}_tY_3)_{1i}(Z_2)_{i1}}$$

and thus, by virtue of condition (7.60),

$$C^{-1} \sum_{j=1}^k (Z_1)_{j1} \leqslant \frac{({}_tY_1)_{11}}{({}_tY_2)_{11}} \leqslant C \sum_{j=1}^k (Z_1)_{j1},$$

from which we obtain

$$|x_{0t} - x_{1t}| \leqslant |\log(Z_1)_{11}| + \log kC^2,$$

and hence (7.55) holds.

The condition (7.56) can be checked by direct examination of $x_{0,\,m+n} - x_{1,\,m+n}$. The rather laborious details are given in Ishitani (1977), building on results of Furstenberg and Kesten (1960). It is shown that

$$x_{0,\,m+n} - x_{1,m+n} = -\log\alpha + O(1 - C^{-3})^n$$

uniformly in m and ω where

$$\alpha = \sum_{i=1}^k \frac{\sum_{r=1}^k ({}_nY_2)_{ri}}{\sum_{r=1}^k ({}_nY_2)_{r1}} (Z_1)_{i1}.$$

Then, since $\log\alpha$ is $\mathscr{M}^n_{-\infty}$-measurable, we obtain

$$|(x_{0t} - x_{1t}) - E(x_{0t} - x_{1t}|\mathscr{M}^n_{-\infty})| = \begin{cases} 0, & t \leqslant n \\ O(1 - C^{-3})^n, & t > n \end{cases}$$

uniformly in t and ω. This shows that the condition (7.56) applies. Thus, if the condition (7.54) is satisfied, then Theorem 7.7 is applicable.

The subadditive process which we have used gives limit results for $\log({}_tY_1)_{11}$, when suitably normed. However, corresponding results for $\log({}_tY_1)_{ij}$ also follow since

$$|\log({}_tY_1)_{ij} - \log({}_tY_1)_{11}| \leqslant 2\log C,$$

as can readily be deduced with the aid of condition (7.60).

7.4. The Hawkins Random Sieve

The random sieve introduced by Hawkins (1958) is analogous to the classical sieve of Eratosthenes and produces a random sequence with asymptotic properties in many ways analogous to the prime numbers. It is constructed as follows. Let $A_1 = \{2,3,4,\ldots\}$.

Stage 1. Put $X_1 = \min A_1$. From the set $A_1 \backslash \{X_1\}$ each number in turn is (independently of the others) deleted with probability X_1^{-1} or not deleted with probability $1 - X_1^{-1}$. The set of elements of $A_1 \backslash \{X_1\}$ which remain is denoted by A_2.

Stage n. Put $X_n = \min A_n$. From the set $A_n \backslash \{X_n\}$ each number in turn is (independently of the others) deleted with probability X_n^{-1} or not deleted with probability $1 - X_n^{-1}$. The set of elements of $A_n \backslash \{X_n\}$ which remain is denoted by A_{n+1}.

Define

$$Y_n = \prod_{k=1}^{n} (1 - X_k^{-1})^{-1}.$$

The basic analogy between the prime numbers and the random sequence $\{X_n\}$ is clear from the following theorem, the results of which are due to Wunderlich (1974, 1976).

Theorem 7.8. (i) $\lim_{n\to\infty}(n \log n)^{-1} X_n = 1$ *a.s.* (ii) $\lim_{n\to\infty}(\log n)^{-1} Y_n = 1$ *a.s.*

The result (i) is an analog of the prime number theorem which, if p_n denotes the nth prime, asserts that $(n \log n)^{-1} p_n \to 1$ as $n \to \infty$. The result (ii) is an analog of Mertens' theorem which, for

$$y_n = \prod_{k=1}^{n} (1 - p_k^{-1})^{-1},$$

asserts that $(\log n)^{-1} y_n \to \exp \gamma$, γ being Euler's constant. See, for example, Ingham (1932).

Proof. We begin by observing that the process $\{(X_n, Y_n), n \geqslant 1\}$ is Markovian with $X_1 = Y_1 = 2$ and

$$P(X_{n+1} - X_n = j | \mathscr{F}_n) = Y_n^{-1}(1 - Y_n^{-1})^{j-1}, \qquad j \geqslant 1, \tag{7.63}$$

$\mathscr{F}_n$ being the σ-field generated by X_j, Y_j, $j \leqslant n$.

Now set

$$Z_n = X_n - 1 \quad \text{and} \quad U_{n+1} = (Z_{n+1} - Z_n) Y_n^{-1}, \qquad n \geqslant 1, \tag{7.64}$$

and, using (7.63), note that

$$E[(U_{n+1} - 1) | \mathscr{F}_n] = 0 \quad \text{a.s.}, \tag{7.65}$$

so that the $U_{n+1} - 1$, $n \geqslant 0$, are martingale differences, while

$$E[(U_{n+1} - 1)^2 | \mathscr{F}_n] = 1 - Y_n^{-1} \leqslant 1 \quad \text{a.s.} \tag{7.66}$$

An application of Theorem 2.18 then gives

$$\sum_{k=1}^{n} (U_{k+1} - 1) = o(n^{(1/2)+\varepsilon}) \quad \text{a.s.} \tag{7.67}$$

for any $\varepsilon > 0$ and, in particular,

$$n^{-1} \sum_{k=1}^{n} U_{k+1} \to 1 \quad \text{a.s.} \tag{7.68}$$

If $Y_n \uparrow Y < \infty$ a.s., we would conclude from the relation

$$Y_{n+1} = Y_n(1 - X_{n+1}^{-1})^{-1} = Y_n(1 + Z_{n+1}^{-1})$$

that $\sum Z_n^{-1} < \infty$ a.s., while (7.68) would yield, via Toeplitz' lemma, the contradictory result $n^{-1}Z_{n+1} \to Y$ a.s. as $n \to \infty$. Thus $Y_n \uparrow \infty$ a.s. and

$$Z_{n+1}Y_{n+1}^{-1} - Z_nY_n^{-1} = U_{n+1} - Y_{n+1}^{-1} = U_{n+1} + o(1) \quad \text{a.s.,}$$

which taken together with (7.68) gives

$$(nY_n)^{-1}Z_n \xrightarrow{\text{a.s.}} 1 \qquad \text{as} \quad n \to \infty. \tag{7.69}$$

Now, with a view to later results we set $H_n = \log Z_n - Y_n, n \geqslant 1, \alpha_n = Y_nZ_n^{-1}$, $\beta_n = Y_nZ_{n+1}^{-1}$. Then from (7.69),

$$\alpha_n = O(n^{-1}) \quad \text{a.s.,} \qquad \beta_n = O(n^{-1}) \quad \text{a.s.} \tag{7.70}$$

as $n \to \infty$. Thus,

$$H_{n+1} - H_n = \log(1 + \alpha_nU_{n+1}) - \beta_n = \alpha_nU_{n+1} - \beta_n + R_n,$$

where

$$|R_n| \leqslant \alpha_n^2U_{n+1}^2 \quad \text{a.s.}$$

Further,

$$|\alpha_n - \beta_n| = |\alpha_n\beta_nU_{n+1}| \leqslant \alpha_n^2|U_{n+1}|,$$

so that

$$H_{n+1} - H_n = \alpha_n(U_{n+1} - 1) + S_n, \tag{7.71}$$

where

$$|S_n| \leqslant \alpha_n^2(|U_{n+1}| + U_{n+1}^2). \tag{7.72}$$

Also, since

$$P(U_{n+1} > x|\mathscr{F}_n) \leqslant (1 - Y_n^{-1})^{xY_n} \leqslant e^{-x},$$

we have from the Borel–Cantelli lemma that

$$U_{n+1} = O(\log n) \quad \text{a.s.} \tag{7.73}$$

and hence, using (7.70) and (7.73) in (7.72),

$$|S_n| = O(n^{-2}(\log n)^2) \quad \text{a.s.} \tag{7.74}$$

We note that the α_n are $\mathscr{F}_n$-measurable, so that the $\alpha_k(U_{k+1} - 1)$, $k \geqslant 1$, are still martingale differences and indeed, by virtue of (7.66), (7.70), and Theorem 2.15,

$$\sum_1^\infty \alpha_n(U_{n+1} - 1) \qquad \text{converges a.s.} \tag{7.75}$$

Thus, partial summation in (7.71), together with (7.74) and (7.75), ensures the a.s. convergence of H_n to some (random) C as $n \to \infty$ and, consequently,

$$\lim_{n\to\infty} X_n e^{-Y_n} = L(=e^C) \quad \text{a.s.} \tag{7.76}$$

for some (random) L. The results of Theorem 7.8 follow immediately from (7.69) and (7.76). ▮

To press further with the analogy between the sequence $\{X_n\}$ and the primes we recall that the famous Riemann hypothesis about the zeros of the Riemann zeta function is equivalent to this statement:

$$\text{li}(p_n) = n + O(n^{(1/2)+\varepsilon}) \tag{7.77}$$

for any $\varepsilon > 0$, where p_n denotes the nth prime and li(x), $x > 1$, is defined by

$$\text{li}(x) = \lim_{\delta\downarrow 0}\left\{\int_0^{1-\delta} + \int_{1+\delta}^x\right\}\frac{dz}{\log z} \sim \frac{x}{\log x} \qquad \text{as} \quad x \to \infty.$$

A form corresponding to (7.77) can be found for the X_n, namely,

$$L\,\text{li}(X_n L^{-1}) = n + O(n^{(1/2)+\varepsilon}) \quad \text{a.s.} \tag{7.78}$$

for any $\varepsilon > 0$, where L denotes the (random) limit defined by (7.76), namely,

$$L = \lim_{n\to\infty} X_n e^{-Y_n}.$$

This result is due to Neudecker and Williams (1974), but (7.78) can be sharpened to a log log form as given in the following theorem due to Foster and Williams (1978); see also Heyde (1978a).

Theorem 7.9.

$$\limsup_{n\to\infty}(2n \log\log n)^{-1/2}[L\,\text{li}(X_n L^{-1}) - n] = 1 \qquad \textit{a.s.},$$

$$\liminf_{n\to\infty}(2n \log\log n)^{-1/2}[L\,\text{li}(X_n L^{-1}) - n] = -1 \quad \textit{a.s.},$$

In addition,

$$n^{-1/2}[L\,\mathrm{li}(X_nL^{-1}) - n] \xrightarrow{\mathrm{d}} N(0,1)$$

as $n \to \infty$.

Proof. We have for $x > 1$ and $h > 0$,

$$\mathrm{li}(x+h) - \mathrm{li}(x) = \int_x^{x+h} (\log z)^{-1}\,dz$$

and hence

$$\begin{aligned}|\mathrm{li}(x+h) - \mathrm{li}(x) - h(\log x)^{-1}| &= \int_x^{x+h} [(\log x)^{-1} - (\log z)^{-1}]\,dz \\ &\leqslant h[(\log x)^{-1} - (\log(x+h))^{-1}] \\ &\leqslant h^2(\log x)^{-2}x^{-1}.\end{aligned}$$

Consequently,

$$L\,\mathrm{li}(Z_{n+1}/L) - L\,\mathrm{li}(Z_n/L) = (\log Z_n - C)^{-1}(Z_{n+1} - Z_n) + T_n, \tag{7.79}$$

where $C = \log L$ and

$$|T_n| \leqslant (Z_{n+1} - Z_n)^2 Z_n^{-1}(\log Z_n - C)^{-2} = O(n^{-1}\log n), \tag{7.80}$$

using Theorem 7.8, (7.73), and (7.76).

Our next task will be to strengthen the result (7.76) to show that

$$H_n = \log Z_n - Y_n = C + \delta(n)(2n^{-1}\log\log n)^{1/2} \quad \text{a.s.}, \tag{7.81}$$

where $\limsup_{n\to\infty} \delta(n) = +1$ a.s., $\liminf_{n\to\infty} \delta(n) = -1$ a.s.

Using (7.71), we have

$$\begin{aligned}H_{n+1} &= \sum_{k=1}^{n} (H_{k+1} - H_k) - 2 \\ &= \sum_{k=1}^{n} \alpha_k(U_{k+1} - 1) + \sum_{k=1}^{n} S_k - 2\end{aligned}$$

and, since $H_n \xrightarrow{\text{a.s.}} C$ (finite a.s.) as $n \to \infty$ and $\sum_1^\infty |S_k| < \infty$ a.s. in view of (7.74),

$$C - H_n = \sum_{k=n}^{\infty} \alpha_k(U_{k+1} - 1) + \sum_{k=n}^{\infty} S_k. \tag{7.82}$$

Now from (7.74),

$$\begin{aligned}\left|(2n^{-1}\log\log n)^{-1/2} \sum_{k=n}^{\infty} S_k\right| &= O\left(n^{1/2}(\log\log n)^{-1/2} \sum_{k=n}^{\infty} k^{-2}(\log k)^2\right) \\ &= o(1) \quad \text{a.s.}\end{aligned} \tag{7.83}$$

Furthermore,

$$\sum_{k=n}^{\infty} \alpha_k(U_{k+1} - 1) = \sum_{k=n}^{\infty} k^{-1}(U_{k+1} - 1) + \sum_{k=n}^{\infty} (\alpha_k - k^{-1})(U_{k+1} - 1) \quad (7.84)$$

and we shall proceed to show that

$$\lim_{n\to\infty} n^{1/2}(\log\log n)^{-1/2} \sum_{k=n}^{\infty} (\alpha_k - k^{-1})(U_{k+1} - 1) = 0 \quad \text{a.s.} \quad (7.85)$$

The variables $(\alpha_k - k^{-1})(U_{k+1} - 1)$, $k \geqslant 1$, are still martingale differences and, using Theorem 2.15,

$$\sum_{k=1}^{n} k^{1/2}(\alpha_k - k^{-1})(U_{k+1} - 1) \qquad \text{converges a.s.} \quad (7.86)$$

provided

$$\sum_{k=1}^{\infty} k(\alpha_k - k^{-1})^2 E[(U_{k+1} - 1)^2 | \mathscr{F}_k] < \infty \quad \text{a.s.,}$$

and hence if

$$\sum_{k=1}^{\infty} k(\alpha_k - k^{-1})^2 < \infty \quad \text{a.s.} \quad (7.87)$$

in view of (7.66). Now,

$$\alpha_k - k^{-1} = k^{-1} Z_k^{-1} Y_k (k - Z_k Y_k^{-1}). \quad (7.88)$$

Furthermore, note that

$$Z_{n+1} Y_{n+1}^{-1} - Z_n Y_n^{-1} = U_{n+1} - Y_{n+1}^{-1}, \qquad n \geqslant 1,$$

from which we obtain by summation

$$Z_n Y_n^{-1} = \sum_{k=2}^{n} (U_k - 1) + n - \sum_{k=1}^{n} Y_k^{-1},$$

and using (7.67),

$$n^{-1}\log n\left(Z_n Y_n^{-1} - n + \sum_{k=1}^{n} Y_k^{-1}\right) = n^{-1}\log n \sum_{k=2}^{n} (U_k - 1) \xrightarrow{\text{a.s.}} 0 \quad (7.89)$$

as $n \to \infty$. But

$$\begin{aligned} n^{-1}\log n \sum_{k=1}^{n} Y_k^{-1} &\sim n^{-1}\log n\left[\sum_{k=2}^{n}(Y_k^{-1} - (\log k)^{-1}) + \sum_{k=2}^{n}(\log k)^{-1}\right] \\ &= n^{-1}\log n\left[\sum_{k=2}^{n}(\log k)^{-1}(Y_k^{-1}\log k - 1) + \sum_{k=2}^{n}(\log k)^{-1}\right] \\ &\to 1 \quad \text{a.s.} \end{aligned}$$

as $n \to \infty$, using Theorem 7.8 and Toeplitz' lemma. Thus, from Theorem 7.8 together with (7.88) and (7.89),

$$\alpha_k - k^{-1} = O((k \log k)^{-1}) \quad \text{a.s.},$$

and hence (7.87) and consequently (7.86) hold.

To establish (7.85) we set

$$K_n = \sup_{M \geqslant n} \left| \sum_{k=n}^{M} k^{1/2}(\alpha_k - k^{-1})(U_{k+1} - 1) \right|,$$

so that $K_n \to 0$ a.s. as $n \to \infty$. Then from Abel's inequality [e.g., see Whittaker and Watson (1927, p. 16)],

$$\left| \sum_{k=n}^{\infty} (\alpha_k - k^{-1})(U_{k+1} - 1) \right| \leqslant K_n n^{-1/2} \quad \text{a.s.},$$

and the result (7.85) follows.

We now apply the iterated logarithm result of Theorem 4.9 to the tail sum $\sum_{k=n}^{\infty} k^{-1}(U_{k+1} - 1)$ of martingale differences. We have

$$s_n^2 = \sum_{k=n}^{\infty} k^{-2} E[E\{(U_{k+1} - 1)^2 | \mathscr{F}_k\}] = \sum_{k=n}^{\infty} k^{-2} E(1 - Y_k^{-1}) \sim n^{-1}$$

as $n \to \infty$ and hence

$$s_n^{-2} \sum_{k=n}^{\infty} k^{-2} E\{(U_{k+1} - 1)^2 | \mathscr{F}_k\} \sim n \sum_{k=n}^{\infty} k^{-2}(1 - Y_k^{-1}) \to 1 \quad \text{a.s.}$$

Furthermore, since $E(U_{n+1} - 1)^4 \leqslant 25$,

$$E[|U_j - 1| I(j^{-1}|U_j - 1| > \varepsilon s_j)] \leqslant 25\varepsilon^{-3} j^{-3} s_j^{-3}$$

and

$$E[(U_j - 1)^2 I(j^{-1}|U_j - 1| > \varepsilon s_n)] \leqslant 25\varepsilon^{-2} j^{-2} s_n^{-2},$$

so that

$$\sum_{j=1}^{\infty} s_j^{-1} E[j^{-1}|U_j - 1| I(j^{-1}|U_j - 1| > \varepsilon s_j)] < \infty$$

and

$$s_n^{-2} \sum_{j=n}^{\infty} E[j^{-2}(U_j - 1)^2 I(j^{-1}|U_j - 1| > \varepsilon s_n)] \to 0$$

as $n \to \infty$. The conditions of Theorem 4.9 are thus satisfied and so

$$\limsup_{n \to \infty} (2n^{-1} \log \log n)^{-1/2} \sum_{k=n}^{\infty} k^{-1}(U_{k+1} - 1) = +1 \quad \text{a.s.} \qquad (7.90)$$

and

$$\liminf_{n\to\infty}(2n^{-1}\log\log n)^{-1/2}\sum_{k=n}^{\infty}k^{-1}(U_{k+1}-1)=-1 \quad \text{a.s.} \tag{7.90}$$

The result (7.81) follows from (7.82)–(7.85) and (7.90).

Returning to (7.79) and using (7.81), we have

$$\begin{aligned}(\log Z_n - C)^{-1}(Z_{n+1}-Z_n) &= (Y_n^{-1}+W_n)(Z_{n+1}-Z_n)\\ &= U_{n+1}+W_n(Z_{n+1}-Z_n),\end{aligned} \tag{7.91}$$

where $W_n = O((n^{-1}\log\log n)^{1/2}(\log n)^{-2})$ a.s. Also, since $Z_{k+1}-Z_k$ is always positive we have, upon writing $\phi_n = (n^{-1}\log\log n)^{1/2}(\log n)^{-2}$,

$$\sum_{k=3}^{n} W_k(Z_{k+1}-Z_k) = O\left(\sum_{k=3}^{n}\phi_k(Z_{k+1}-Z_k)\right) \quad \text{a.s.} \tag{7.92}$$

and

$$\begin{aligned}\sum_{k=3}^{n}\phi_k(Z_{k+1}-Z_k) &= \phi_n Z_{n+1} - \phi_3 Z_3 + \sum_{k=4}^{n}(\phi_{k-1}-\phi_k)Z_k\\ &= O(n\phi_n\log n) + O\left(\sum_{k=4}^{n}\phi_k\log k\right)\\ &= O(n\phi_n\log n) \quad \text{a.s.,}\end{aligned} \tag{7.93}$$

so that, from (7.79) and (7.91)–(7.93),

$$L\,\mathrm{li}(Z_n/L) - n = \sum_{k=2}^{n}(U_k-1) + O((n\log\log n)^{1/2}(\log n)^{-1}). \tag{7.94}$$

Finally, we shall show that the martingale law of the iterated logarithm (Corollary 4.2) and central limit theorem (Corollary 3.2) apply to $\sum_{k=2}^{n}(U_k-1)$ to give

$$\begin{aligned}&\limsup_{n\to\infty}(2n\log\log n)^{-1/2}\sum_{k=2}^{n}(U_k-1) = +1 \quad \text{a.s.,}\\ &\liminf_{n\to\infty}(2n\log\log n)^{-1/2}\sum_{k=2}^{n}(U_k-1) = -1 \quad \text{a.s.,}\end{aligned} \tag{7.95}$$

and

$$n^{-1/2}\sum_{k=2}^{n}(U_k-1) \xrightarrow{\mathrm{d}} N(0,1).$$

The result of Theorem 7.9 then follows immediately from (7.94) and (7.95).

To apply Corollaries 4.2 and 3.2 we note that

$$\sum_{k=1}^{n} E(U_k - 1)^2 = \sum_{k=1}^{n} (1 - EY_k^{-1}) \sim n \qquad \text{as} \quad n \to \infty$$

and

$$\sum_{n=1}^{\infty} n^{-1}[(U_{n+1} - 1)^2 - E\{(U_{n+1} - 1)^2 | \mathscr{F}_n\}] \qquad \text{converges a.s.,} \tag{7.96}$$

using Theorem 2.15, since

$$\begin{aligned} E(U_{n+1} - 1)^4 &= E[E\{(U_{n+1} - 1)^4 | \mathscr{F}_n\}] \\ &= E[E\{(Y_n^{-1}(X_{n+1} - X_n) - 1)^4 | \mathscr{F}_n\}] \\ &= E\left[\sum_{j=1}^{\infty} (jY_n^{-1} - 1)^4 Y_n^{-1}(1 - Y_n^{-1})^{j-1}\right] \\ &\leqslant E\left[\sum_{j=1}^{Y_n} Y_n^{-1}(1 - Y_n^{-1})^{j-1} + \sum_{j=Y_n+1}^{\infty} (jY_n^{-1})^4 Y_n^{-1}(1 - Y_n^{-1})^{j-1}\right] \\ &\leqslant 1 + E\left[Y_n^{-5} \sum_{j=1}^{\infty} j^4(1 - Y_n^{-1})^{j-1}\right] \\ &\leqslant 1 + 24E[Y_n^{-5}\{1 - (1 - Y_n^{-1})\}^{-5}] \leqslant 25. \end{aligned} \tag{7.97}$$

Furthermore, from (7.96) in view of (7.66), and using Kronecker's lemma,

$$n^{-1} \sum_{k=1}^{n} (U_k - 1)^2 \xrightarrow{\text{a.s.}} 1 \qquad \text{as} \quad n \to \infty.$$

In addition, using (7.97),

$$\begin{aligned} \sum_1^{\infty} n^{-1/2} E[|U_{n+1} - 1| I(|U_{n+1} - 1| > \varepsilon n^{1/2})] &\leqslant \sum_1^{\infty} \varepsilon^{-3} n^{-2} E[U_{n+1} - 1]^4 \\ &\leqslant 25\varepsilon^{-3} \sum_1^{\infty} n^{-2} < \infty \end{aligned}$$

and

$$\sum_1^{\infty} n^{-2} E[(U_{n+1} - 1)^4 I(|U_{n+1} - 1| \leqslant \delta n^{1/2})] \leqslant 25 \sum_1^{\infty} n^{-2} < \infty,$$

while

$$n^{-1} \sum_{k=1}^{n} E[(U_{k+1} - 1)^2 I(|U_{k+1} - 1| > \varepsilon n^{1/2})] \leqslant 25\varepsilon^{-2} n^{-1} \to 0.$$

This completes the verification of the conditions of Corollaries 3.2 and 4.2 and hence the proof of Theorem 7.9. ∎

7.5. Genetic Balance When the Population Size is Varying

A standard problem in mathematical genetics concerns a finite population of haploid individuals in which the successive generations are nonoverlapping. Suppose that there are M individuals in the population which are either of genotype a or A. Suppose also that a has a small selective advantage σ ($\geqslant 0$) over A.

In the classical Sewall Wright formulation each generation is produced by binomial sampling with probabilities proportional to the numbers of a and A individuals in the previous generation multiplied by their respective selective values. Thus, if there are i individuals of type a in a particular generation, the probability that there will be j in the next is

$$P_{ij} = \binom{M}{j} \left\{\frac{(1+\sigma)i}{M+\sigma i}\right\}^j \left\{\frac{M-i}{M+\sigma i}\right\}^{M-j}, \qquad 0 \leqslant j \leqslant M. \tag{7.98}$$

If X_n denotes the number of a alleles in the nth generation, then $\{X_n\}$ is a finite homogeneous Markov chain with transition probabilities given by (7.98). The states 0 and M are absorbing and the others are transient, so fixation of one allele or the other (i.e., $X_n = 0$ or M) is eventually attained with probability one. This classical model has been discussed by many authors. See, for example, Iosifescu and Tăutu (1973, Chapter 4), Ewens (1969, Chapter 4), Moran (1962, Chapters 4, 6), and references therein.

One of the major criticisms leveled at the above model concerns the assumption of a constant population size M, which must, in a sense, restrict modes of long-term behavior. We can, however, relax this assumption in a general way while retaining the essential features of the Wright model. This then allows consideration of, for example, a fluctuating environment.

Let M_n be a random variable denoting the population size at time n (with $M_n > 0$ for all n) and assume further that

(a) the process $\{X_n, M_n, n \geqslant 0\}$ is bivariate Markov, and

(b) the distribution of X_{n+1} given M_{n+1}, M_n, X_n is binomial with probability given by

$$\binom{M_{n+1}}{j} \left\{\frac{(1+\sigma)X_n}{M_n+\sigma X_n}\right\}^j \left\{\frac{M_n - X_n}{M_n+\sigma X_n}\right\}^{M_{n+1}-j}, \qquad 0 \leqslant j \leqslant M_{n+1}. \tag{7.99}$$

The question now is to consider the behavior of X_n in this general case. Write $\mathscr{F}_n$ for the σ-field generated by $\{X_0, M_0, \ldots, X_n, M_n\}$, $n \geqslant 0$, that is, the history of the process up to time n, and set $Y_n = X_n/M_n$. Then

$$E(Y_{n+1}|\mathscr{F}_n) = \frac{(1+\sigma)Y_n}{1+\sigma Y_n} = Y_n + \frac{\sigma Y_n(1-Y_n)}{1+\sigma Y_n} \geqslant Y_n, \tag{7.100}$$

so that $\{Y_n, \mathscr{F}_n, n \geqslant 1\}$ is a submartingale and, since $0 \leqslant Y_n \leqslant 1$, the martingale convergence theorem (Theorem 2.5) assures the existence of a random variable Y such that $Y_n \to Y$ a.s. as $n \to \infty$.

In order to identify Y we need to consider separately the cases $\sigma = 0$ and $\sigma > 0$. The case $\sigma > 0$ is simple and we shall dispose of this first.

Taking expectations in (7.100) we obtain

$$EY_{n+1} = EY_n + \sigma E\left[\frac{Y_n(1 - Y_n)}{1 + \sigma Y_n}\right],$$

so that letting $n \to \infty$ and using dominated convergence,

$$\sigma E\left[\frac{Y(1 - Y)}{1 + \sigma Y}\right] = 0.$$

Therefore, since $0 \leqslant Y \leqslant 1$ and $\sigma > 0$, it follows that $Y(1 - Y) = 0$ a.s. Consequently, the presence of selection assures ultimate homozygosity with probability one. This result is due to Heyde (1977b).

When there is no selection ($\sigma = 0$) the situation is much more complicated. For this case we deduce from (7.99) that

$$E(Y_{n+1} | M_{n+1}, \mathscr{F}_n) = Y_n, \qquad \operatorname{var}(Y_{n+1} | M_{n+1}, \mathscr{F}_n) = M_{n+1}^{-1} Y_n(1 - Y_n),$$

so that

$$E(Y_{n+1}(1 - Y_{n+1}) | \mathscr{F}_n) = (1 - E(M_{n+1}^{-1} | \mathscr{F}_n)) Y_n(1 - Y_n). \tag{7.101}$$

Suppose firstly that M_{n+1} is independent of $\mathscr{F}_n$ for each n. That is, we have a situation of deterministically, or at least independently, fluctuating environments. Then, taking expectations in (7.101),

$$\begin{aligned} EY_{n+1}(1 - Y_{n+1}) &= (1 - EM_{n+1}^{-1}) EY_n(1 - Y_n) \\ &= \prod_{i=1}^{n+1} (1 - EM_i^{-1}) EY_0(1 - Y_0). \end{aligned} \tag{7.102}$$

Write $h = \lim_{n \to \infty} EY_n(1 - Y_n) = EY(1 - Y)$. Of course $h = 0$ is equivalent to $Y(1 - Y) = 0$ a.s., and hence corresponds to the case where ultimate homozygosity is certain. Furthermore, from (7.102) we see that a necessary and sufficient condition for $h = 0$ is $\sum_{i=1}^{\infty} EM_i^{-1} = \infty$. If $\sum_{i=1}^{\infty} EM_i^{-1} < \infty$, we have $h > 0$ and the system may settle down to a state of balanced polymorphism rather than one of ultimate homozygosity.

In the case of a completely general population size process $\{M_n\}$ such a comprehensive discussion of $Y(1 - Y)$ has not been achieved. We can obtain some results by noting, with the aid of (7.101), that

$$\left\{ Z_n = \frac{Y_n(1 - Y_n)}{\prod_{i=1}^{n} (1 - E(M_i^{-1} | \mathscr{F}_{i-1}))}, \mathscr{F}_n, n \geqslant 1 \right\} \tag{7.103}$$

is a nonnegative martingale with mean $EY_0(1 - Y_0)$. Thus, from the martingale convergence theorem, we have the a.s. convergence of Z_n as $n \to \infty$. The following result, due to Heyde and Seneta (1975) following on work of Seneta (1974), is simply deduced.

Theorem 7.10. (i) *If $\sum_{n=1}^{\infty} E(M_n^{-1}|\mathscr{F}_{n-1}) = \infty$ a.s. on some set A of histories of the process (or equivalently $\sum_{n=0}^{\infty} M_n^{-1} = \infty$ a.s. on A), then fixation occurs a.s. on A (i.e., $Y(1 - Y) = 0$ a.s. on A).*

(ii) *If $E(M_n^{-1}|\mathscr{F}_{n-1}) \leqslant \alpha_n$ a.s. for all n, where $\{\alpha_n\}$ is a sequence of positive constants with $\sum \alpha_n < \infty$, or if $M_n \geqslant \beta_n$ a.s. for all n, where $\{\beta_n\}$ is a sequence of positive constants with $\sum \beta_n^{-1} < \infty$, then $P(Y(1 - Y) > 0) > 0$.*

Proof. (i) If $\sum_1^{\infty} E(M_n^{-1}|\mathscr{F}_{n-1}) = \infty$ a.s. on A, then from the a.s. convergence of Z_n as $n \to \infty$ we deduce immediately that $Y_n(1 - Y_n) \to 0$ a.s. on A. The equivalence of $\sum_1^{\infty} E(M_n^{-1}|\mathscr{F}_{n-1}) = \infty$ a.s. on A and $\sum_0^{\infty} M_n^{-1} = \infty$ a.s. on A follows from Corollary 2.3.

(ii) If $E(M_n^{-1}|\mathscr{F}_{n-1}) \leqslant \alpha_n$ a.s. for constants α_n with $\sum \alpha_n < \infty$, then we obtain from (7.103) that $EY(1 - Y) > 0$ and hence $P(Y(1 - Y) > 0) > 0$. Similar reasoning applies to the case $M_n \geqslant \beta_n$ a.s. for constants β_n with $\sum \beta_n^{-1} < \infty$, and this completes the proof. ∎

In the case where M_{n+1} is independent of $\mathscr{F}_n$ for each n and $\sum_1^{\infty} EM_n^{-1} < \infty$, it is possible to obtain precise results on the rate of approach to balanced polymorphism in the case where it obtains. These results are given in the following theorem in which it is interesting to observe the strong dependence of the rates on the growth of M_n.

Theorem 7.11. *Suppose that M_{n+1} is independent of $\mathscr{F}_n$ for each n, and $\sum_1^{\infty} E(M_n^{-1}) < \infty$. Then, conditional on $Y(1 - Y) > 0$,*

$$\frac{Y_n - Y}{\left(\sum_{j=n}^{\infty} E(M_j^{-1})\right)^{1/2}} \xrightarrow{d} \eta^{1/2}(1 - \eta)^{1/2} N(0,1),$$

where η has the distribution of Y and is independent of the standard normal $N(0,1)$. Furthermore, on the set $\{Y(1 - Y) > 0\}$,

$$\limsup_{n\to\infty} \frac{Y_n - Y}{\left[2\sum_{j=n}^{\infty} E(M_j^{-1}) \log\log\left(\sum_{j=n}^{\infty} E(M_j^{-1})\right)^{-1}\right]^{1/2}} = Y^{1/2}(1 - Y)^{1/2} \quad \text{a.s.},$$

$$\liminf_{n\to\infty} \frac{Y_n - Y}{\left[2\sum_{j=n}^{\infty} E(M_j^{-1}) \log\log\left(\sum_{j=n}^{\infty} E(M_j^{-1})\right)^{-1}\right]^{1/2}} = -Y^{1/2}(1 - Y)^{1/2} \quad \text{a.s.}$$

The results follow as a simple application of Theorem 4.9 to the martingale $\{Y_n, \mathscr{F}_n, n \geqslant 1\}$. The details are omitted.

7.6. Stochastic Approximation

7.6.1. Introduction

In certain statistical applications such as bioassay, sensitivity testing, or fatigue trials, problems are obtained which can be conveniently attacked using stochastic approximation methods with a minimum of distributional assumptions. Typical examples of such problem types are as follows.

(a) In testing insecticides it is observed how an insect responds to a certain dose. Let $Y(x)$ be the quantitative response to the dose x. The problem is to determine the critical dose y for a given quantitative response $EY(y)$.

(b) In an agricultural experiment let $Y(x)$ be the crop yield on an experimental plot on which x units of fertilizer have been applied. Generally $Y(x)$ will increase with x for x small and decrease with x for x large. The problem is to find the level y for which the maximum yield $EY(y)$ will be attained.

Quite a large number of stochastic approximation schemes have been discussed in the literature, but they essentially amount to modifications of two basic schemes. These are the Robbins–Monro procedure (appropriate for problem (a)) and the Kiefer–Wolfowitz procedure (appropriate for problem (b)). We shall discuss these procedures in turn, but first we comment on the use of martingale theory in this context.

It is standard practice to iterate the approximation procedure and obtain as the end product a martingale plus other terms which are asymptotically negligible. The asymptotic behavior is then obtained from suitable martingale results. The procedures in essence consist of a convergent deterministic component with a superimposed random element.

Stochastic approximation methods are closely related to recursive least squares [as pointed out by Ho (1962)], while Albert and Gardner (1967) applied stochastic approximation techniques to the estimation of parameters of a nonlinear regression. A comprehensive discussion of both stochastic approximation and recursive estimation and their relationship is provided by Nevel'son and Has'minskii (1973/76). The control engineering literature contains many applications of stochastic approximation methods in the system identification (i.e., parameter estimation) of time-series [e.g., Saridis (1974, 1977, Chapters 3, 6); Tsypkin (1971, Chapter 3)]. Traditional stochastic approximation is treated in the book of Wasan (1967) and the more practical side of the subject is discussed in Wetherill (1966, Chapter 9).

We shall give strong laws and also central limit and iterated logarithm results for the basic Robbins–Monro and Kiefer–Wolfowitz procedures. Similar analysis can also be given for various modifications of these schemes [e.g., those of Venter (1967) and Fabian (1968)]. The strong law results come from Dvoretzky (1956). The other work follows, at least in spirit, Sacks (1958). The iterated logarithm results are given in Heyde (1974b).

7.6.2. The Robbins–Monro Procedure

For the Robbins–Monro (R–M) scheme, the observable is a random variable $Y(x)$ having a distribution depending on the level x at which the observation is made. We have $EY(x) = M(x)$, where the function M is such that the equation $M(x) = \alpha$ has a unique solution $x = \theta$. The R–M scheme for finding θ starts with an initial guess X_1, and successive observations $Y_r(X_r)$ are taken at levels X_r chosen according to the formula

$$X_{n+1} = X_n - a_n\{Y_n(X_n) - \alpha\}. \tag{7.104}$$

Here $\{a_n\}$ is a decreasing sequence of positive numbers chosen so that

$$\sum a_n = \infty, \qquad \sum a_n^2 < \infty,$$

and $Y_r(X_r)$ is a random variable whose conditional distribution, given $X_1 = x_1, \ldots, X_r = x_r$, is the same as the distribution of $Y(x_r)$. Under the additional conditions (A1) and (A2) below and $\sup_x \operatorname{var} Y(x) \to \infty$, it will be shown in Theorem 7.12 that X_n converges a.s. to θ as $n \to \infty$. The relevance of martingale limit theory to the study of the limit behavior of this approximation scheme may be made rather transparent by rewriting (7.104) as

$$X_{n+1} = X_n - a_n\{M(X_n) - \alpha\} - a_n Z(X_n),$$

where

$$Z(X_n) = Y_n(X_n) - M(X_n)$$

and $\lim_{n\to\infty} M(X_n) = \alpha$ a.s. Clearly

$$E(Z(X_n)|X_1, \ldots, X_n) = 0 \quad \text{a.s.},$$

so that the $Z(X_k)$ are martingale differences.

We shall use the following assumptions (A1) through (A4) about $M(x)$ and $Z(x)$.

(A1) M is a Borel-measurable function; $M(\theta) = \alpha$ and

$$(x - \theta)(M(x) - \alpha) > 0$$

for all $x \neq \theta$.

(A2) For some positive constants K_1 and K_2, and for all x,

$$K_1|x - \theta| \leqslant |M(x) - \alpha| \leqslant K_2|x - \theta|.$$

(A3), (A3′) For all x,

$$M(x) = \alpha + \alpha_1(x - \theta) + \delta(x,\theta),$$

where $\alpha_1 > 0$ and for (A3)

$$\delta(x,\theta) = o(|x - \theta|),$$

as $x - \theta \to 0$, while for (A3′)

$$\delta(x,\theta) = \alpha_2(x-\theta)^2 + o(x-\theta)^2$$

as $x - \theta \to 0$ for some finite constant α_2.

(A4) (a) $\sup_x E|Z(x)|^{2+\eta} < \infty$ for some $\eta > 0$, and (b) $\lim_{x\to\theta} EZ^2(x) = \sigma^2$.

We shall obtain the following results on the convergence of X_n to θ.

Theorem 7.12. *Suppose that assumptions* (A1) *and* (A2) *are satisfied and* $\sup_x \operatorname{var} Y(x) < \infty$. *Then X_n converges a.s. to θ.*

Theorem 7.13. *Suppose that assumptions* (A1) *through* (A4) *are satisfied. Let $a_n = An^{-1}$, where A is such that $2K_1A > 1$. Then*

$$n^{1/2}(X_n - \theta) \xrightarrow{\text{d}} N(0, A^2\sigma^2(2A\alpha_1 - 1)^{-1}).$$

In addition, if (A3) *is strengthened to* (A3′), *then*

$$X_n = \theta + A\sigma(2A\alpha_1 - 1)^{1/2}\zeta_1(n)(2n^{-1}\log\log n)^{1/2},$$

where $\zeta_1(n)$ has its set of a.s. limit points confined to $[-1,1]$ with

$$\limsup_{n\to\infty} \zeta_1(n) = 1 \quad a.s., \qquad \liminf_{n\to\infty} \zeta_1(n) = -1 \quad a.s.$$

These results are only illustrative and of course more general sequences $\{a_n\}$ may be treated in Theorem 7.13. Invariance principles for the CLT for the case $a_n = An^{-1}$ have been provided by McLeish (1976) and Kersting (1977). A detailed discussion for more general $\{a_n\}$ is given by Solo (1978). Some situations where $2A\alpha_1 \leqslant 1$ may also be treated; see Major and Révész (1973).

7.6.3. The Kiefer–Wolfowitz Procedure

For the Kiefer–Wolfowitz (K–W) scheme, again we have as the observable a random variable $Y(x)$ having a distribution depending on the level x at which the observation is made. Here $EY(x) = M(x)$, where M has a unique maximum at $x = \theta$. The K–W scheme for finding θ operates as follows. Let $\{a_n\}$, $\{c_n\}$ be two sequences of positive numbers such that

$$\sum a_n = \infty, \qquad c_n \to 0, \qquad \sum a_n^2 c_n^{-2} < \infty. \tag{7.105}$$

Starting with an initial guess X_1, successive observations $Y_r(X_r + c_r)$, $Y_r(X_r - c_r)$ are taken for levels determined from the recursion formula

$$X_{n+1} = X_n - a_n c_n^{-1}\{Y_n(X_n - c_n) - Y_n(X_n + c_n)\}. \tag{7.106}$$

Here $Y_n(X_n \pm c_n)$ is a random variable whose conditional distribution, given $X_1 = x_1, \ldots, X_n = x_n$, is the same as the distribution of $Y(x_n \pm c_n)$. For this scheme and under the additional conditions (B1) below and $\sup_x \operatorname{var} Y(x) < \infty$, it will be shown in Theorem 7.14 that X_n converges a.s. to θ.

To see the relevance of martingale limit theory here we again use $Z(x) = Y(x) - M(x)$ and rewrite (7.106) as

$$X_{n+1} = X_n - a_n c_n^{-1}(M_n + Z_n),$$

where

$$M_n = M(X_n - c_n) - M(X_n + c_n), \qquad Z_n = Z(X_n - c_n) - Z(X_n + c_n).$$

Then $\{Z_n\}$ is a sequence of martingale differences while $M_n \to 0$ a.s. as $n \to \infty$.

We shall use the following assumptions (B1) through (B5).

(B1) M is a Borel-measurable function, has a unique maximum at $x = \theta$, and, for all $0 < t_0 < t_1 < t_2 < \infty$,

$$\inf_{\substack{t_1 \leqslant |x-\theta| \leqslant t_2 \\ 0 < \varepsilon < t_0}} (x - \theta)\{M(x - \varepsilon) - M(x + \varepsilon)\}\varepsilon^{-1} > 0.$$

In addition, for suitable positive constants D_1, D_2 and all x,

$$|M(x + 1) - M(x)| < D_1 + D_2|x|.$$

(B2), (B2′) For all x,

$$M(x) = \alpha_0 - \alpha(x - \theta)^2 + \delta(x,\theta),$$

where α_0 is some real number, $\alpha > 0$ and for (B2),

$$\delta(x,\theta) = o(|x - \theta|^2),$$

as $x - \theta \to 0$, while for (B2′),

$$\delta(x,\theta) = O(|x - \theta|^3)$$

as $x - \theta \to 0$.

(B3) For some $c_0 > 0$ there exist positive constants K_1 and K_2 such that for all c for which $0 < c \leqslant c_0$,

$$K_1(x - \theta)^2 \leqslant (x - \theta)\{M(x - c) - M(x + c)\}c^{-1} \leqslant K_2(x - \theta)^2.$$

(B4) For every $\varepsilon > 0$ there exists $c_\varepsilon > 0$ such that, for all c satisfying $0 < c \leqslant c_\varepsilon$ and all x satisfying $|x - \theta| < c$,

$$|\delta(x - c, \theta) - \delta(x + c, \theta)|c^{-1} \leqslant \varepsilon|x - \theta|.$$

(B5) (a) $\sup_x E|Z(x)|^{2+\eta} < \infty$ for some $\eta > 0$, and
(b) $\lim_{x\to\theta,\, a\to 0} E\{Z(x - a) - Z(x + a)\}^2 = \sigma^2$.

We shall obtain the following results on the convergence of X_n to θ.

Theorem 7.14. *Suppose that assumption* (B1) *is satisfied and* $\sup_x \operatorname{var} Y(x) < \infty$. *Then* X_n *converges a.s. to* θ.

Theorem 7.15. *Suppose that assumptions* (B1) *through* (B5) *are satisfied. Let* $a_n = An^{-1}$, *where* A *is such that* $2K_1A > 1$ *and* $c_n = n^{-\gamma}$, *where* $0 < \gamma < \frac{1}{2}$. *Then*

$$n^{(1/2)-\gamma}(X_n - \theta) \xrightarrow{d} N(0, A^2\sigma^2(8\alpha A + 2\gamma - 1)^{-1}).$$

In addition, if (B2) *is strengthened to* (B2′) *and* $\frac{1}{6} < \gamma < \frac{1}{3}$, *then*

$$X_n = \theta + A\sigma(8\alpha A + 2\gamma - 1)^{1/2}\zeta_2(n)(2n^{-1+2\gamma}\log\log n)^{1/2},$$

where $\zeta_2(n)$ *has its set of a.s. limit points confined to* $[-1,1]$ *with* $\limsup_{n\to\infty} \zeta_2(n) = 1$ *a.s.,* $\liminf_{n\to\infty} \zeta_2(n) = -1$ *a.s.*

The constants $\{a_n\}$ and $\{c_n\}$ should in principle be chosen with a view to obtaining an optimal rate of convergence of X_n to θ in some appropriate sense. It follows from work of Sacks (1958) that under the above assumptions, but with $\{c_n\}$ satisfying (7.105) and $c_n c_{n+1}^{-1} = 1 + o(n^{-1})$ as $n \to \infty$ (e.g., $c_n = \log n$, $n > 1$), $n^{1/3}c_n(X_n - \theta)$ is asymptotically normally distributed with mean 0 and variance $A^2\sigma^2(8\alpha A - 1)^{-1}$. The analysis is similar to that given below. This result provides a better convergence rate than our choice of c_n, as evidenced by Theorem 7.15. However, in the interests of clarity and simplicity we have chosen $c_n = n^{-\gamma}$. This yields results which are direct analogs of those for the R–M scheme and the proofs can be conducted simultaneously. More general results have been provided by Solo (1978). For extensions of Theorems 7.12 and 7.14 see Ljung (1978).

7.6.4. Proofs of Theorems 7.12 and 7.14

Almost sure convergence results for both the R–M and K–W procedures are obtained with the aid of the following general result of Dvoretzky (1956).

Theorem 7.16. *Let* $\{X_n\}$, $\{T_n = T_n(X_1, \ldots, X_n)\}$, *and* $\{Y_n = Y_n(X_1, \ldots, X_n)\}$ *be sequences of random variables with* X_1 *arbitrary and*

$$X_{n+1} = T_n + Y_n, \tag{7.107}$$

where

$$E(Y_n | X_1, \ldots, X_n) = 0 \quad a.s. \tag{7.108}$$

and

$$\sum EY_n^2 < \infty. \tag{7.109}$$

Suppose also that

$$|T_n| \leqslant \max(\alpha_n, (1 + \beta_n)|X_n| - \gamma_n), \tag{7.110}$$

where α_n, β_n, *and* γ_n *are nonnegative random variables such that*

$$\alpha_n \xrightarrow{a.s.} 0 \qquad \sum \beta_n < \infty \quad a.s., \qquad and \qquad \sum \gamma_n = \infty \quad a.s. \tag{7.111}$$

Then $X_n \xrightarrow{a.s.} 0$ *as* $n \to \infty$.

As a prelude to the proof of this result, in which we follow Derman and Sacks (1959), we need the following lemma.

Lemma 7.2. *Let* $\{a_n\}$, $\{b_n\}$, $\{c_n\}$, $\{d_n\}$, *and* $\{e_n\}$ *be sequences of real numbers satisfying*

(i) $\{a_n\}$, $\{b_n\}$, $\{c_n\}$, *and* $\{e_n\}$ *are nonnegative,*

(ii) $\lim_{n\to\infty} a_n = 0$, $\sum b_n < \infty$, $\sum c_n = \infty$, $\sum d_n$ *converges and, for all n larger than some* N_0,

(iii) $e_{n+1} \leqslant \max(a_n, (1 + b_n)e_n + d_n - c_n)$.

Then $\lim_{n\to\infty} e_n = 0$.

Proof. Let $N > N_0$ and write

$$B_n = \prod_{i=1}^{n} (1 + b_i). \tag{7.112}$$

Take $n > N$ and iterate (iii) back to N. This yields

$$e_{n+1} \leqslant \max\Bigg(B_n B_{N-1}^{-1} e_N + B_n \sum_{j=N}^{n} (d_j - c_j) B_j^{-1}, \\ \max_{N \leqslant k \leqslant n} \Bigg[B_n B_k^{-1} a_k + B_n \sum_{j=k+1}^{n} (d_j - c_j) B_j^{-1} \Bigg] \Bigg). \tag{7.113}$$

Now (i) and (ii) imply that B_n increases to B, say, which is finite. It can then be shown that $\sum_{j=1}^{\infty} d_j B_j^{-1} < \infty$ and $\sum_{j=1}^{\infty} c_j B_j^{-1} = \infty$. Since $B_n B_{N-1}^{-1} e_N$ is bounded, we see that the first term on the right-hand side of (7.113) must be negative for large enough n and can therefore be ignored. Thus, for n large enough,

$$e_{n+1} \leqslant \max_{N \leqslant k \leqslant n} \Bigg(B_n B_k^{-1} a_k + B_n \sum_{j=k+1}^{n} (d_j - c_j) B_j^{-1} \Bigg) \\ \leqslant B \Bigg(\max_{k \geqslant N} B_k^{-1} a_k + \max_{N \leqslant k \leqslant n} \Bigg| \sum_{j=k+1}^{n} d_j B_j^{-1} \Bigg| \Bigg). \tag{7.114}$$

Since $\sum d_j B_j^{-1}$ converges and $a_k \to 0$, the right member of (7.114) can be made arbitrarily small by choosing N large enough. This completes the proof of the lemma. ▮

Proof of Theorem 7.16. We may assume that $\{\alpha_n\}$ is such that

$$\sum E(\alpha_n^{-2} Y_n^2) < \infty. \tag{7.115}$$

For, if this is not the case, there is always a (nonstochastic) sequence $\{\alpha_n^*\}$ which satisfies (7.111) and (7.115). Taking $A_n = \max(\alpha_n, \alpha_n^*)$, we obtain a sequence which satisfies (7.110), (7.111), and (7.115).

Define $Z_n = Y_n \operatorname{sgn} T_n$. Then (7.108), (7.109), and (7.115) hold with Y_n replaced by Z_n. From (7.108), (7.109), and Theorem 2.18 we see that $\sum Z_n$ converges a.s. Further, from (7.115), Chebyshev's inequality, and the Borel–Cantelli lemma we conclude that with probability tending to one as $N \to \infty$,

$$|Z_n| \leqslant \alpha_n \tag{7.116}$$

for all $n \geqslant N$. Then, using (7.107) and (7.116), we can write with probability tending to one as $N \to \infty$,

$$\begin{aligned} |X_{n+1}| &\leqslant 2\alpha_n && \text{if} \quad |T_n| \leqslant \alpha_n \\ &= |T_n| + Z_n && \text{if} \quad |T_n| > \alpha_n \end{aligned}$$

for all $n \geqslant N$. Hence

$$\begin{aligned} |X_{n+1}| &\leqslant \max(2\alpha_n, |T_n| + Z_n) \\ &\leqslant \max(2\alpha_n, (1 + \beta_n)|X_n| + Z_n - \gamma_n) \end{aligned}$$

with probability tending to one. The required result then follows from Lemma 7.2 with $e_n = |X_n|$, $a_n = 2\alpha_n$, $b_n = \beta_n$, $d_n = Z_n$, and $c_n = \gamma_n$. ▮

Proofs of Theorems 7.12 and 7.14. For the R–M case we may suppose that $\theta = 0$ and $\alpha = 0$ without loss of generality, so that

$$X_{n+1} = X_n - a_n M(X_n) - a_n Z(X_n).$$

We apply Theorem 7.16 with

$$T_n = X_n - a_n M(X_n), \qquad Y_n = -a_n Z(X_n).$$

The required conditions on Y_n are seen to be satisfied since $\sum a_n^2 < \infty$. Furthermore, letting $\{\rho_n\}$ be a sequence of positive constants with $\rho_n \to 0$ as $n \to \infty$ and $\sum \rho_n a_n = \infty$, we have from (A1) and (A2) that for $|X_n| > \rho_n$ and n sufficiently large,

$$|T_n| = |X_n - a_n M(X_n)| < |X_n| K_1 a_n \rho_n,$$

while for $|X_n| \leqslant \rho_n$,

$$|T_n| = |X_n - a_n M(X_n)| \leqslant \rho_n(1 + K_2 a_n).$$

Now take $\alpha_n = \rho_n(1 + K_2 a_n)$, $\beta_n = 0$, and $\gamma_n = K_1 a_n \rho_n$, and the result of Theorem 7.12 follows. The proof for the K–W case is quite similar and is omitted. ▌

7.6.5. Proofs of Theorems 7.13 and 7.15

We conduct both proofs in the one exercise. Take without loss of generality $\theta = 0$ and $\alpha = 0$ in the R–M case, $\theta = 0$ and $\alpha_0 = 0$ in the K–W case. We have to investigate the approximation scheme

$$X_{n+1} = (1 - Acn^{-1})X_n - An^{-1+\gamma}\delta_n - An^{-1+\gamma}Z_n. \tag{7.117}$$

For the R–M scheme we have $c = \alpha_1$, $\gamma = 0$, $\delta_n = \delta(X_n, 0)$, and $Z_n = Z(X_n)$, and for the K–W scheme we have $c = 4\alpha$, $0 < \gamma < \frac{1}{2}$ or $\frac{1}{6} < \gamma < \frac{1}{3}$, $\delta_n = \delta(X_n - c_n, 0) - \delta(X_n + c_n, 0)$, and $Z_n = Z(X_n - c_n) - Z(X_n + c_n)$.

Write

$$\beta_{mn} = \begin{cases} \prod_{j=m+1}^{n}(1 - Acj^{-1}), & 0 \leqslant m < n \\ 1, & m = n. \end{cases}$$

Then if

$$\gamma_n = \prod_{j=1}^{n}(1 - Acj^{-1}), \qquad n > 0, \quad \gamma_0 = 1,$$

we have

$$\beta_{mn} = \gamma_n \gamma_m^{-1}, \qquad 0 \leqslant m \leqslant n. \tag{7.118}$$

Using $[x]$ to denote the integer part of x, we have for $n \geqslant [Ac] + 1$,

$$\begin{aligned} \gamma_n &= \gamma_{[Ac]} \frac{[Ac]!}{n!} \prod_{j=[Ac]+1}^{n} (j - Ac) \\ &= \gamma_{[Ac]} \frac{[Ac]!}{n!} \frac{\Gamma(n + 1 - Ac)}{\Gamma(1 + [Ac] - Ac)} \\ &\sim \frac{\gamma_{[Ac]}[Ac]!}{\Gamma(1 + [Ac] - Ac)} n^{-Ac} \end{aligned} \tag{7.119}$$

as $n \to \infty$ from an application of Stirling's formula. Of course $\gamma_{[Ac]}$ need not be positive.

In the case of the CLT we must prove that

$$n^{(1/2)-\gamma}X_n \xrightarrow{d} N(0, A^2\sigma^2(2Ac + 2\gamma - 1)^{-1}),$$

and for the LIL,

$$\limsup_{n\to\infty} n^{(1/2)-\gamma}(2\log\log n)^{-1/2}X_n = A\sigma(2Ac + 2\gamma - 1)^{-1/2} \quad \text{a.s.},$$

$$\liminf_{n\to\infty} n^{(1/2)-\gamma}(2\log\log n)^{-1/2}X_n = -A\sigma(2Ac + 2\gamma - 1)^{-1/2} \quad \text{a.s.}$$

Iterating in (7.117) gives

$$X_{n+1} = \beta_{0n} X_1 - A \sum_{m=1}^{n} m^{-1+\gamma} \beta_{mn} \delta_m - A \sum_{m=1}^{n} m^{-1+\gamma} \beta_{mn} Z_m. \quad (7.120)$$

Now, from (7.118) and (7.119), $n^{(1/2)-\gamma}\beta_{0n} = O(n^{(1/2)-\gamma-Ac})$ and in the R–M case $\gamma = 0$, $c = \alpha_1$, and $2A\alpha_1 > 1$ (from (A2), (A3), and $2K_1 A > 1$), while in the K–W case $c = 4\alpha$ and $8\alpha A > 1$ (from (B2), (B3), (B4), and $2K_1 A > 1$). Thus,

$$\lim_{n\to\infty} n^{(1/2)-\gamma}\beta_{0n} X_1 = 0 \quad \text{a.s.}$$

and, in view of (7.118), (7.119), and (7.120), in order to prove the CLT it suffices to show that

$$n^{(1/2)-\gamma}|\gamma_n| \sum_{m=1}^{n} m^{-1+\gamma} |\gamma_m|^{-1} \delta_m \xrightarrow{\text{p}} 0 \quad (7.121)$$

and

$$n^{(1/2)-\gamma}|\gamma_n| \sum_{m=1}^{n} m^{-1+\gamma} |\gamma_m|^{-1} Z_m \xrightarrow{\text{d}} N(0, \sigma^2(2Ac + 2\gamma - 1)^{-1}), \quad (7.122)$$

while to establish the LIL it suffices to show that

$$\lim_{n\to\infty} n^{(1/2)-\gamma}|\gamma_n| (\log\log n)^{-1/2} \sum_{m=1}^{n} m^{-1+\gamma} |\gamma_m|^{-1} \delta_m = 0 \quad \text{a.s.} \quad (7.123)$$

and

$$\limsup_{n\to\infty} n^{(1/2)-\gamma}|\gamma_n| (\log\log n)^{-1/2} \sum_{m=1}^{n} m^{-1+\gamma} |\gamma_m|^{-1} Z_m = \sigma(2Ac + 2\gamma - 1)^{-1/2} \quad \text{a.s.,}$$

$$\liminf_{n\to\infty} n^{(1/2)-\gamma}|\gamma_n| (\log\log n)^{-1/2} \sum_{m=1}^{n} m^{-1+\gamma} |\gamma_m|^{-1} Z_m = -\sigma(2Ac + 2\gamma - 1)^{-1/2} \quad \text{a.s.} \quad (7.124)$$

We shall first dispose of (7.121) and (7.123) in the R–M case. To do this we need the estimate $EX_n^2 = O(n^{-1})$, which we now proceed to obtain.

Squaring both sides of (7.104), taking expected values, and using (A4), we obtain

$$EX_{n+1}^2 = E(X_n - An^{-1}M(X_n))^2 + O(n^{-2}).$$

Then, by (A1) and (A2), for ε sufficiently small that $2K_1A - \varepsilon > 1$, and for n sufficiently large,

$$\begin{aligned} EX_{n+1}^2 &\leqslant (1 - 2K_1An^{-1} + A^2K_2^2n^{-2})EX_n^2 + O(n^{-2}) \\ &\leqslant (1 - (2K_1A - \varepsilon)n^{-1})EX_n^2 + D_1n^{-2}. \end{aligned} \tag{7.125}$$

Iteration of (7.125) yields $EX_n^2 = O(n^{-1})$.

To deal with (7.121), we note that $\delta(x,0) = O(|x|)$ from (A3), so that given $\lambda > 0$ we can find $\varepsilon > 0$ such that

$$|\delta(x,0)| \leqslant \lambda^2|x| \qquad \text{for} \quad |x| < \varepsilon.$$

Since $X_n \to 0$ a.s. by Theorem 7.12, we have

$$\lim_{n\to\infty} n^{1/2}|\gamma_n| \sum_{m=1}^{n} m^{-1}|\gamma_m|^{-1}|\delta_m|I(|X_m| > \varepsilon) = 0 \quad \text{a.s.}$$

and

$$\begin{aligned} n^{1/2}|\gamma_n| \sum_{m=1}^{n} m^{-1}|\gamma_m|^{-1}E|\delta_m|I(|X_m| \leqslant \varepsilon) &\leqslant \lambda^2 n^{1/2}|\gamma_n| \sum_{m=1}^{n} m^{-1}|\gamma_m|^{-1}E|X_m| \\ &\leqslant \lambda^2 C n^{1/2}|\gamma_n| \sum_{m=1}^{n} m^{-3/2}|\gamma_m|^{-1} \end{aligned}$$

for some constant $C > 0$. Now λ is arbitrarily small and

$$n^{1/2}|\gamma_n| \sum_{m=1}^{n} m^{-3/2}|\gamma_m|^{-1} \to (Ac - \tfrac{1}{2})^{-1},$$

using (7.119). The result (7.121) follows.

In order to prove (7.123) we note that, using (A3′), it suffices to show that

$$\lim_{n\to\infty} n^{1/2}|\gamma_n|(\log\log n)^{-1/2} \sum_{m=1}^{n} m^{-1}|\gamma_m|^{-1}X_m^2 = 0 \quad \text{a.s.,}$$

which holds via the Kronecker lemma if

$$\sum (n \log\log n)^{-1/2} X_n^2 < \infty \quad \text{a.s.,}$$

and hence if

$$\sum (n \log\log n)^{-1/2} EX_n^2 < \infty.$$

This last result is of course satisfied since $EX_n^2 = O(n^{-1})$ as $n \to \infty$.

Next we shall deal with (7.121) and (7.123) for the K–W case. Here we need the estimate $EX_n^2 = O(n^{-1+2\gamma})$, which we now proceed to obtain.

Squaring both sides of (7.106), taking expected values, and using (B3) and (B5), we obtain

$$EX_{n+1}^2 \leqslant (1 - 2K_1An^{-1} + K_2^2A^2n^{-2})EX_n^2 + O(n^{-2+2\gamma}).$$

Let $\varepsilon > 0$ be such that $2K_1A - \varepsilon > 1$. Then for sufficiently large n,

$$EX_{n+1}^2 \leqslant (1 - 2K_1A - \varepsilon)n^{-1}EX_n^2 + O(n^{-2+2\gamma}). \tag{7.126}$$

Iteration of (7.126) yields $EX_n^2 = O(n^{-1+2\gamma})$.

We shall deal with (7.121) by showing that

$$n^{(1/2)-\gamma}|\gamma_n| \sum_{m=1}^{n} m^{-1+\gamma}|\gamma_m|^{-1}|\delta_m|I(|X_m| \leqslant (\log m)^{-1}) \xrightarrow{\text{P}} 0 \tag{7.127}$$

and

$$n^{(1/2)-\gamma}|\gamma_n| \sum_{m=1}^{n} m^{-1+\gamma}|\gamma_m|^{-1}|\delta_m|I(|X_m| > (\log m)^{-1}) \xrightarrow{\text{P}} 0. \tag{7.128}$$

The result (7.127) follows since (B4) yields

$$m^{\gamma}|\delta_m|I(|X_m| \leqslant (\log m)^{-1}) < \varepsilon|X_m|$$

for m sufficiently large, and

$$n^{(1/2)-\gamma}|\gamma_n| \sum_{m=1}^{n} m^{-1}|\gamma_m|^{-1}(EX_m^2)^{1/2} = O(1)$$

as $n \to \infty$ when $EX_n^2 = O(n^{-1+2\gamma})$. On the other hand, (B3) gives $|\delta_m|m^{\gamma} = O(|X_m|)$. Also,

$$E(|X_m|I(|X_m| > (\log m)^{-1}) \leqslant [P(|X_m| > (\log m)^{-1})EX_m^2]^{1/2},$$

and $P(|X_m| > (\log m)^{-1}) = o(1)$ using Chebyshev's inequality, so that (7.128) is implied by

$$n^{(1/2)-\gamma}|\gamma_n| \sum_{m=1}^{n} m^{-1}|\gamma_m|^{-1}[P(|X_m| > (\log m)^{-1})EX_m^2]^{1/2} = o(1)$$

as $n \to \infty$.

To prove (7.123) we note that it suffices to show, via the Kronecker lemma, that

$$\sum (n \log\log n)^{-1/2}|\delta_n| < \infty \quad \text{a.s.}$$

and hence, in view of (B2′), that

(i) $\sum (n \log\log n)^{-1/2}|X_n|^3 < \infty$ a.s.,
(ii) $\sum (n \log\log n)^{-1/2}n^{-\gamma}X_n^2 < \infty$ a.s.,
(iii) $\sum (n \log\log n)^{-1/2}n^{-2\gamma}|X_n| < \infty$ a.s., and
(iv) $\sum (n \log\log n)^{-1/2}n^{-3\gamma} < \infty$.

Condition (iv) is obvious since $\gamma > \frac{1}{6}$ for the case in question, and (ii) holds since $EX_n^2 = O(n^{-1+2\gamma})$, so that $\sum n^{-\gamma}(n \log\log n)^{-1/2}EX_n^2 < \infty$. Furthermore, $\sum(n \log\log n)^{-1/2} n^{-2\gamma} E|X_n| < \infty$ by virtue of

$$(\log\log n)^{-1/2} E|X_n| = o((EX_n^2)^{1/2}) = o(n^{-(1/2)+\gamma}),$$

and so (iii) holds. To prove (i) we first show that $n^{\varepsilon} X_n = o(1)$ a.s. as $n \to \infty$ for any $0 < \varepsilon < \frac{1}{2} - \gamma$. This is obtained with the aid of the following version of a final value theorem.

Lemma 7.3. *Suppose that $\{e_n\}$ is a real sequence satisfying*

$$e_{n+1} = (1 - a_n)e_n + b_n,$$

where $a_n \geqslant 0$, $a_n \to 0$, and $\sum a_n = \infty$. If $b_0 < \infty$,

$$\lim_{n\to\infty} e_n = \lim_{n\to\infty} b_n/a_n \tag{7.129}$$

if the right-hand limit exists.

Proof. Without loss of generality each $a_n < 1$. We write

$$A_{n+1} = \prod_{k=1}^{n} (1 - a_k)$$

and observe that $A_n \downarrow 0$ as $n \to \infty$. Then

$$e_{n+1} = A_{n+1} A_n^{-1} e_n + b_n,$$

and iteration gives

$$e_{n+1} = A_{n+1} \sum_{k=1}^{n} A_{k+1}^{-1} b_k + b_0 A_{n+1}.$$

The result (7.129) is then immediate from the Toeplitz lemma since

$$\lim_{n\to\infty} A_{n+1} \sum_{k=1}^{n} A_{k+1}^{-1} a_k = 1. \quad \blacksquare$$

To apply Lemma 7.3 we multiply both sides of (7.117) by $(n+1)^{\varepsilon}$. Then the lemma certainly gives $\lim_{n\to\infty} n^{\varepsilon} X_n = 0$ a.s. if $\sum n^{-1+\gamma+\varepsilon}\delta_n$ and $\sum n^{-1+\gamma+\varepsilon} Z_n$ both converge a.s. The former holds, however, since $\sum n^{-1+\gamma+\varepsilon} E|\delta_n| < \infty$ by virtue of (B3), which gives for n sufficiently large,

$$\begin{aligned} n^{-1+\gamma+\varepsilon} E|\delta_n| &= O(n^{-1+\varepsilon} E|X_n|) \\ &= O(n^{-1+\varepsilon}(EX_n^2)^{1/2}) = O(n^{-(3/2)+\varepsilon+\gamma}). \end{aligned}$$

The latter holds, via the martingale convergence theorem, since the Z_n are martingale differences and

$$\begin{aligned} E(\textstyle\sum n^{-1+\gamma+\varepsilon} Z_n)^2 &= \textstyle\sum n^{-2+2\gamma+2\varepsilon} EZ_n^2 \\ &< C \textstyle\sum n^{-2+2\gamma+2\varepsilon} < \infty \end{aligned}$$

for some $C > 0$ using (B5). Then, since $n^{\varepsilon} X_n = o(1)$ a.s., we have

$$|X_n|^3 (n \log\log n)^{-1/2} = o(n^{-(1/2) - 3\varepsilon}) \quad \text{a.s.}$$

for any $0 < \varepsilon < \frac{1}{2} - \gamma$, so that (i) holds for $\gamma < \frac{1}{3}$. This completes the proof of (7.123).

To prove (7.122) and (7.124) we first write

$$U_m = m^{-1+\gamma} |\gamma_m|^{-1} Z_m, \qquad m \geqslant 1,$$

and note that the U_m are martingale differences. We set

$$s_n^2 = E\left(\sum_{m=1}^{n} U_m \right)^2 = \sum_{m=1}^{n} m^{-2+2\gamma} \gamma_m^{-2} E Z_m^2.$$

Also, $EZ_n^2 \to \sigma^2$ as $n \to \infty$ via (A4) and (B5), so that using the Toeplitz lemma and the relation (7.119),

$$s_n^2 \sim \sigma^2 \gamma_n^{-2} (2Ac + 2\gamma - 1)^{-1} n^{2\gamma - 1}$$

as $n \to \infty$. Thus, we need to show that

$$s_n^{-1} \sum_{m=1}^{n} U_m \xrightarrow{\text{d}} N(0,1)$$

in the case of (7.122), and

$$\limsup_{n \to \infty} (2 s_n^2 \log\log s_n^2)^{-1/2} \sum_{m=1}^{n} U_m = 1 \qquad \text{a.s.,}$$

$$\liminf_{n \to \infty} (2 s_n^2 \log\log s_n^2)^{-1/2} \sum_{m=1}^{n} U_m = -1 \quad \text{a.s.}$$

in the case of (7.124). These results follow from Theorem 4.1 and Corollary 4.2, provided that

(I) $\sum_1^{\infty} s_n^{-4} E(U_n^4 I(|U_n| < \delta s_n)) < \infty$, some $\delta > 0$,
(II) $\sum_1^{\infty} s_n^{-2} E(U_n^2 I(|U_n| \geqslant \varepsilon s_n)) < \infty$, all $\varepsilon > 0$, and
(III) $s_n^{-2} \sum_{m=1}^{n} U_m^2 \xrightarrow{\text{a.s.}} 1$ as $n \to \infty$

hold.

To check (I) we need to show that

$$\sum_1^{\infty} n^{-2} E(Z_n^4 I(|Z_n| < \delta s_n n^{1-\gamma} |\gamma_n|)) < \infty$$

for some $\delta > 0$, and this holds if

$$\sum_1^{\infty} n^{-2} E(Z_n^4 I(|Z_n| < \delta' n^{1/2})) < \infty$$

for some $\delta' > 0$, since

$$s_n n^{1-\gamma}|\gamma_n| \sim \sigma(2Ac + 2\gamma - 1)^{-1/2} n^{1/2} \tag{7.130}$$

as $n \to \infty$. That (I) holds then follows since

$$E(Z_n^4 I(|Z_n| < \delta' n^{1/2})) < E|Z_n|^{2+\eta}(\delta' n^{1/2})^{2-\eta} = O(n^{1-\eta/2})$$

from (A4) and (B5).

For (II) we need to show that

$$\sum_1^\infty n^{-1} E(Z_n^2 I(|Z_n| \geqslant \varepsilon n^{1/2})) < \infty$$

for every $\varepsilon > 0$. This holds, however, since, using Hölder's inequality and then Chebyshev's inequality together with (A4) and (B5),

$$E(Z_n^2 I(|Z_n| \geqslant \varepsilon n^{1/2})) \leqslant [E|Z_n|^{2+\eta}]^{2/(2+\eta)}[P(|Z_n| > \varepsilon n^{1/2})]^{\eta/(2+\eta)} = O(n^{-\eta/2}).$$

In order to check (III), we start by proving

$$s_n^{-2} \sum_{m=1}^n \{U_m^2 - E(U_m^2|\mathscr{F}_{m-1})\} \xrightarrow{\text{a.s.}} 0 \tag{7.131}$$

as $n \to \infty$, where $\mathscr{F}_k$ is the σ-field generated by the Z_j, $j \leqslant k$. That is,

$$n^{1-2\gamma}\gamma_n^2 \sum_{m=1}^n m^{-2+2\gamma}\gamma_m^{-2}\{Z_m^2 - E(Z_m^2|\mathscr{F}_{m-1})\} \xrightarrow{\text{a.s.}} 0. \tag{7.132}$$

Set

$$Z_m' = Z_m I(|Z_m| \leqslant m^{1/2}), \qquad m \geqslant 1.$$

Then

$$\sum_{n=1}^\infty n^{-2} E\{(Z_n')^2 - E((Z_n')^2|\mathscr{F}_{n-1})\}^2 \leqslant \sum_{n=1}^\infty n^{-2} E(Z_n')^4 < \infty$$

since

$$E(Z_n')^4 < E|Z_n|^{2+\eta} n^{1-\eta/2} = O(n^{1-\eta/2}).$$

Thus, using Theorem 2.18,

$$n^{1-2\gamma}\gamma_n^2 \sum_{m=1}^n m^{-2+2\gamma}\gamma_m^{-2}\{(Z_m')^2 - E((Z_m')^2|\mathscr{F}_{m-1})\} \xrightarrow{\text{a.s.}} 0.$$

Furthermore, $P(Z_m' \neq Z_m \text{ i.o.}) = 0$ by the Borel–Cantelli lemma since

$$\sum_1^\infty P(Z_m' \neq Z_m) = \sum_1^\infty P(|Z_m| > m^{1/2}) \leqslant \sup_y E|Z(y)|^{2+\eta} \sum_1^\infty m^{-(2+\eta)/2} < \infty,$$

and hence

$$n^{1-2\gamma}\gamma_n^2 \sum_{m=1}^{n} m^{-2+2\gamma}\gamma_m^{-2}\{Z_m^2 - E((Z_m')^2|\mathscr{F}_{m-1})\} \xrightarrow{\text{a.s.}} 0.$$

The final step in proving (7.132) is thus to show that

$$n^{1-2\gamma}\gamma_n^2 \sum_{m=1}^{n} m^{-2+2\gamma}\gamma_m^{-2} E(Z_m^2 I(|Z_m| > m^{1/2})|\mathscr{F}_{m-1}) \xrightarrow{\text{a.s.}} 0.$$

This holds, using the Kronecker lemma, if

$$\sum_{1}^{\infty} n^{-1} E(Z_n^2 I(|Z_n| > n^{1/2})|\mathscr{F}_{n-1}) < \infty \quad \text{a.s.},$$

which in turn holds if

$$\sum_{1}^{\infty} n^{-1} E(Z_n^2 I(|Z_n| > n^{1/2})) < \infty,$$

which was established in the course of proving (II).

In order to obtain (III) it now suffices to prove that

$$s_n^{-2} \sum_{m=1}^{n} E(U_m^2|\mathscr{F}_{m-1}) \xrightarrow{\text{a.s.}} 1,$$

or equivalently,

$$n^{1-2\gamma}\gamma_n^2 \sum_{m=1}^{n} m^{-2+2\gamma}\gamma_m^{-2} E(Z_m^2|\mathscr{F}_{m-1}) \xrightarrow{\text{a.s.}} \sigma^2(2Ac + 2\gamma - 1)^{-1}.$$

This, however, follows from the relation (7.119) and an application of the Toeplitz lemma, since

$$E(Z_n^2|\mathscr{F}_{n-1}) \xrightarrow{\text{a.s.}} \sigma^2$$

as $n \to \infty$, again via (A4) and (B5). This completes the proofs. ∎

7.7. On Adaptive Control of Linear Systems

In this section we shall deal with the control of an evolutionary system whose state is observed at discrete time intervals. At each observation time the unknown parameters of the system are estimated and a control policy which is optimal in a certain sense is computed as if the estimates were the exact values of the parameters. This is used to control the system in the next time period. The results given below are a simplified version of those of Mandl (1974). They are less general but have the bonus of avoiding some rather obscure regularity conditions. For a review of the extensive related literature on stochastic control theory see Hinderer (1975).

We consider a discrete time linear system

$$X_{j+1} = \Phi X_j + \Psi U_j + e_{j+1}, \qquad j = 0, 1, 2, \ldots.$$

Here X_j denotes the state of the system and U_j the control signal at time j, these being vectors of dimension r and s, respectively. Also,

$$U_j = K_j X_j, \qquad j = 0, 1, 2, \ldots,$$

where the $K_j, j = 0, 1, \ldots$, are $s \times r$ matrices with

$$K_j = K_j(X_0, \ldots, X_j),$$

and Φ and Ψ are constant matrices of size $r \times r$ and $r \times s$, respectively. Finally, the disturbances $\{e_j, j \geqslant 1\}$ are i.i.d. random vectors of dimension r with zero mean, covariance matrix $\sigma^2 I$, and with finite fourth moment. The sequence of matrices $\{K_j, j = 0,1, \ldots\}$ is the *control policy*, and if $K_j = K, j = 0, 1, \ldots$, the policy is called a *stationary control* K.

For comparing control policies we introduce the cost

$$C_n = \sum_{j=0}^{n-1} (X_j' R X_j + U_j' S U_j),$$

where R and S are symmetric positive definite $r \times r$ and $s \times s$ matrices, respectively. The most common approach is to assume n fixed and to minimize the expected cost EC_n. An alternative approach, which we adopt here, is to minimize $\lim_{n\to\infty} n^{-1} C_n$.

Before proceeding, we mention some notation and conventions which will be adopted. We use E_x to denote expectation conditional on the event $X_0 = x$. A prime denotes the transpose of a matrix and the norm $\|M\|$ denotes the square root of the sum of squares of the elements of M. The terminology that M is positive definite (nonnegative definite) is used only for symmetric matrices and is written as $M > 0$ ($M \geqslant 0$). We write $M_0 \geqslant M_1$ to mean $M_0 - M_1 \geqslant 0$, and the minimum of a set of matrices is taken with respect to this ordering.

We shall say that the process $\{X_j, j = 0,1, \ldots\}$ is *stable* under the control policy $\{K_j, j = 0,1, \ldots\}$ if

$$\limsup_{n\to\infty} n^{-1} \sum_{j=0}^{n-1} E_x \|X_j\|^2 \leqslant C < \infty,$$

where C is independent of x. Stability is a kind of asymptotic stationarity and it is a straightforward matter to check that a stationary policy $K_j = K$, $j = 0, 1, \ldots$, is stable if and only if all the eigenvalues of $\Phi + \Psi K$ have modulus less than unity. Considerable emphasis will be given to stationary controls. We shall say that the stationary control K is *optimal* if it is stable and if

$$\lim_{n\to\infty} n^{-1}C_n = \theta \qquad \text{a.s. under } K,$$

where θ is a constant with the property that under an arbitrary stable control $\{K_j, j = 0,1, \ldots\}$,

$$\limsup_{n\to\infty} n^{-1}C_n \geqslant \theta \quad \text{a.s.}$$

The ultimate objective of the theory is to deal with systems posessing unknown parameters, but for the time being we shall suppose that Φ and Ψ are known. Within this framework we shall show how to find an optimal control using a version of the policy improvement algorithm of dynamic programming originally suggested by Howard.

The aim is to find a symmetric $r \times r$ matrix $W(\Phi)$ satisfying

$$x'W(\Phi)x = \min_u \{x'Rx + u'Su + (\Phi x + \Psi u)'W(\Phi)(\Phi x + \Psi u)\}. \quad (7.133)$$

Let

$$u = K(\Phi)x$$

denote the value of u which minimizes the expression on the right-hand side of (7.133). It will be shown in Theorem 7.17 that $K(\Phi)$ is an optimal stationary control.

Before proceeding we make the following assumption, which is important to the proof of the stability of the system under the controls constructed during the iteration procedure.

Assumption 1. *There exists an $s \times r$ matrix $K^{(0)}$ such that $\Phi + \Psi K^{(0)}$ has all its eigenvalues in the unit circle.*

We introduce

$$W(K|\Phi) = \sum_{k=0}^{\infty} ((\Phi + \Psi K)^k)'(R + K'SK)(\Phi + \Psi K)^k$$

whenever K is a stable stationary control. It is not difficult to verify that $W(K|\Phi)$ is the only symmetric matrix satisfying

$$W = R + K'SK + (\Phi + \Psi K)'W(\Phi + \Psi K). \quad (7.134)$$

The iterative solution of (7.133) proceeds as follows. Starting with a $K^{(0)}$ which satisfies Assumption 1, a sequence $\{K^{(0)}, K^{(1)}, \ldots\}$ is constructed with the property that

$$(K^{(m+1)})'SK^{(m+1)} + (\Phi + \Psi K^{(m+1)})'W(K^{(m)}|\Phi)(\Phi + \Psi K^{(m+1)})$$
$$= \min_K \{K'SK + (\Phi + \Psi K)'W(K^{(m)}|\Phi)(\Phi + \Psi K)\}. \tag{7.135}$$

We note that, since

$$\frac{\partial}{\partial K}\{K'SK + (\Phi + \Psi K)'W(K^{(m)}|\Phi)(\Phi + \Psi K)\}$$
$$= 2SK + 2\Psi'W(K^{(m)}|\Phi)\Phi + 2\Psi'W(K^{(m)}|\Phi)\Psi K,$$

the result (7.135) holds for

$$K^{(m+1)} = -(S + \Psi'W(K^{(m)}|\Phi)\Psi)^{-1}\Psi'W(K^{(m)}|\Phi)\Phi. \tag{7.136}$$

Lemma 7.4. *If $\{X_j, j = 0,1, \ldots\}$ is stable under $K^{(m)}$, then it is also stable under $K^{(m+1)}$. Furthermore,*

$$W(K^{(m+1)}|\Phi) \leqslant W(K^{(m)}|\Phi), \qquad m = 0, 1, \ldots. \tag{7.137}$$

Proof. We first remark that the process $\{X_j, j = 0,1, \ldots\}$ is stable under $K^{(0)}$, which is chosen to satisfy Assumption 1.

Now consider $\{X_j, j = 0,1, \ldots\}$ under the stationary control $K^{(m+1)}$. For convenience we write

$$\Phi + \Psi K^{(m+1)} = A, \qquad W(K^{(m)}|\Phi) = W,$$

and we note that

$$x'A'WAx = E_x X_1' W X_1 - \sigma^2 \operatorname{tr} W.$$

From (7.134) and (7.135) we have

$$W \geqslant R + (K^{(m+1)})'SK^{(m+1)} + A'WA, \tag{7.138}$$

and hence

$$x'Wx \geqslant x'(R + (K^{(m+1)})'SK^{(m+1)})x + E_x X_1' W X_1 - \sigma^2 \operatorname{tr} W. \tag{7.139}$$

Furthermore, (7.139) is easily generalized to

$$E_x X_k' W X_k \geqslant E_x X_k'(R + (K^{(m+1)})'SK^{(m+1)})X_k$$
$$+ E_x X_{k+1}' W X_{k+1} - \sigma^2 \operatorname{tr} W, \qquad k = 0, 1, \ldots. \tag{7.140}$$

Then, adding (7.140) for $k = 0, 1, \ldots, n-1$, we obtain

$$E_x X_0' W X_0 + \sigma^2 n \operatorname{tr} W \geqslant \sum_{k=0}^{n-1} E_x X_k'(R + (K^{(m+1)})'SK^{(m+1)})X_k + E_x X_n' W X_n$$

and, since $R + (K^{(m+1)})'SK^{(m+1)} > 0$ in view of $R > 0$ and $S > 0$,

$$\sum_{k=0}^{n} E_x\|X_k\|^2 = O(E_x X_0' W X_0 + \sigma^2 n \operatorname{tr} W),$$

which implies the stability of $\{X_j, j = 0,1, \ldots\}$ under $K^{(m+1)}$.

To establish (7.137) we first conclude from (7.138) that

$$(A^k)'WA^k \geqslant (A^k)'(R + (K^{(m+1)})'SK^{(m+1)})A^k + (A^{k+1})'WA^{k+1}, \qquad k = 0, 1, \ldots,$$

and summation gives

$$W(K^{(m)}|\Phi) \geqslant \sum_{k=0}^{\infty} (A^k)'(R + (K^{(m+1)})'SK^{(m+1)})A^k = W(K^{(m+1)}|\Phi),$$

which completes the proof. ▮

Next we show that the above iterative procedure leads to the solution of (7.133). Since $\{W(K^{(m)}|\Phi), m = 0,1, \ldots\}$ is a nonincreasing sequence of nonnegative matrices,

$$\lim_{m\to\infty} W(K^{(m)}|\Phi) = W(\Phi)$$

exists and (7.136) implies the existence of

$$\lim_{m\to\infty} K^{(m)} = K(\Phi).$$

Furthermore, we obtain from (7.134) and (7.136) the relations

$$W(\Phi) = R + (K(\Phi))'SK(\Phi) + (\Phi + \Psi K(\Phi))'W(\Phi)(\Phi + \Psi K(\Phi)) \qquad (7.141)$$

and

$$K(\Phi) = -(S + \Psi' W(\Phi)\Psi)^{-1}\Psi' W(\Phi)\Phi. \qquad (7.142)$$

The stability of $\{X_j, j = 0,1, \ldots\}$ under $K(\Phi)$ follows straightforwardly from (7.142) and the proof of Lemma 7.4. Also, the validity of (7.133) is ensured by (7.141) and (7.142). The value of u at which the minimum in (7.133) occurs can be found as in (7.136). Note that $W(\Phi)$ is the unique symmetric matrix satisfying (7.133). (To see this observe that $\tilde{W} \geqslant W(\Phi)$ and $\tilde{W} \leqslant W(\Phi)$ must hold wherever $\tilde{W}$ satisfies (7.133).) Consequently, $W(\Phi)$ and $K(\Phi)$ are uniquely determined by (7.141) and (7.142).

We now proceed to consider a system with an unknown parameter. The matrix Ψ in the model is assumed to be known while

$$\Phi = \Phi(\alpha) = F_0 + \alpha F_1, \qquad (7.143)$$

F_0 and F_1 being known matrices and α an unknown parameter with true value α_0. The set I of possible values of α is assumed to be an open interval. For convenience we set $\Phi_0 = \Phi(\alpha_0)$ and $k_0 = K(\Phi_0)$.

We shall consider the least squares estimation of α_0 which proceeds simultaneously with the evolution of the system. For $n = 1, 2, \ldots$ an estimate α_n^* of α_0 is obtained by minimizing the sum of squares

$$L_n(\alpha) = \sum_{j=0}^{n-1} (X_{j+1} - \Phi(\alpha)X_j - \Psi U_j)'(X_{j+1} - \Phi(\alpha)X_j - \Psi U_j) \quad (7.144)$$

for $\alpha \in I$. First we establish the consistency of the estimation method. Later we shall link the identification and control using

$$K_j^* = K(\Phi(\alpha_j^*)), \qquad j = 0, 1, \ldots. \quad (7.145)$$

The following regularity condition is assumed in order to ensure that, for each $\alpha \in I$, the optimal stationary control $K(\Phi(\alpha))$ can be calculated.

Assumption 1.′ *For all $\alpha \in I$, $\Phi(\alpha)$ fulfills Assumption* 1.

Lemma 7.5. *For any control policy $\{K_j, j = 0,1, \ldots\}$, the least squares estimator $\overline{\alpha}_n$ obtained by minimizing the sum of squares in* (7.144) *is strongly consistent for α_0. That is, $\overline{\alpha}_n \xrightarrow{\text{a.s.}} \alpha_0$ as $n \to \infty$.*

Proof. From (7.143) and (7.144) we have

$$L_n(\alpha) = \sum_{j=0}^{n-1} e'_{j+1}e_{j+1} + (\alpha - \alpha_0)^2 \sum_{j=0}^{n-1} X'_jF'_1F_1X_j - 2(\alpha - \alpha_0) \sum_{j=0}^{n-1} e'_{j+1}F_1X_j,$$

and by differentiating with respect to α and equating to zero we find that $L_n(\alpha)$ is minimized by

$$\overline{\alpha}_n = \alpha + \sum_{j=0}^{n-1} e'_{j+1}F_1X_j \Big/ \sum_{j=0}^{n-1} X'_jF'_1F_1X_j. \quad (7.146)$$

Set

$$Y_j = e'_{j+1}F_1X_j. \quad (7.147)$$

Then, writing $\mathscr{F}_n$ for the σ-field generated by $X_j, j \leqslant n$,

$$E(Y_j|\mathscr{F}_j) = E(e'_{j+1}|\mathscr{F}_j)F_1X_j = 0 \quad \text{a.s.}, \qquad j = 0, 1, \ldots,$$

so that the $Y_j, j = 0, 1, \ldots$, are martingale differences. Furthermore,

$$\begin{aligned} E(Y_j^2|\mathscr{F}_j) &= X'_jF'_1E(e_{j+1}e'_{j+1}|\mathscr{F}_j)F_1X_j \\ &= \sigma^2X'_jF'_1F_1X_j \quad \text{a.s.} \end{aligned} \quad (7.148)$$

Therefore, setting

$$a_n = \sum_{j=1}^{n} X'_jF'_1F_1X_j, \qquad n \geqslant 1, \quad a_0 = 0,$$

we have

$$\sum_{n=1}^{\infty} a_n^{-2}(a_n - a_{n-1}) = \sum_{n=1}^{\infty} a_n(a_n^{-2} - a_{n+1}^{-2})$$
$$= \sum_{n=1}^{\infty} a_n(a_n^{-1} - a_{n+1}^{-1})(a_n^{-1} + a_{n+1}^{-1})$$
$$\leqslant 2 \sum_{n=1}^{\infty} (a_n^{-1} - a_{n+1}^{-1}) \leqslant 2a_1^{-1} < \infty \quad \text{a.s.},$$

so that Theorem 2.18 gives

$$\sum_{j=1}^{n-1} Y_j \bigg/ \sum_{j=1}^{n-1} E(Y_j^2|\mathscr{F}_j) \xrightarrow{\text{a.s.}} 0 \tag{7.149}$$

as $n \to \infty$. The required result then follows from (7.146)–(7.149). ∎

The $\bar{\alpha}_n$ of Lemma 7.5 are not necessarily all elements of I, but minor modification will cope with this difficulty.

We shall meet a variety of quadratic forms like (7.148) in the analysis below. Their asymptotic behavior for suitably convergent control policies is specified by the following lemma.

Lemma 7.6. *Let $\{K_j, j = 0,1, \ldots\}$ be a stable control policy for which*

$$K_n \xrightarrow{\text{a.s.}} k_0 = K(\Phi_0)$$

as $n \to \infty$. Then

$$\limsup_{n\to\infty} n^{-1} \sum_{j=0}^{n-1} \|X_j\|^4 < \infty \quad \textit{a.s.}, \tag{7.150}$$

and in particular,

$$\lim_{n\to\infty} n^{-1}\|X_n\|^2 = 0 \quad \textit{a.s.} \tag{7.151}$$

Furthermore, for any symmetric $r \times r$ matrix A,

$$\lim_{n\to\infty} n^{-1} \sum_{j=0}^{n-1} X_j'AX_j = \sigma^2 \operatorname{tr} V \quad \textit{a.s.}, \tag{7.152}$$

where V is the unique symmetric matrix satisfying

$$V = A + (\Phi_0 + \Psi k_0)'V(\Phi_0 + \Psi k_0).$$

Proof. First we shall establish (7.150). Set

$$A = \Phi_0 + \Psi k_0, \qquad B_j = \Psi(K_j - k_0), \qquad j = 0, 1, \ldots,$$

and recall that stability ensures that all the eigenvalues of A must lie within the unit circle. Then for $j \geqslant m$,

$$\begin{aligned} X_j &= e_j + (A + B_{j-1})e_{j-1} + \cdots + (A + B_{j-1}) \cdots (A + B_{j-m+1})e_{j-m+1} \\ &\quad + (A + B_{j-1}) \cdots (A + B_{j-m})X_{j-m} \\ &= \tilde{e}_{jm} + (A + B_{j-1}) \cdots (A + B_{j-m})X_{j-m}, \end{aligned}$$

say. If $\|B_{j-1}\| \leqslant b, \ldots, \|B_{j-m}\| \leqslant b$, then

$$\|(A + B_{j-1}) \cdots (A + B_{j-m})\| \leqslant \|A^m\| + mb(\|A\| + b)^{m-1}. \tag{7.153}$$

Denote the right-hand side of (7.153) by Δ, and fix m and $b > 0$ so that $8\Delta^4 < 1$. Further, let $\varepsilon > 0$ be arbitrary and choose a $j_0 = j_0(\varepsilon)$ so large that

$$P\left(\bigcap_{j=j_0}^{\infty} \{\|B_j\| < b\}\right) \geqslant 1 - \varepsilon.$$

Write

$$D = \bigcap_{j=j_0}^{\infty} \{\|B_j\| < b\}.$$

Then, for $j \geqslant j_0 + m$,

$$I(D)\|X_j\| \leqslant I(D)(\|\tilde{e}_{jm}\| + \Delta\|X_{j-m}\|),$$

so that

$$I(D)\|X_j\|^4 \leqslant I(D)(8\|\tilde{e}_{jm}\|^4 + 8\Delta^4\|X_{j-m}\|^4). \tag{7.154}$$

Further,

$$I(D)\|\tilde{e}_{jm}\|^4 \leqslant I(D)\left[\sum_{k=0}^{m-1} (\|A\| + b)^{2k}\right]^2 m \sum_{k=0}^{m-1} \|e_{j-k}\|^4, \tag{7.155}$$

and, from (7.154) and (7.155) we have for n sufficiently large,

$$\begin{aligned} I(D) \sum_{j=j_0+m}^{n-1} \|X_j\|^4 &\leqslant (1 - 8\Delta^4)^{-1}(\|X_{j_0}\|^4 + \cdots + \|X_{j_0+m-1}\|^4) \\ &\quad + C \sum_{j=j_0}^{n-1} \|e_j\|^4, \end{aligned} \tag{7.156}$$

C denoting a positive constant. The strong law of large numbers holds for the sequence $\{\|e_j\|^4, j = 1,2, \ldots\}$ and hence

$$\limsup_{n\to\infty} I(D)n^{-1} \sum_{j=0}^{n-1} \|X_j\|^4 \leqslant \text{const.} \quad \text{a.s.},$$

from which the result (7.150) follows since $P(D)$ can be made arbitrarily close to unity.

Let $A_n = \Sigma_{j=1}^{n} \|X_j\|^4$. Using (7.150) we have for a constant C,

$$\begin{aligned}\sum_{n=1}^{\infty} n^{-2}\|X_n\|^2 &= \left[\sum_{n=1}^{\infty} n^{-2}(A_n - A_{n-1})^{1/2}\right]^2 \\ &\leqslant \left[\sum_{n=1}^{\infty} n^{-2}\right]\left[\sum_{n=1}^{\infty} n^{-2}(A_n - A_{n-1})\right] \\ &\leqslant C \sum_{n=1}^{\infty} A_n(n^{-2} - (n+1)^{-2}) \\ &\leqslant 3C \sum_{n=1}^{\infty} n^{-3} A_n < \infty \quad \text{a.s.,}\end{aligned}$$

and (7.151) follows immediately.

To establish (7.152) we write

$$Z_j = e'_{j+1} V e_{j+1} - \sigma^2 \operatorname{tr} V + 2e'_{j+1} V(\Phi_0 + \Psi K_j) X_j$$

and note that

$$E(Z_j | \mathscr{F}_j) = 0 \quad \text{a.s.,}$$

so that $\{Z_j, j = 0,1, \ldots\}$ is a sequence of martingale differences. Further,

$$E(Z_j^2 | \mathscr{F}_j) = O(1 + \|X_j\|^2) \quad \text{a.s.}$$

and, in view of the stability of $\{K_j, j = 0,1, \ldots\}$,

$$\sum_{n=1}^{\infty} n^{-2} E Z_n^2 < \infty,$$

so that, from Theorem 2.18,

$$n^{-1} \sum_{j=0}^{n-1} Z_j \xrightarrow{\text{a.s.}} 0. \tag{7.157}$$

But, after some algebra,

$$\begin{aligned}\sum_{j=0}^{n-1} Z_j = \sum_{j=0}^{n-1} X'_j A X_j - n\sigma^2 \operatorname{tr} V + X'_n A X_n - X'_0 A X_0 \\ + \sum_{j=0}^{n-1} X'_j\{(\Phi_0 + \Psi k_0)' V(\Phi_0 + \Psi k_0) - (\Phi_0 + \Psi K_j)' V(\Phi_0 + \Psi K_j)\} X_j.\end{aligned}$$

Also, since $K_n \to k_0$ a.s. as $n \to \infty$,

$$T_n = (\Phi_0 + \Psi k_0)' V(\Phi_0 + \Psi k_0) - (\Phi_0 + \Psi K_n)' V(\Phi_0 + \Psi K_n) \xrightarrow{\text{a.s.}} 0$$

and, using Schwarz's inequality,

$$\begin{aligned}0 &\leqslant \limsup_{n\to\infty} n^{-1} \sum_{j=0}^{n-1} X_j' T_j X_j \\ &\leqslant \limsup_{n\to\infty} n^{-1} \sum_{j=0}^{n-1} \|T_j\| \, \|X_j\|^2 \\ &\leqslant \limsup_{n\to\infty} \left(n^{-1} \sum_{j=0}^{n-1} \|T_j\|^2 \right)^{1/2} \left(n^{-1} \sum_{j=0}^{n-1} \|X_j\|^4 \right)^{1/2} = 0 \quad \text{a.s.}\end{aligned}$$

Consequently, (7.152) follows from (7.157) in view of (7.151). ∎

Our assumptions imply that the iterative procedure described earlier can be applied for each $\alpha \in I$ to calculate the optimal stationary control $K(\Phi(\alpha))$, which we denote more compactly by $K(\alpha)$. We introduce the set

$$\mathscr{K} = \{K(\alpha), \alpha \in I\}.$$

Only matrices from $\mathscr{K}$ will be used in our control. We shall in addition make the following assumption which guarantees the stability of the system, whatever rule is used to choose controls from $\mathscr{K}$.

Assumption 2. *For each $\alpha \in I$ an $r \times r$ matrix $V > 0$ can be found so that*

$$V \geqslant I + (\Phi(\alpha) + \Psi K)' V (\Phi(\alpha) + \Psi K), \qquad K \in \mathscr{K}. \tag{7.158}$$

Lemma 7.7. *The process $\{X_j, j = 0,1, \ldots\}$ is stable under a control policy $\{K_j, j = 0,1, \ldots\}$ taking values in $\mathscr{K}$.*

Proof. Suppose that V satisfies Assumption 2 with $\alpha = \alpha_0$. We have

$$\begin{aligned}\|X_j\|^2 + E_x\{X_{j+1}' V X_{j+1} | \mathscr{F}_j\} &= \|X_j\|^2 + X_j'(\Phi_0 + \Psi K_j)' V (\Phi_0 + \Psi K_j) \\ &\quad + \sigma^2 \operatorname{tr} V \leqslant X_j' V X_j + \sigma^2 \operatorname{tr} V,\end{aligned}$$

so that

$$E_x\|X_j\|^2 + E_x X_{j+1}' V X_{j+1} \leqslant E_x X_j' V X_j + \sigma^2 \operatorname{tr} V, \qquad j = 0, 1, \ldots,$$

and hence

$$\sum_{j=0}^{n-1} E_x\|X_j\|^2 + E_x X_n' V X_n \leqslant E_x X_0' V X_0 + n\sigma^2 \operatorname{tr} V, \qquad n = 1, 2, \ldots,$$

from which stability follows. This completes the proof. ∎

In the remainder of this section we shall write, for the sake of brevity, $W_0 = W(\Phi_0)$ and $\theta = \sigma^2 \operatorname{tr} W_0$.

Theorem 7.17. *Under any stable control policy* $\{K_j, j = 0,1, \ldots\}$,

$$\limsup_{n\to\infty} n^{-1}C_n \geqslant \theta \quad a.s. \tag{7.159}$$

Furthermore, if $\{K_j, j = 0,1, \ldots\}$ *is any stable control policy such that*

$$\lim_{n\to\infty} K_n = k_0 \quad a.s.,$$

then under $\{K_j, j = 0,1, \ldots\}$ *we have*

$$\lim_{n\to\infty} n^{-1}C_n = \theta \quad a.s. \tag{7.160}$$

Thus, in particular, k_0 *provides an optimal stationary control.*

Proof. The proof is closely analogous to that of Lemma 7.6. First we show that under any stable control policy $\{K_j, j = 0,1, \ldots\}$,

$$\begin{aligned} M_n = {} & C_n - n\theta + X_n'W_0X_n - X_0'W_0X_0 \\ & - \sum_{j=0}^{n-1} X_j'(K_j - k_0)'(S + \Psi'W_0\Psi)(K_j - k_0)X_j \end{aligned} \tag{7.161}$$

is a martingale with respect to $\{\mathscr{F}_n, n = 1,2, \ldots\}$.

Using (7.141) and (7.142) we obtain, after some algebra,

$$\begin{aligned} &(K_j - k_0)'(S + \Psi'W_0\Psi)(K_j - k_0) + W_0 \\ &\quad = R + K_j'SK_j + (\Phi_0 + \Psi K_j)'W_0(\Phi_0 + \Psi K_j). \end{aligned}$$

Further, writing

$$M_n = \sum_{j=0}^{n-1} Y_j, \qquad n \geqslant 1,$$

we find that

$$\begin{aligned} Y_j &= X_j'RX_j + U_j'SU_j - \theta + X_{j+1}'W_0X_{j+1} - X_j'W_0X_j \\ &\quad - X_j'(K_j - k_0)'(S + \Psi'W_0\Psi)(K_j - k_0)X_j \\ &= e_{j+1}'W_0e_{j+1} - \theta + 2e_{j+1}'W_0(\Phi_0X_j + \Psi U_j), \end{aligned} \tag{7.162}$$

from which we deduce that $E(Y_j|\mathscr{F}_j) = 0$ a.s., $j = 0, 1, \ldots$, which gives the martingale property.

In addition,

$$\begin{aligned} E(Y_j^2|\mathscr{F}_j) &= E(e_{j+1}'W_0e_{j+1} - \theta)^2 \\ &\quad + 4(\Phi_0X_j + \Psi U_j)'E[W_0e_{j+1}(e_{j+1}'W_0e_{j+1} - \theta)] \\ &\quad + 4\sigma^2(\Phi_0X_j + \Psi U_j)'W_0^2(\Phi_0X_j + \Psi U_j) \\ &= O(1 + \|\Phi_0X_j + \Psi U_j\|^2) \quad \text{a.s.}, \end{aligned} \tag{7.163}$$

so that

$$EY_j^2 = E[E(Y_j^2|\mathscr{F}_j)] = O(1 + E\|X_j\|^2). \tag{7.164}$$

Then, since the control $\{K_j, j = 0, 1, \ldots\}$ is stable,

$$\sum_{n=1}^{\infty} n^{-2} EY_n^2 < \infty,$$

which implies, using Theorem 2.18, that

$$\lim_{n\to\infty} n^{-1} M_n = 0 \quad \text{a.s.} \tag{7.165}$$

From (7.161) we see that

$$n^{-1}C \geqslant \theta + n^{-1}M_n - n^{-1}X_n' W_0 X_n$$

and then, using (7.165),

$$\limsup_{n\to\infty} n^{-1}C_n \geqslant \theta - \liminf_{n\to\infty} n^{-1}X_n' W_0 X_n \quad \text{a.s.} \tag{7.166}$$

However, using Fatou's lemma,

$$\begin{aligned} 0 \leqslant E\left[\liminf_{n\to\infty} n^{-1}X_n' W_0 X_n\right] &\leqslant \liminf_{n\to\infty} n^{-1}EX_n' W_0 X_n \\ &\leqslant \|W_0\| \liminf_{n\to\infty} n^{-1}E\|X_n\|^2 = 0 \end{aligned}$$

since the control is stable. This forces

$$\liminf_{n\to\infty} n^{-1}X_n' W_0 X_n = 0 \quad \text{a.s.,}$$

and (7.159) follows from (7.166).

That (7.160) holds when $K_n \to k_0$ a.s. follows simply from (7.161) and Lemma 7.6. We have

$$\lim_{n\to\infty} n^{-1}X_n' W_0 X_n = 0 \quad \text{a.s.}$$

from (7.151) and

$$\lim_{n\to\infty} n^{-1} \sum_{j=0}^{n-1} X_j'(K_j - k_0)'(S + \Psi' W_0 \Psi)(K_j - k_0)X_j$$

$$\leqslant \limsup_{n\to\infty} n^{-1}\|S + \Psi' W_0 \Psi\| \sum_{j=0}^{n-1} \|K_j - k_0\|^2 \|X_j\|^2$$

$$\leqslant \limsup_{n\to\infty} \|S + \Psi' W_0 \Psi\| \left(n^{-1} \sum_{j=0}^{n-1} \|K_j - k_0\|^4\right)^{1/2} \left(n^{-1} \sum_{j=0}^{n-1} \|X_j\|^4\right)^{1/2} = 0 \quad \text{a.s.}$$

in view of (7.150) and $\|K_n - k_0\| \xrightarrow{\text{a.s.}} 0$. This completes the proof. ∎

Finally, we investigate the control policy based on estimation of α_0. At time 0 the controller selects an $\hat{\alpha} \in I$ and sets

$$\alpha_0^* = \hat{\alpha}, \qquad K_0 = K(\alpha_0^*).$$

For $n = 1, 2, \ldots$ he observes $\{X_0, X_1, \ldots, X_n\}$, find $\bar{\alpha}_n$, the least squares estimator of α_0, and sets

$$K_n = K(\alpha_n^*),$$

where

$$\alpha_n^* = \begin{cases} \bar{\alpha}_n & \text{if } \bar{\alpha}_n \in I \\ \hat{\alpha} & \text{otherwise.} \end{cases}$$

From Lemma 7.5 we see that $\bar{\alpha}_n$ must ultimately lie in I with probability one since α_0 is an interior point of I and hence $\alpha_n^* \xrightarrow{\text{a.s.}} \alpha_0$ as $n \to \infty$. The policy $\{K_n = K(\alpha_n^*), n = 0, 1, \ldots\}$ is an estimated version of the optimal stationary control and its properties are described in the following theorem.

Theorem 7.18. *Under the control policy $\{K_n = K(\alpha_n^*), n = 0,1, \ldots\}$, we have*

$$\lim_{n\to\infty} n^{-1} C_n = \theta \quad \text{a.s.} \tag{7.167}$$

and

$$n^{-1/2}(C_n - n\theta) \xrightarrow{\text{d}} N(0, E(e_1' W_0 e_1 - \theta)^2 + 4\sigma^4 \operatorname{tr} H), \tag{7.168}$$

where H is the unique symmetric matrix satisfying

$$H = (\Phi_0 + \Psi k_0)'(W_0^2 + H)(\Phi_0 + \Psi k_0). \tag{7.169}$$

Proof. The differentiability of $K(\alpha)$, $\alpha \in I$, with respect to α can be straightforwardly established, and Taylor expansion gives

$$K(\alpha_n^*) = K(\alpha_0) + (\alpha_n^* - \alpha_0)\frac{\partial K(\alpha_0)}{\partial \alpha}(1 + o(1)) \quad \text{a.s.} \tag{7.170}$$

as $n \to \infty$, from which we deduce that $K_n = K(\alpha_n^*) \xrightarrow{\text{a.s.}} K(\alpha_0) = k_0$ as $n \to \infty$. The result (7.167) then follows from Theorem 7.17.

To establish (7.168) we first show that a central limit result holds for the martingale M_n defined by (7.161). We shall show that

$$n^{-1/2}(M_n - n\theta) \xrightarrow{\text{d}} N(0, E(e_1' W_0 e_1 - \theta)^2 + 4\sigma^4 \operatorname{tr} H), \tag{7.171}$$

and (7.168) follows provided that

$$n^{-1/2}(X_n' W_0 X_n - X_0' W_0 X_0) \xrightarrow{\text{p}} 0 \tag{7.172}$$

and

$$n^{-1/2} \sum_{j=0}^{n-1} X_j'(K_j - k_0)'(S + \Psi' W_0 \Psi)(K_j - k_0) X_j \xrightarrow{p} 0. \tag{7.173}$$

That (7.172) holds is immediate from (7.151). To establish (7.173) we first note from (7.170) that

$$K_n - k_0 = O(\alpha_n^* - \alpha_0) = O\left(\sum_{j=0}^{n-1} e_{j+1}' F_1 X_j \Big/ \sum_{j=0}^{n-1} X_j' F_1' X_j \right) \quad \text{a.s.,} \tag{7.174}$$

using (7.146). But from Lemma 7.6,

$$\lim_{n\to\infty} n^{-1} \sum_{j=0}^{n-1} X_j' F_1' F_1 X_j = \sigma^2 \operatorname{tr} V \quad \text{a.s.,} \tag{7.175}$$

where V is the unique symmetric matrix satisfying

$$V = F_1' F_1 + (\Phi_0 + \Psi k_0)' V (\Phi_0 + \Psi k_0).$$

Also, writing

$$Z_j = e_{j+1}' F_1 X_j, \qquad j = 0, 1, \ldots,$$

we have from (7.148) that

$$EZ_j^2 = O(E\|X_j\|^2),$$

and hence, using the stability condition,

$$\sum_{n=1}^{\infty} n^{-5/4} EZ_n^2 < \infty. \tag{7.176}$$

Thus, using Theorem 2.18 we obtain

$$n^{-5/8} \sum_{j=0}^{n-1} Z_j \xrightarrow{\text{a.s.}} 0,$$

that is,

$$n^{-5/8} \sum_{j=0}^{n-1} e_{j+1}' F_1 X_j \xrightarrow{\text{a.s.}} 0, \tag{7.177}$$

and from (7.174), (7.175), and (7.177),

$$K_n - k_0 = o(n^{-3/8}) \quad \text{a.s.} \tag{7.178}$$

as $n \to \infty$. The result (7.173) then follows by observing that the basic expression in (7.173) is

$$O\left(n^{-1/2} \sum_{j=0}^{n-1} \|K_j - k_0\|^2 \|X_j\|^2 \right) = O\left(n^{-1/2} \sum_{j=0}^{n-1} j^{-3/4} \|X_j\|^2 \right) = o(1) \quad \text{a.s.}$$

as $n \to \infty$, provided that

$$\sum_{n=1}^{\infty} n^{-5/4} \|X_n\|^2 < \infty \quad \text{a.s.} \tag{7.179}$$

However, (7.179) holds using the reasoning that led to (7.176), and hence (7.173) is established.

To obtain (7.171) it certainly suffices, using Corollary 3.1, to show that the martingale $M_n = \sum_{j=0}^{n-1} Y_j$ satisfies

$$\lim_{n\to\infty} n^{-1} \sum_{j=0}^{n-1} E(Y_j^2 | \mathscr{F}_j) = E(e'W_0 e - \theta)^2 + 4\sigma^4 \operatorname{tr} H \quad \text{a.s.} \tag{7.180}$$

and

$$\limsup_{n\to\infty} n^{-1} \sum_{j=0}^{n-1} E(Y_j^2 I(|Y_j| > \varepsilon n^{1/2}) | \mathscr{F}_j) = 0 \quad \text{a.s.} \tag{7.181}$$

From (7.163) we see that (7.180) holds provided that

$$\lim_{n\to\infty} n^{-1} \sum_{j=0}^{n-1} X_j = 0 \quad \text{a.s.} \tag{7.182}$$

and

$$\lim_{n\to\infty} n^{-1} \sum_{j=0}^{n-1} X_j'(\Phi_0 + \Psi K_j)' W_0^2 (\Phi_0 + \Psi K_j) X_j = \sigma^2 \operatorname{tr} H \quad \text{a.s.} \tag{7.183}$$

The result (7.182) follows from setting

$$T = \sum_{j=0}^{\infty} (\Phi_0 + \Psi k_0)^j$$

and applying the law of large numbers to

$$T \sum_{j=0}^{n-1} e_{j+1} = \sum_{j=0}^{n-1} X_j + T\Big(X_n - X_0 + \sum_{j=0}^{n-1} \Psi(k_0 - K_j) X_j\Big),$$

since $K_n \xrightarrow{\text{a.s.}} k_0$ as $n \to \infty$.

To deal with (7.183) we apply Lemma 7.6 to show that

$$\lim_{n\to\infty} n^{-1} \sum_{j=0}^{n-1} X_j'(\Phi_0 + \Psi k_0)' W_0^2 (\Phi_0 + \Psi k_0) X_j = \sigma^2 \operatorname{tr} H \quad \text{a.s.,}$$

and hence (7.183) holds if

$$\lim_{n\to\infty} n^{-1} \sum_{j=0}^{n-1} X_j' T_j X_j = 0 \quad \text{a.s.,} \tag{7.184}$$

where

$$T_n = (\Phi_0 + \Psi k_0)' W_0^2 (\Phi_0 + \Psi k_0) - (\Phi_0 + \Psi K_n)' W_0^2 (\Phi_0 + \Psi K_n) \xrightarrow{\text{a.s.}} 0$$

as $n \to \infty$. The result (7.184) follows, however, using the same reasoning as in the final part of the proof of Lemma 7.6.

Finally, to establish (7.181) we note that

$$E(Y_j^4 | \mathscr{F}_j) = O(1 + \|\Phi_0 X_j + \Psi U_j\|^4) = O(1 + \|X_j\|^4) \quad \text{a.s.},$$

so that, using (7.150),

$$\lim_{n\to\infty} n^{-2} \sum_{j=0}^{n-1} E(Y_j^4 | \mathscr{F}_j) = 0 \quad \text{a.s.},$$

and (7.181) holds since

$$E(Y_j^2 I(|Y_j| > \varepsilon n^{1/2}) | \mathscr{F}_j) \leqslant \varepsilon^{-2} n^{-1} E(Y_j^4 | \mathscr{F}_j) \quad \text{a.s.}$$

This completes the proof. ∎

It should be noted from the above proof that the result

$$n^{-1/2}(C_n - n\theta) \xrightarrow{\text{d}} N(0, E(e_1' W_0 e_1 - \theta)^2 + 4\sigma^4 \operatorname{tr} H)$$

also holds under the stationary control k_0.

Under rather stronger regularity conditions, Mandl (1977) has used the martingale LIL to show that

$$\limsup_{n\to\infty} (C_n - n\theta)/(2n \log\log n)^{1/2} = [E(e_1' W_0 e_1 - \theta)^2 + 4\sigma^4 \operatorname{tr} H]^{1/2} \quad \text{a.s.},$$

$$\liminf_{n\to\infty} (C_n - n\theta)/(2n \log\log n)^{1/2} = -[E(e_1' W_0 e_1 - \theta)^2 + 4\sigma^4 \operatorname{tr} H]^{1/2} \quad \text{a.s.}$$

for both the estimated control $\{K(\alpha_n^*), n = 0,1,\ldots\}$ and the optimal stationary control k_0. These results can be obtained straightforwardly by an application of Corollary 4.2 to the martingale M_n of (7.161), but we shall omit the details.

Appendix

I. The Skorokhod Representation

Many important stochastic processes approximate Brownian motion when suitably scaled, and their asymptotics can conveniently be studied by embedding them in Brownian motion. Such an embedding was introduced by Skorokhod (1965) for sums of independent r.v., and this was later extended to martingales; see Dambis (1965), Dubins (1968), Dubins and Schwartz (1965), Monroe (1972), and Strassen (1967). The result for martingales has been used extensively in Chapters 3 and 4 in connection with establishing both central limit and iterated logarithm results. For a discussion of the general question of which processes can be embedded in Brownian motion see Drogin (1973) and Monroe (1978).

Theorem A.1. *Let $\{S_n = \sum_1^n X_i, \mathscr{F}_n, n \geqslant 1\}$ be a zero-mean, square-integrable martingale. Then there exists a probability space supporting a (standard) Brownian motion W and a sequence of nonnegative variables $\tau_1, \tau_2, \ldots$ with the following properties. If $T_n = \sum_1^n \tau_i$, $S_n' = W(T_n)$, $X_1' = S_1'$, $X_n' = S_n' - S_{n-1}'$ for $n \geqslant 2$, and $\mathscr{G}_n$ is the σ-field generated by $S_1', \ldots, S_n'$ and by $W(t)$ for $0 \leqslant t \leqslant T_n$, then*

(i) $\{S_n, n \geqslant 1\} \overset{\mathrm{d}}{=} \{S_n', n \geqslant 1\}$,
(ii) T_n is *$\mathscr{G}_n$-measurable,*
(iii) *for each real number $r \geqslant 1$,*

$$E(\tau_n^r|\mathscr{G}_{n-1}) \leqslant C_r E(|X_n'|^{2r}|\mathscr{G}_{n-1}) = C_r E(|X_n'|^{2r}|X_1', \ldots, X_{n-1}') \qquad a.s.,$$

where $C_r = 2(8/\pi^2)^{r-1}\Gamma(r+1)$, and

(iv) $E(\tau_n|\mathscr{G}_{n-1}) = E(X_n'^2|\mathscr{G}_{n-1}) \qquad a.s.$

(Here $\overset{\mathrm{d}}{=}$ means "equals in distribution.")

Proof. Our proof is based on Skorokhod (1965) and Sawyer (1967). Note that if $\mathscr{F}_n^*$ equals the σ-field generated by $S_1, \ldots, S_n$, then $\{S_n, \mathscr{F}_n^*\}$ is also a zero-mean martingale, and so there is no loss of generality in assuming that $\mathscr{F}_n = \mathscr{F}_n^*$.

Embedding a two-point distribution. Suppose that

$$P(X = -u) + P(X = v) = 1$$

for $u, v > 0$, and that $E(X) = 0$. Let $a = (u + v)/2$ and $b = (u - v)/(u + v)$, and define

$$\begin{aligned} T &= \inf\{t | W(t) = -u \text{ or } v\} \\ &= \inf\{t | a^{-1} W(t) + b = \pm 1\} \\ &\overset{\mathrm{d}}{=} a^2 \inf\{t | W(t) + b = \pm 1\}. \end{aligned} \tag{A.1}$$

Since T is a stopping time and $\{W(t), t \geqslant 0\}$ and $\{W(t)^2 - t, t \geqslant 0\}$ are (continuous time) martingales, then

$$E(W(T)) = 0, \tag{A.2}$$

$$E(W(T)^2 - T) = 0 \tag{A.3}$$

[see, e.g., Breiman (1968, pp. 302, 303) or Freedman (1971, p. 193)]. Clearly $W(T)$ has its distribution concentrated on $\{-u,v\}$, and from (A.2) we see that $W(T)$ must have the same distribution as X. Equation (A.3) implies that $E(T) = \operatorname{var}(X)$. To evaluate higher-order moments, note that for $r \geqslant 1$ we have from (A.1) that

$$E(T^r) = a^{2r} E(\tau^r) = a^{2r} r \int_0^\infty x^{r-1} P(\tau > x)\, dx, \tag{A.4}$$

where $\tau = \inf\{t | W(t) + b = \pm 1\}$. Now, $P(\tau > x)$ is the probability that a particle moving with Brownian motion and starting at the origin will not have reached the points $1 - b$ or $-(1 + b)$ by time x. From Feller (1971, p. 342), we see that this probability equals

$$\lambda_2(x, 1 + b) = \frac{4}{\pi} \sum_{n=0}^\infty \frac{(-1)^n \cos[(n + \frac{1}{2})\pi b]}{2n + 1} \exp\left[-\frac{(2n + 1)^2 \pi^2 x}{8}\right].$$

Therefore,

$$r \int_0^\infty x^{r-1} P(\tau > x)\, dx = \frac{4}{\pi} \Gamma(r + 1) \left(\frac{8}{\pi^2}\right)^r \sum_{n=0}^\infty \frac{(-1)^n \cos[(n + \frac{1}{2})\pi b]}{(2n + 1)^{2r+1}}. \tag{A.5}$$

Let $f_r(b) = E(\tau^r)$. From (A.4) and (A.5) we deduce that $f_r(1) = 0$ and $f_r'(0) = 0$, and that $f_r''(b) = -2rf_{r-1}(b)$. Therefore

$$\begin{aligned} f_r(b) &= -2r \int_1^b dx \int_0^x f_{r-1}(y)\,dy \\ &= 2r \iint_{\substack{b<x<1,\\ 0<y<x}} f_{r-1}(y)\,dx\,dy \qquad (\text{if } b \geqslant 0) \\ &\leqslant r(1-b^2) \sup_{|y|\leqslant 1} f_{r-1}(y). \end{aligned} \tag{A.6}$$

By symmetry this inequality must hold for all b. Since $f_r''(b) \leqslant 0$ for all b and $f_r'(0) = 0$, then

$$\sup_{|y|\leqslant 1} f_{r-1}(y) = f_{r-1}(0) = \frac{4}{\pi}\Gamma(r)\left(\frac{8}{\pi^2}\right)^{r-1} \sum_{n=0}^{\infty} \frac{(-1)^n}{(2n+1)^{2r-1}}.$$

By taking the terms in pairs, the series on the right is easily shown to be dominated by $\frac{3}{2}$ for all $r \geqslant 1$, and so from (A.6),

$$f_r(b) \leqslant \frac{6\Gamma(r+1)}{\pi}\left(\frac{8}{\pi^2}\right)^{r-1}(1-b^2) \leqslant C_r uv/a^2,$$

where C_r is as in Theorem A.1. The equations (A.4) now imply that

$$E(T^r) \leqslant C_r a^{2r-2} uv,$$

and we have proved the following result.

Lemma A.1. *There exists a stopping time T such that $X \stackrel{\mathrm{d}}{=} W(T)$, $E(T) = \operatorname{var}(X)$, and for $r \geqslant 1$,*

$$E(T^r) \leqslant C_r 2^{2-2r}(u+v)^{2r-2}uv.$$

Embedding a continuous distribution. Suppose that Y has a continuous distribution F, with zero mean and finite variance. Define the function G so that $G(x) \geqslant 0$, for $x \leqslant 0$, $G(x) < 0$ for $x > 0$, and

$$\int_x^{G(x)} y\,dF(y) = 0, \qquad -\infty < x < \infty. \tag{A.7}$$

Let $W(t)$, $t \geqslant 0$, be a Brownian motion independent of Y. Define the distribution of the r.v. X, given $Y\ (\neq 0)$, by

$$\begin{aligned} P(X = Y|Y) &= |G(Y)|/(|Y| + |G(Y)|), \\ P(X = G(Y)|Y) &= |Y|/(|Y| + |G(Y)|). \end{aligned}$$

Conditional on Y, X has a two-point distribution with mean zero, and by embedding X in W according to Lemma A.1, we obtain a stopping time

$T = T(Y)$ such that $X \stackrel{d}{=} W(T)$. We show that, unconditionally, X and Y have the same distribution. Since $G(G(x)) = x$, then for all $x > 0$,

$$P(0 < X < x) = \int_0^x \frac{-G(y)}{y - G(y)}\, dF(y) + \int_{G(x)}^0 \frac{-y}{-y + G(y)}\, dF(y)$$

$$= \int_0^x \frac{-G(y)}{y - G(y)}\, dF(y) + \int_0^x \frac{G(z)}{z - G(z)}\, dF(G(z)).$$

Condition (A.7) implies that $x\, dF(x) = G(x)\, dF(G(x))$, and so

$$P(0 < X < x) = \int_0^x \frac{-G(y)}{y - G(y)}\, dF(y) + \int_0^x \frac{z}{z - G(z)}\, dF(z)$$

$$= \int_0^x dF(y) = P(0 < Y < x).$$

Similarly $P(x < X < 0) = P(x < Y < 0)$ for $x < 0$, and so $X \stackrel{d}{=} Y$.

It follows from Lemma A.1 that

$$E(T) = E[E(T|Y)] = E[\mathrm{var}(X|Y)],$$

and since $\mathrm{var}(X) = E[\mathrm{var}(X|Y)] + \mathrm{var}[E(X|Y)]$ and $E(X|Y) = 0$ a.s., it follows that $E(T) = \mathrm{var}(X) = \mathrm{var}(Y)$. Also, for $r \geqslant 1$,

$$E(T^r) \leqslant C_r 2^{2-2r} E\{|Y - G(Y)|^{2r-2}|YG(Y)|\}$$

$$\leqslant C_r E\left\{|YG(Y)| \frac{|Y|^{2r-1} + |G(Y)|^{2r-1}}{|Y| + |G(Y)|}\right\}$$

$$= C_r E|X|^{2r} = C_r E|Y|^{2r},$$

and we have established the following result.

Lemma A.2. *There exists a stopping time T such that $Y = W(T)$, $E(T) = \mathrm{var}(Y)$, and for $r \geqslant 1$,*

$$E(T^r) \leqslant C_r E|Y|^{2r}.$$

Since any distribution may be approximated arbitrarily closely by a continuous distribution, it is plain that Lemma A.2 holds for *any* r.v. Y with zero mean and finite variance.

Embedding a martingale. Let $\{S_n = \sum_1^n X_i, \mathscr{F}_n, n \geqslant 1\}$ be a zero-mean, square-integrable martingale with $\mathscr{F}_n$ equal to the σ-field generated by $S_1, \ldots, S_n$, and let $W(t), t \geqslant 0$, be a standard Brownian motion. Let $Y_1 = X_1$ and let $Y_{n+1}(x_1, \ldots, x_n)$ denote an r.v. with the distribution of X_{n+1} given $X_i = x_i$, $1 \leqslant i \leqslant n$ $(n \geqslant 1)$. Since $E(X_{n+1}|X_1, \ldots, X_n) = 0$ a.s., $E(Y_{n+1}) = 0$ for almost all $x_1, \ldots, x_n$. We consider only such values of x_i.

Let τ_1 be the stopping time of Lemma A.2 applied to $Y = X_1$ (which we may take to be independent of W), so that $X_1 \overset{d}{=} W(\tau_1) = Z_1$, say, $E(\tau_1) = E(X_1^2)$, and

$$E(\tau_1^r) \leqslant C_r E|X_1|^{2r}, \qquad r \geqslant 1.$$

Now set $W_1(t) = W(t + \tau_1) - W(\tau_1)$, $t \geqslant 0$, and let $\mathscr{G}_1$ be the σ-field generated by Z_1 and by $W(t)$ for $t \leqslant \tau_1$. The r.v. τ_1 is $\mathscr{G}_1$-measurable. Conditional on $\mathscr{G}_1$, W_1 is a Brownian motion, and we may apply Lemma A.2 with $Y = Y_2(x_1)$ (taken independent of W_1, given $\mathscr{G}_1$), obtaining a stopping time $\tau_2 = \tau_2(x_1)$ such that the distribution of $Z_2 = W_1(\tau_2)$ (conditional on $\mathscr{G}_1$ and $W(\tau_1) = x_1$) equals that of $Y_2(x_1)$,

$$E(\tau_2|\mathscr{G}_1;\, W(\tau_1) = x_1) = E(Y_2^2(x_1)) = E(X_2^2|X_1 = x_1)$$

for almost all x_1, and for $r \geqslant 1$,

$$E(\tau_2^r|\mathscr{G}_1;\, W(\tau_1) = x_1) \leqslant C_r E|Y_2(x_1)|^{2r} = C_r E(|X_2|^{2r}|X_1 = x_1).$$

Continuing in this way we define a σ-field $\mathscr{G}_n$ generated by $Z_1, \ldots, Z_n$ and by $W(t)$ for $t \leqslant \sum_1^n \tau_i$, and a process $W_n(t) = W(t + \sum_1^n \tau_i) - W(\sum_1^n \tau_i)$. The r.v. $\tau_n = \tau_n(x_1, \ldots, x_{n-1})$ is $\mathscr{G}_n$-measurable. Conditional on $\mathscr{G}_n$, W_n is Brownian motion, and there exists a stopping time $\tau_{n+1} = \tau_{n+1}(x_1, \ldots, x_n)$ such that the distribution of $Z_{n+1} = W_n(\tau_{n+1})$ (conditional on $\mathscr{G}_n$ and $Z_j = x_j$, $1 \leqslant j \leqslant n$) equals that of $Y_{n+1}(x_1, \ldots, x_n)$,

$$\begin{aligned} E(\tau_{n+1}|\mathscr{G}_n;\, Z_j = x_j,\, 1 \leqslant j \leqslant n) &= E(Y_{n+1}^2(x_1, \ldots, x_n)) \\ &= E(X_{n+1}^2|X_1 = x_1, \ldots, X_n = x_n) \end{aligned}$$

for almost all $\{x_i\}$, and for $r \geqslant 1$,

$$E(\tau_{n+1}^r|\mathscr{G}_n;\, Z_j = x_j,\, 1 \leqslant j \leqslant n) \leqslant C_r E(|X_{n+1}|^{2r}|X_1 = x_1, \ldots, X_n = x_n).$$

Since Z_{n+1} (conditional on $\mathscr{G}_n$ and $Z_j = x_j$, $1 \leqslant j \leqslant n$) $\overset{d}{=} Y_{n+1}(x_1, \ldots, x_n)$,

$$P(Z_{n+1} \leqslant z|\mathscr{G}_n;\, Z_j = x_j,\, 1 \leqslant j \leqslant n) = P(X_{n+1} \leqslant z|X_1 = x_1, \ldots, X_n = x_n),$$

and so $P(Z_{n+1} \leqslant z|\mathscr{G}_n)$ can be expressed as a function of z and $Z_1, \ldots, Z_n$ alone. Therefore $\{Z_n\} \overset{d}{=} \{X_n\}$ and

$$E(|Z_{n+1}|^{2r}|\mathscr{G}_n) = E(|Z_{n+1}|^{2r}|Z_1, \ldots, Z_n) \quad \text{a.s.,} \qquad n \geqslant 1.$$

This proves the theorem. ∎

II. Weak Convergence on Function Spaces

The modern theory of weak convergence on function spaces grew from a need to understand the behavior of functions of sums of independent random variables. It was natural to think of the functions as elements of a metric space, and to derive the desired results from limit theorems for measures on metric spaces. This abstraction has developed into an elegant and cohesive theory which has considerable importance in its own right. However, its main interest from the point of view of applications is still as a tool for proving limit laws for functions of random variables. We shall consider the theory with this in mind, and present here only those few results which are needed in this book. We refer the reader to Billingsley (1968) for a more detailed and rigorous exposition.

Let $C = C[0,1]$ denote the metric space of continuous functions on $[0,1]$, where C has the uniform topology obtained by defining the distance between x and y as

$$\rho(x,y) = \sup_{t\in[0,1]} |x(t) - y(t)|.$$

Let $\mathscr{C}$ be the class of Borel sets in C—that is, the σ-field generated by the open sets—and let $(\Omega,\mathscr{F},P)$ be a probability space. A *random element* is a measurable function $\xi:\Omega \to C$ (i.e., $\xi^{-1}(\mathscr{C}) \subseteq \mathscr{F}$). Under this mapping, $\omega \mapsto \xi(\cdot)(\omega)$, a continuous function on $[0,1]$. For each fixed $t \in [0,1]$, $\xi(t)(\cdot)$ is an r.v. on $(\Omega,\mathscr{F},P)$. We say that ξ is a random element *of* C, defined *on* $(\Omega,\mathscr{F},P)$. The random element ξ determines a probability measure π on the pair $(C,\mathscr{C})$ by the relation

$$\pi(A) = P(\xi \in A) = P(\xi^{-1}(A)), \qquad A \in \mathscr{C}.$$

Two random elements have the same distribution if and only if they induce the same measure π.

Consider now a sequence $\{\xi_n, n \geqslant 1\}$ of random elements on $(\Omega,\mathscr{F},P)$. This induces a sequence of probability measures $\{\pi_n\}$ on $(C,\mathscr{C})$, and we say that π_n converges weakly to π (written $\pi_n \xrightarrow{w} \pi$) if $\pi_n(A) \to \pi(A)$ for all π-continuity sets A. (The set A is a π-continuity set if the boundary of A has π-measure zero, i.e., $\pi(\partial A) = 0$.) There are a number of equivalent definitions of weak convergence, perhaps the most important being that $\pi_n \xrightarrow{w} \pi$ if and only if

$$\text{for all bounded, continuous} \quad f: C \to \mathbb{R}, \qquad \int f d\pi_n \to \int f d\pi. \tag{A.8}$$

Returning to our sequence $\{\xi_n\}$, suppose that the measure π is induced by a random element ξ. We say that $\xi_n \xrightarrow{d} \xi$(in the sense (C,ρ)) if $\pi_n \xrightarrow{w} \pi$. This definition of convergence in distribution is a natural analog of the definition of convergence in distribution of random variables, but it is not entirely satisfying since it does not give us much understanding of the behavior of the variables $\xi_n(t)$. We would expect that if $\xi_n \xrightarrow{d} \xi$, then for each $t \in [0,1]$,

$\xi_n(t) \xrightarrow{d} \xi(t)$ (convergence of r.v.). Furthermore, we would hope that for each finite k and each collection $0 \leqslant t_1 < t_2 < \cdots < t_k \leqslant 1$,

$$(\xi_n(t_1), \ldots, \xi_n(t_k)) \xrightarrow{d} (\xi(t_1), \ldots, \xi(t_k)) \tag{A.9}$$

(convergence of random vectors). That is, the *finite-dimensional distributions* of ξ_n converge to those of ξ.

Condition (A.9) is necessary for $\xi_n \xrightarrow{d} \xi$ in the sense (C,ρ), but not sufficient. We must know in addition that the sequence $\{\xi_n\}$ is *tight*; that is,

(i) for each $\varepsilon > 0$, $P(\sup_{|s-t|<\delta}|\xi_n(s) - \xi_n(t)| > \varepsilon) \to 0$ uniformly in n as $\delta \to 0$, and (A.10)

(ii) $P(|\xi_n(0)| > \lambda) \to 0$ uniformly in n as $\lambda \to \infty$.

There exist sequences $\{\xi_n\}$ for which (A.9) holds but not (A.10), and sequences for which (A.10) holds but not (A.9) (see Example 1 of Chapter 4).

Theorem A.2. $\xi_n \xrightarrow{d} \xi$ *in the sense* (C,ρ) *if and only if both* (A.9) *and* (A.10) *hold.*

We have had occasion in this book to mention the joint convergence of random elements and r.v. Let ξ_n and ξ be random elements of C, and X_n and X be r.v., all defined on the same probability space. We say that

$$(\xi_n, X_n) \xrightarrow{d} (\xi, X) \tag{A.11}$$

if for all continuity sets A of ξ and E of X,

$$P(\xi_n \in A \text{ and } X_n \in E) \to P(\xi \in A \text{ and } X \in E).$$

(Note that $A \subseteq C[0,1]$ while $E \subseteq \mathbb{R}$.) Theorem A.2 has an obvious extension, which states that (A.11) holds if and only if $\{\xi_n\}$ is tight and for all $0 \leqslant t_1 < t_2 < \cdots < t_k \leqslant 1$,

$$(\xi_n(t_1), \ldots, \xi_n(t_k), X_n) \xrightarrow{d} (\xi(t_1), \ldots, \xi(t_k), X).$$

The limit random element ξ is frequently standard Brownian motion, which we designate by W. We can completely determine W by specifying its finite dimensional distributions: for each $0 \leqslant t_1 < t_2 < \cdots < t_k \leqslant 1$, $(W(t_1), W(t_2) - W(t_1), \ldots, W(t_k) - W(t_{k-1}))$ has a multivariate normal distribution with zero mean and diagonal covariance matrix equal to $\operatorname{diag}(t_1, t_2 - t_1, \ldots, t_k - t_{k-1})$. The distribution of any random element is completely determined by its finite-dimensional distributions, although a given set of finite-dimensional distributions may not come from any random element. The existence of Brownian motion is proved in, for example, Billingsley (1968, Section 9).

Suppose that $\xi_n \xrightarrow{d} \xi$, and let $\mathscr{B}$ denote the Borel σ-field of $\mathbb{R}$. Let $h: C \to \mathbb{R}$ be a continuous mapping. If ξ_n induces the measure π_n then each $\pi_n \circ h^{-1}$ is a probability measure on $(\mathbb{R},\mathscr{B})$. Condition (A.8) implies that for all bounded, continuous $g: \mathbb{R} \to \mathbb{R}$,

$$\int (g \circ h)\, d\pi_n \to \int (g \circ h)\, d\pi;$$

that is,

$$\int g\, d(\pi_n \circ h^{-1}) \to \int g\, d(\pi \circ h^{-1}).$$

This is the so-called continuous mapping theorem.

Theorem A.3. *If $\xi_n \xrightarrow{d} \xi$ (convergence of random elements) and $h: C \to \mathbb{R}$ is continuous, then $h(\xi_n) \xrightarrow{d} h(\xi)$ (convergence of r.v.).*

The requirement that h be continuous can be relaxed. For example, it is sufficient to have h measurable and $P(\xi \in D_h) = 0$, where D_h is the set of discontinuities of h.

Let S and S' be arbitrary metric spaces with Borel fields $\mathscr{S}$ and $\mathscr{S}'$, and let $(\Omega,\mathscr{F},P)$ be a probability space. Suppose that $\{\xi_n, n \geqslant 1\}$ are random elements of $(S,\mathscr{S})$; that is, each $\xi_n: \Omega \to S$ is a measurable mapping. The theory of weak convergence on general metric spaces can be set up along the lines described above, and the analog of Theorem A.3 states that if $\xi_n \xrightarrow{d} \xi$ (convergence on $(S,\mathscr{S})$) and $h: S \to S'$ is continuous, then $h(\xi_n) \xrightarrow{d} h(\xi)$ (convergence on $(S',\mathscr{S}')$). As a special case, let $C_\lambda = C[0,\lambda]$ denote the space of continuous functions on $[0,\lambda]$, and let $S = C_\lambda \times [0,\lambda]$ and $S' = C_1 \times [0,\lambda]$ have the usual product topologies. The following result was used in Chapter 4.

Theorem A.4. *Let $\{\xi_n, n \geqslant 1\}$ be random elements of $C[0,\lambda]$ and let $\{X_n, n \geqslant 1\}$ be r.v. satisfying $0 \leqslant X_n \leqslant \lambda$. If $(\xi_n, X_n) \xrightarrow{d} (\xi, X)$ and $h: C_\lambda \times [0,\lambda] \to C_1 \times [0,\lambda]$ is continuous, then $h(\xi_n, X_n) \xrightarrow{d} h(\xi, X)$.*

III. Mixing Inequalities

Let $(\Omega,\mathscr{F},P)$ be a basic probability space and $\mathscr{G}$ and $\mathscr{H}$ denote two sub-σ-fields of $\mathscr{F}$. We define two measures of dependence between $\mathscr{G}$ and $\mathscr{H}$ by

$$\phi(\mathscr{G},\mathscr{H}) = \sup_{G \in \mathscr{G},\, H \in \mathscr{H},\, P(G) > 0} |P(H|G) - P(H)|$$

and

$$\alpha(\mathscr{G},\mathscr{H}) = \sup_{G \in \mathscr{G},\, H \in \mathscr{H}} |P(G \cap H) - P(G)P(H)|.$$

Note that $\alpha(\mathscr{G},\mathscr{H}) = 0$ implies $\phi(\mathscr{G},\mathscr{H}) = 0$ and that both α and ϕ are zero if $\mathscr{G}$ and $\mathscr{H}$ are independent. Our object is to obtain a variety of Hölder-type inequalities which are used in various places in the book, particularly in Chapter 5. We shall denote $(E|X|^p)^{1/p}$ by $\|X\|_p$.

Theorem A.5. *Suppose that X and Y are random variables which are $\mathscr{G}$- and $\mathscr{H}$-measurable, respectively, and $|X| \leqslant C_1$, $|Y| \leqslant C_2$ a.s. Then*

$$|EXY - EXEY| \leqslant 4C_1C_2\alpha(\mathscr{G},\mathscr{H}).$$

This result, but with a constant 16 rather than 4, is due to Volkonskii and Rozanov (1959). In the present form it is due to Ibragimov (1962) and also appears in the book of Ibragimov and Linnik (1971).

Proof. Set $\eta = \operatorname{sgn}\{E(Y|\mathscr{G}) - EY\}$ and note that η is $\mathscr{G}$-measurable. Then, using the properties of conditional expectation, we have

$$\begin{aligned} |EXY - EXEY| &= |E[X\{E(Y|\mathscr{G}) - EY\}]| \\ &\leqslant C_1 E|E(Y|\mathscr{G}) - EY| \\ &= C_1 E[\eta\{E(Y|\mathscr{G}) - EY\}] \leqslant C_1|E(\eta Y) - E\eta EY|. \end{aligned} \tag{A.12}$$

Furthermore, if $\xi = \operatorname{sgn}\{E(X|\mathscr{H}) - EX\}$, a similar reduction can be effected to give

$$|E\eta Y - E\eta EY| \leqslant C_2|E\xi\eta - E\xi E\eta|. \tag{A.13}$$

Then, setting $A_1 = \{\xi = 1\}$, $A_2 = \{\xi = -1\}$, $B_1 = \{\eta = 1\}$, and $B_2 = \{\eta = -1\}$, we have

$$\begin{aligned} |E\xi\eta - E\xi E\eta| &= |P(A_1 \cap B_1) + P(A_2 \cap B_2) - P(A_2 \cap B_1) - P(A_1 \cap B_2) \\ &\quad - P(A_1)P(B_1) - P(A_2)P(B_2) + P(A_2)P(B_1) + P(A_1)P(B_2)| \\ &\leqslant 4\alpha(\mathscr{G},\mathscr{H}) \end{aligned} \tag{A.14}$$

since $A_1, A_2 \in \mathscr{G}$ and $B_1, B_2 \in \mathscr{H}$. The required result then follows from (A.12), (A.13), and (A.14). ∎

Corollary A.1. *Suppose that X and Y are random variables which are $\mathscr{G}$- and $\mathscr{H}$-measurable, respectively, and that $E|X|^p < \infty$ for some $p > 1$, while $|Y| \leqslant C$ a.s. Then*

$$|EXY - EXEY| \leqslant 6C\|X\|_p[\alpha(\mathscr{G},\mathscr{H})]^{1-p^{-1}}.$$

Proof. Suppose that $\alpha(\mathscr{G},\mathscr{H}) > 0$, since otherwise the inequality is trivial. We set $\alpha = \alpha(\mathscr{G},\mathscr{H})$, $C_1 = \alpha^{-1/p}\|X\|_p$, $X_1 = XI(|X| \leqslant C_1)$, and $X_2 = X - X_1$.

Then, using Theorem A.5,

$$|EXY - EXEY| \leqslant |EX_1Y - EX_1EY| + |EX_2Y - EX_2EY| \leqslant 4CC_1\alpha + 2CE|X_2|. \tag{A.15}$$

But

$$E|X_2| = E[|X|I(|X| > C_1)] \leqslant C_1^{-p+1}E|X|^p. \tag{A.16}$$

The required result follows from (A.15) and (A.16) in view of the definition of C_1. ▮

Corollary A.1 and the subsequent Corollary A.2 are due to Davydov (1968).

Corollary A.2. *Suppose that X and Y are random variables which are $\mathscr{G}$- and $\mathscr{H}$-measurable, respectively, and that $E|X|^p < \infty$, $E|Y|^q < \infty$, where p, $q > 1$, $p^{-1} + q^{-1} < 1$. Then*

$$|EXY - EXEY| \leqslant 8\|X\|_p\|Y\|_q[\alpha(\mathscr{G},\mathscr{H})]^{1-p^{-1}-q^{-1}}.$$

Proof. Again suppose that $\alpha = \alpha(\mathscr{G},\mathscr{H}) > 0$. We now set

$$C = \alpha^{-1/q}\|Y\|_q, \qquad Y_1 = YI(|Y| \leqslant C), \qquad \text{and} \qquad Y_2 = Y - Y_1.$$

Then, using Corollary A.1 and Hölder's inequality,

$$|EXY - EXEY| \leqslant |EXY_1 - EXEY_1| + |EXY_2 - EXEY_2| \leqslant 6C\|X\|_p\alpha^{1-p^{-1}} + 2\|X\|_p\|Y_2\|_{p(p-1)^{-1}}, \tag{A.17}$$

while

$$E|Y_2|^{p(p-1)^{-1}} = E[|Y|^{p(p-1)^{-1}}I(|Y| > C)] \leqslant C^{-q+p(p-1)^{-1}}E|Y|^q. \tag{A.18}$$

The result follows from (A.17) and (A.18) in view of the definition of C. ▮

Theorem A.6. *Suppose that X and Y are random variables which are $\mathscr{G}$- and $\mathscr{H}$-measurable, respectively, and that $E|X|^p < \infty$, $E|Y|^q < \infty$, where p, $q > 1$, $p^{-1} + q^{-1} = 1$. Then*

$$|EXY - EXEY| \leqslant 2[\phi(\mathscr{G},\mathscr{H})]^{p^{-1}}\|X\|_p\|Y\|_q.$$

Furthermore, the result continues to hold for $p = 1$, $q = \infty$, where

$$\|Y\|_\infty = \operatorname{ess\,sup}|Y| = \inf\{C|P(|Y| > C) = 0\}.$$

Proof. First we shall deal with the case $p, q > 1$. Since we can approximate to X and Y by simple random variables, we begin by supposing that

$$X = \sum_i x_iI(A_i), \qquad Y = \sum_j y_jI(B_j), \tag{A.19}$$

where $\{A_i\}$ and $\{B_j\}$ are, respectively, finite decompositions of the sample space Ω into disjoint elements of $\mathscr{G}$ and $\mathscr{H}$. Then, using Hölder's inequality twice,

$$\begin{aligned}
|EXY - EXEY| &= \left|\sum_i \sum_j x_i y_j [P(A_i \cap B_j) - P(A_i)P(B_j)]\right| \\
&= \left|\sum_i x_i (P(A_i))^{1/p} \sum_j [P(B_j|A_i) - P(B_j)] y_j (P(A_i))^{1/q}\right| \\
&\leqslant \left\{\sum_i |x_i|^p P(A_i)\right\}^{1/p} \left\{\sum_i P(A_i) \left|\sum_j y_j [P(B_j|A_i) - P(B_j)]\right|^q\right\}^{1/q} \\
&\leqslant \|X\|_p \left\{\sum_i P(A_i) \left[\sum_j |y_j| \, |P(B_j|A_i)\right.\right. \\
&\qquad \left.\left. - P(B_j)|^{1/q} |P(B_j|A_i) - P(B_j)|^{1/p}\right]^q\right\}^{1/q} \\
&\leqslant \|X\|_p \left\{\sum_i P(A_i) \sum_j |y_j|^q [P(B_j|A_i)\right. \\
&\qquad \left. + P(B_j)] \sum_j |P(B_j|A_i) - P(B_j)|^{q/p}\right\}^{1/q} \\
&\leqslant 2^{1/q} \|X\|_p \|Y\|_q \sup_i \left\{\sum_j |P(B_j|A_i) - P(B_j)|\right\}^{1/p}. \qquad \text{(A.20)}
\end{aligned}$$

Further, if C_i^+ (respectively C_i^-) is the union of those B_j for which $P(B_j|A_i) - P(B_j)$ is positive (nonpositive), then C_i^+ and C_i^- are elements of $\mathscr{H}$ and hence

$$\begin{aligned}
\sum_j |P(B_j|A_i) - P(B_j)| &= [P(C_i^+|A_i) - P(C_i^+)] + [P(C_i^-|A_i) - P(C_i^-)] \\
&\leqslant 2\phi(\mathscr{G},\mathscr{H}). \qquad \text{(A.21)}
\end{aligned}$$

The required result for simple random variables then follows from (A.20) and (A.21).

To deal with the general case it suffices to note that

$$E|X - X_n|^p \to 0, \qquad E|Y - Y_n|^q \to 0$$

as $n \to \infty$, where X_n and Y_n are simple random variables defined by

$$X_n = n^{-1} \sum_{-n^2 \leqslant k < n^2} kI(k/n < X \leqslant (k+1)/n),$$

$$Y_n = n^{-1} \sum_{-n^2 \leqslant k < n^2} kI(k/n < Y \leqslant (k+1)/n).$$

The proof for the case $p = 1$, $q = \infty$ is very similar and we may suppose that (A.19) is satisfied with $|y_j| \leqslant \|Y\|_\infty < \infty$. Then

$$\begin{aligned}|EXY - EXEY| &= \left|\sum_i x_i P(A_i) \sum_j [P(B_j|A_i) - P(B_j)] y_j\right| \\ &\leqslant \|Y\|_\infty \sum_i |x_i| P(A_i) \sum_j |P(B_j|A_i) - P(B_j)| \\ &\leqslant 2\phi(\mathscr{G},\mathscr{H}) \|X\|_1 \|Y\|_\infty\end{aligned}$$

using (A.21), and this is the required form. ∎

Theorem A.6 has been widely reported in the literature and appears, for example, in the books of Billingsley (1968), Ibragimov and Linnik (1971), and Stout (1974). Quite a number of variants of the above results also appear in the literature, particularly ones concerning

$$\|E(X|\mathscr{G}) - EY\|_p$$

[e.g., in Serfling (1968), Dvoretzky (1972), and McLeish (1975b)], but these can readily be derived from the forms given herein by introducing $\operatorname{sgn}\{E(Y|\mathscr{G}) - EY\}$ as in the proof of Theorem A.5.

IV. Stationarity and Ergodicity

A process $\{X_n, n \geqslant 0\}$ is called (strictly) *stationary* if for each $k \geqslant 0$, the sequence $(X_{k+1}, X_{k+2}, \ldots)$ has the same distribution as $(X_0, X_1, \ldots)$. That is, any sequence of consecutive observations of the process $\{X_n\}$ has the same probabilistic structure. Many processes exhibit the properties of stationarity as they approach their equilibrium state.

Stationary processes first achieved prominence in studies of the laws of large numbers, although their general importance in probability theory has led to the study of many of their other properties. Birkhoff (1931) discovered that an integrable, stationary sequence $\{X_n\}$ obeys a strong law of large numbers, in the sense that $n^{-1} \sum_0^{n-1} X_j \xrightarrow{\text{a.s.}} Z$ for an r.v. Z. von Neumann (1932) considered the convergence of functions in a Hilbert space, showing that if $\{X_n\}$ is square integrable, then $n^{-1} \sum_0^{n-1} X_j \to Z$ in L^2 as well as with probability 1. These laws are referred to as *ergodic theorems*, a term which arose from much older averaging problems in physics and statistical mechanics. Today the field of ergodic theory is more widely embracing, containing many limit theorems for averages $n^{-1} \sum_1^n X_j$. Thus, Birkhoff and von Neumann's ergodic theorems are predated by those of Bernoulli and Chebyshev for sums of independent r.v. However, Birkhoff's result (as improved and generalized by Khintchine (1933), Hopf (1937), Riesz (1945) and others) is generally referred to as *the* ergodic theorem.

If $(\Omega,\mathscr{F},P)$ is a probability space, a measurable map $T:\Omega \to \Omega$ is said to be *measure preserving* if $P(T^{-1}A) = P(A)$ for all $A \in \mathscr{F}$. Any stationary process $\{X_n\}$ may be thought of as being generated by a measure-preserving transformation, in the sense that there exists a variable X defined on a probability space $(\Omega,\mathscr{F},P)$, and a measure-preserving map $T:\Omega \to \Omega$, such that the sequence $\{X'_n\}$ defined by $X'_0 = X$ and $X'_n(\omega) = X(T^n\omega)$, $n \geqslant 1$, $\omega \in \Omega$, has the same distribution as $\{X_n\}$. If T is measure preserving, a set $A \in \mathscr{F}$ is called *invariant* if $T^{-1}(A) = A$. The class $\mathscr{T}$ of all invariant sets is a sub-σ-field of $\mathscr{F}$, called the *invariant σ-field*, and T is said to be *ergodic* if all the sets in $\mathscr{T}$ have probability zero or one.

Ergodic Theorem. *If T is a measure-preserving transformation and X is an r.v. with $E|X| < \infty$, then*

$$n^{-1} \sum_{0}^{n-1} X(T^k\omega) \to E(X|\mathscr{T})(\omega)$$

for almost all $\omega \in \Omega$, and the convergence is also in L^1.

In general $E(X|\mathscr{T})$ may be a nondegenerate r.v. However, if T is ergodic, then $E(X|\mathscr{T}) = E(X)$ a.s.

Garsia (1965) has provided a very simple proof of the ergodic theorem. Short, rigorous accounts of stationarity and ergodicity are contained in many texts—for example, in Doob (1953), Loève (1978), and Breiman (1968). For a more detailed discussion we refer the reader to Dunford and Schwartz (1958) and Billingsley (1965).

V. Miscellanea

We present here two useful and unrelated theorems which have found several applications in this book. The first is a version of the dominated convergence theorem, and in the form given here it is due to Pratt (1960). However, there are several very similar versions of earlier origin; see, for example, Loève (1977, pp. 164–165).

Theorem A.7. *Let X_n, Y_n, Z_n, X, Y, and Z be r.v., and suppose that*

$$X_n \xrightarrow{p} X, \qquad Y_n \xrightarrow{p} Y, \qquad \text{and} \qquad Z_n \xrightarrow{p} Z, \tag{A.22}$$

$$X_n \leqslant Y_n \leqslant Z_n \quad \text{a.s.} \qquad \text{for all} \quad n, \tag{A.23}$$

$$E(X_n) \to E(X) \qquad \text{and} \qquad E(Z_n) \to E(Z), \qquad E(X), E(Z) \text{ finite}. \tag{A.24}$$

Then $E(Y_n) \to E(Y)$ and $E(Y)$ is finite.

Proof. Suppose that $E(Y_n) \nrightarrow E(Y)$. Then there exists a subsequence $\{n_k\}$ and a number l (possibly infinite) such that $E(Y_{n_k}) \to l \neq E(Y)$. Since (A.22) continues to hold along this subsequence, and since any subsequence which converges in measure has a further subsequence which converges a.s., there is no loss of generality in assuming that $X_{n_k} \xrightarrow{\text{a.s.}} X$, $Y_{n_k} \xrightarrow{\text{a.s.}} Y$, and $Z_{n_k} \xrightarrow{\text{a.s.}} Z$. Fatou's lemma states that for any sequence of r.v. $\{W_n, n \geqslant 1\}$ with $W_n \geqslant 0$ a.s. and $W_n \xrightarrow{\text{a.s.}} W$, we have

$$E(W) \leqslant \liminf E(W_n).$$

Hence,

$$\begin{aligned} E(Y - X) = E(\lim(Y_{n_k} - X_{n_k})) &\leqslant \liminf E(Y_{n_k} - X_{n_k}) \\ &= \liminf(EY_{n_k} - EX_{n_k}) \\ &= \liminf E(Y_{n_k}) - E(X). \end{aligned}$$

That is, $E(Y) \leqslant \liminf E(Y_{n_k})$. Similarly,

$$E(Z - Y) \leqslant \liminf E(Z_{n_k} - Y_{n_k}) = E(Z) - \limsup E(Y_{n_k}),$$

and so $\limsup E(Y_{n_k}) \leqslant E(Y)$. Combining these two inequalities we deduce that $\lim E(Y_{n_k})$ exists and equals $E(Y)$, which contradicts our assumption.

The finiteness of $E(Y)$ follows from (A.23) and (A.24), and the finiteness of $E(X)$ and $E(Z)$.

Our second result is a special case of a rather striking inequality for convex functions. For the general inequality, see Theorem III.4.3 of Garsia (1973, p. 106).

Theorem A.8. *Let $Z_1, Z_2, \ldots, Z_n$ be nonnegative r.v. and let $\mathscr{F}_0 \subseteq \mathscr{F}_1 \subseteq \cdots \subseteq \mathscr{F}_{n-1}$ be a nested sequence of σ-fields. For any $p \geqslant 1$,*

$$E\left[\sum_1^n E(Z_i|\mathscr{F}_{i-1})\right]^p \leqslant p^{p+1} E\left[\sum_1^n Z_i\right]^p.$$

Proof. Define $A_0 = 0$ and $A_r = \sum_1^r E(Z_i|\mathscr{F}_{i-1})$, $1 \leqslant r \leqslant n$. Then

$$A_n^p = \int_0^{A_n} px^{p-1}\,dx = \sum_{r=1}^n \int_{A_{r-1}}^{A_r} px^{p-1}\,dx \leqslant \sum_{r=1}^n pA_r^{p-1}(A_r - A_{r-1}).$$

Now set $B_r = A_r^{p-1} - A_{r-1}^{p-1}$, $1 \leqslant r \leqslant n$, and note that

$$A_n^p \leqslant p \sum_{r=1}^n (A_r - A_{r-1}) \sum_{s=1}^r B_s = p \sum_{s=1}^n B_s \sum_{r=s}^n (A_r - A_{r-1}) = p \sum_{s=1}^n B_s(A_n - A_{s-1}).$$

Since

$$\begin{aligned} E[B_s(A_n - A_{s-1})] &= E[B_s E(A_n - A_{s-1}|\mathscr{F}_{s-1})] \\ &= E\left[B_s E\left(\sum_s^n Z_i|\mathscr{F}_{s-1}\right)\right] \\ &= E\left[E\left(B_s \sum_s^n Z_i|\mathscr{F}_{s-1}\right)\right] \\ &\leqslant E\left(B_s \sum_1^n Z_i\right), \end{aligned}$$

it follows that

$$E(A_n^p) \leqslant pE\left[\left(\sum_{s=1}^n B_s\right)\left(\sum_{i=1}^n Z_i\right)\right] = pE\left(A_n^{p-1} \sum_1^n Z_i\right). \tag{A.25}$$

Now consider the function

$$f(x,y) = p^p x^p + (1 - p^{-1})y^p - pxy^{p-1},$$

where x, $y \geqslant 0$. By examining the partial derivitives of f and observing that $f \geqslant 0$ on the boundaries $x = 0$ and $y = 0$ and that $f \to \infty$ as either $x \to \infty$ or $y \to \infty$, we deduce that $f \geqslant 0$ everywhere. Hence

$$pxy^{p-1} \leqslant p^p x^p + (1 - p^{-1})y^p.$$

Using this inequality with (A.25), setting $x = \sum_1^n Z_i$ and $y = A_n$, and taking expectations, we derive

$$E(A_n^p) \leqslant p^p E\left(\sum_1^n Z_i\right)^p + (1 - p^{-1})E(A_n^p);$$

that is,

$$E(A_n^p) \leqslant p^{p+1} E\left(\sum_1^n Z_i\right)^p,$$

which proves Theorem A.8. ∎

References

The numbers in square brackets denote pages on which the references are cited; P denotes Preface.

Adler, R. J. (1978). A martingale central limit theorem without negligibility conditions. *Bull. Austral. Math. Soc.* **18**, 13–19. [52]

Adler, R. J., and Scott, D. J. (1975). Martingale central limit theorems without uniform asymptotic negligibility. *Bull. Austral. Math. Soc.* **13**, 45–56. [52]

Adler, R. J., and Scott, D. J. (1978). Martingale central limit theorems without negligibility conditions: Corrigendum. *Bull. Austral. Math. Soc.* **18**, 311–319. [52]

Albert, A. E., and Gardner, L. A. (1967). "*Stochastic Approximation and Nonlinear Regression.*" M.I.T. Press, Cambridge, Massachusetts. [238]

Aldous, D. J. (1977a). Subsequences of sequences of random variables. *Bull. Amer. Math. Soc.* **83**, 121–123. [207]

Aldous, D. J. (1977b). Limit theorems for subsequences of arbitrarily-dependent sequences of random variables. *Z. Wahrsch. Verw. Gebiete* **40**, 59–82. [207, 208]

Aldous, D. J. (1978). Stopping times and tightness. *Ann. Probab.* **6**, 335–340. [52, 100]

Aldous, D. J. and Eagleson, G. K. (1978). On mixing and stability of limit theorems. *Ann Probab.* **6**, 325–331. [52, 56, 59]

Anderson, B. D. O., and Moore, J. B. (1976). A matrix Kronecker lemma. *Linear Algebra Appl.* **15**, 227–234. [156]

Anderson, T. W. (1959). On asymptotic distributions of estimates of parameters of stochastic difference equations. *Ann. Math. Statist.* **30**, 676–687. [172]

Anderson, T. W. (1970). "*Time Series Analysis.*" Wiley, New York. [134]

Anderson, T. W., and Taylor, J. B. (1979). Strong consistency of least squares estimates in dynamics models. *Ann. Statist.* **7**, 484–489. [194]

Bachelier, L. (1900). "*Théorie de la speculation.*" Gauthier-Villars, Paris. [2]

Basawa, I. V., and Scott, D. J. (1976). Efficient tests for branching processes. *Biometrika* **63**, 531–536. [165]

Basawa, I. V., and Scott, D. J. (1977). Efficient tests for stochastic processes. *Sankhyā Ser. A* **39**, 21–31. [165]

Basawa, I. V., Feigin, P. D., and Heyde, C. C. (1976). Asymptotic properties of maximum likelihood estimators for stochastic processes. *Sankhyā Ser. A.* **38**, 259–270. [158]

Basu, A. K. (1976). On the rate of convergence to normality for sums of dependent variables. *Acta Math. Acad. Sci. Hungar.* **28**, 261–265. [89]

Bergström, H. (1970). A comparison method for distribution functions of sums of independent and dependent random variables. *Theory Probab. Appl.* **15**, 430–457. [7, 51]

Berkes, I. (1973). On Strassen's version of the loglog law for multiplicative systems. *Studia Sci. Math. Hungar.* **8**, 424–431. [7]

Berkes, I. (1974). The law of the iterated logarithm for subsequences of random variables. *Z. Wahrsch. Verw. Gebiete* **30**, 209–215. [7, 206, 208]

Berkes, I. (1975). An almost sure invariance principle for lacunary trigonometric series. *Acta Math. Acad. Sci. Hungar.* **26**, 209–220. [7]

Berkes, I. (1976a). On the asymptotic behavior of $\sum f(n_k x)$. Main theorems. *Z. Wahrsch. Verw. Gebiete* **34**, 319–345. [7]

Berkes, I. (1976b). On the asymptotic behavior of $\sum f(n_k x)$. Applications. *Z. Wahrsch. Verw. Gebiete* **34**, 347–365. [7]

Berman, J. (1976). An example of a weak martingale. *Ann. Probab.* **4**, 107–108. [P]

Bernstein, S. (1927). Sur l'extension du théorème limite du calcul des probabilitiés aux sommes de quantités dépendantes. *Math. Ann.* **85**, 1–59. [2, 51]

Bernstein, S. (1939). Quelques remarques sùr le théorème limite Liapounoff. *Dokl. Akad. Nauk. SSSR* **24**, 3–8. [2, 70]

Bernstein, S. N. (1940). New applications of almost independent quantities. *Izv. Akad. Nauk. SSSR Ser. Math.* **4**, 137–150 (in Russian). [2]

Bernstein, S. N. (1941). On sums of dependent quantities having a joint regression which is almost zero. *Dokl. Akad. Nauk. SSSR* **32**, 303–307 (in Russian). [2]

Berry, A. C. (1941). The accuracy of the Gaussian approximation to the sum of independent variates. *Trans. Amer. Math. Soc.* **49**, 122–136. [81]

Beśka, M. (1978). Poisson distribution as a limit law for sums of dependent random variables. *Preprint No. 6/78, Institute of Mathematics, Nicholas Copernicus Univ. Torun, Poland.* [52]

Bhat, B. R. (1974). On the method of maximum likelihood for dependent observations. *J. Roy. Statist. Soc. Ser. B* **36**, 48–53. [158]

Billingsley, P. (1961a). "*Statistical Inference for Markov Processes.*" Univ. of Chicago Press, Chicago, Illinois. [155, 179, 180, 181]

Billingsley, P. (1961b). The Lindeberg-Lévy theorem for martingales. *Proc. Amer. Math. Soc.* **12**, 788–792. [51]

Billingsley, P. (1965). "*Ergodic Theory and Information.*" Wiley, New York. [281]

Billingsley, P. (1968). "*Convergence of Probability Measure.*" Wiley, New York. [71, 97, 161, 175, 225, 274, 276, 280]

Birkhoff, G. D. (1931). Proof of the ergodic theorem. *Proc. Nat. Acad. Sci. U.S.A.* **17**, 656–660. [280]

Blackwell, D., and Freedman, D. (1973). On the amount of variance needed to escape from a strip. *Ann. Probab.* **1**, 772–787. [54]

Blum, J. R., Hanson, D. L. and Koopmans, L. (1963). On the strong law of large numbers for a class of stochastic processes. *Z. Wahrsch. Verw. Gebiete* **2**, 1–11. [40]

Breiman, L. (1968). "*Probability.*" Addison-Wesley, Reading, Massachusetts. [270, 281]

Brown, B. M. (1969). Moments of a stopping rule related to the central limit theorem. *Ann. Math. Statist.* **40**, 1236–1249. [70]

Brown, B. M. (1970). Characteristic functions, moments and the central limit theorem. *Ann. Math. Statist.* **41**, 658–664. [70, 89, 90]

Brown, B. M. (1971). Martingale central limit theorems. *Ann. Math. Statist.* **42**, 59–66. [9, 16, 51, 59, 70, 99]

Brown, B. M., and Eagleson, G. K. (1971). Martingale convergence to infinitely divisible laws with finite variances. *Trans. Amer. Math. Soc.* **162**, 449–453. [52]

Burkholder, D. L. (1973). Distribution function inequalities for martingales. *Ann. Probab.* **1**, 19–42. [13]

Caines, P. E. (1975). A note on the consistency of maximum likelihood estimates for finite families of stochastic processes, *Ann. Statist.* **3**, 539–546. [158]

Chatterji, S. D. (1970). A general strong law: *Invent, Math.* **9**, 235–245. [206]

Chatterji, S. D. (1974a). A subsequence principle in probability theory. *Bull. Amer. Math. Soc.* **80**, 495–497. [206]

Chatterji, S. D. (1974b). A principle of subsequences in probability theory: the central limit theorem. *Adv. in Math.* **13**, 31–54. [52, 59, 206, 208]

Chow, Y. S. (1965). Local convergence of martingales and the law of large numbers. *Ann. Math. Statist.* **36**, 552–558. [35]

Chow, Y. S. (1971). On the L_p convergence for $n^{-1/p}S_n$, $0 < p < 2$. *Ann. Math. Statist.* **42**, 393–394. [42]

Chow, Y. S., and Teicher, H. (1978). "*Probability Theory: Independence, Interchangeability, Martingales.*" Springer-Verlag, Berlin and New York. [11, 31, 204]

Chung, K. L. (1967). "*Markov Chains with Stationary Transition Probabilities,*" 2nd ed. Springer-Verlag, Berlin and New York. [7]

Chung, K. L. (1974). "*A Course in Probability Theory,*" 2nd ed. Academic Press, New York. [4, 11, 33, 39, 71]

Cocke, W. J. (1972). Central limit theorems for sums of dependent vector variables. *Ann. Math. Statist.* **43**, 968–976. [7]

Cogburn, R. (1960). Asymptotic properties of stationary sequences. *Univ. California Publ. Statist.* **3**, 99–146. [128]

Csörgö, M. (1968). On the strong law of large numbers and the central limit theorem for martingales. *Trans. Amer. Math. Soc.* **131**, 259–275. [31]

Dambis, K. E. (1965). On the decomposition of continuous submartingales. *Theory Probab. Appl.* **10**, 401–410. [269]

Davydov, Yu. A. (1968). Convergence of distributions generated by stationary stochastic processes. *Theory Probab. Appl.* **13**, 691–696. [278]

Derman, C., and Sacks, J. (1959). On Dvoretzky's stochastic approximation theorem. *Ann. Math. Statist.* **30**, 601–606. [243]

Dobrushin, R. L. (1956). Central limit theorem for nonstationary Markov chains I and II. *Theory Probab. Appl.* **1**, 65–80, 329–383.

Donsker, M. (1951). An invariance principle for certain probability limit theorems. *Mem. Amer. Math. Soc.* **6**. [97]

Donsker, M. (1952). Justification and extension of Doob's neuristic approach to the Kolmogorov-Smirnov theorems. *Ann. Math. Statist.* **23**, 277–281. [97]

Doob, J. L. (1953). "*Stochastic Processes.*" Wiley, New York. [2, 20, 22, 31, 51, 54, 132, 281]

Drogin, R. (1972). An invariance principle for martingales. *Ann. Math. Statist.* **43**, 602–620. [52, 989]

Drogin, R. (1973). On covergence in probability to Brownian motion. *Ann. Probab.* **1**, 254–262. [269]

Dubins, L. E. (1968). On a theorem of Skorokhod. *Ann. Math. Statist.* **39**. 2094–2097. [269]

Dubins, L. E., and Schwartz, G. (1965). On continuous martingales. *Proc. Nat. Acad. Sci. U.S.A.* **53**, 913–916. [269]

Dunford, N., and Schwartz, J. T. (1958). "*Linear Operators,*" Part I. Wiley (Interscience), New York. [281]

Dunsmuir, W. (1979). A central limit theorem for parameter estimation in stationary vector time series and its application to models for a signal observed with noise. *Ann. Statist.* **7**, 490–506. [194]

Dunsmuir, W., and Hannan, E. J. (1976). Vector linear time series models. *Adv. Appl. Probab.* **8**, 339–364. [194]

Durrett, R., and Resnick, S. I. (1978). Functional limit theorems for dependent variables. *Ann. Probab.* **6**, 829–846. [52]

Dvoretzky, A. (1956). On stochastic approximation. *Proc 3rd Berkeley Symp. Math. Statist. Prob.* **1**, 39–56. [238, 242]

Dvoretzky, A. (1969). Central limit theorems for dependent random variables and some applications. Abstract. *Ann. Math. Statist.* **40**, 1871. [51]

Dvoretzky, A. (1971). The central limit problem for dependent random variables. *In "Actes du Congres International deś Mathématiciens, Nice,"* pp. 565–570. Gauthier-Villars, Paris. [51]

Dvoretzky, A. (1972). Asymptotic normality for sums of dependent random variables. *Proc. 6th Berkeley Symp. Math. Statist. Prob.*, **2**, 513–535. [44, 51, 280]

Dynkin, E. B. (1953). Classes of equivalent functions. *Uspekhi Mat. Nauk.* **54**, 125–134 (in Russian). [240]

Eagleson, G. K. (1975a). Martingale convergence to mixtures of infinitely divisible laws. *Ann. Probab.* **3**, 557–562. [52, 59, 64]

Eagleson, G. K. (1975b). On Gordin's central limit theorem for stationary processes. *J. Appl. Probab.* **12**, 176–179. [130]

Eagleson, G. K. (1976). Martingale convergence to the Poisson distribution. *Casopis Pest. Mat.* **101**, 271–277. [52]

Eagleson, G. K., and Weber, N. C. (1978). Limit theorems for weakly exchangeable arrays. *Math. Proc. Cambridge Philos. Soc.* **84**, 123–130. [78]

Erdös, P., and Kac, M. (1946). On certain limit theorems in the theory of probability. *Bull. Amer. Math. Soc.* **52**, 292–302. [97]

Erdös, P., and Kac, M. (1947). On the number of positive sums of independent random variables. *Bull. Amer. Math. Soc.* **53**, 1011–1020. [97]

Erickson, R. V., Quine, M. P., and Weber, N. C. (1978). Explicit bounds for the departure from normality of sums of dependent random variables. *Acta. Math. Acad. Sci. Hungar.* **34**, 27–32. [89]

Esseen, C. G. (1942). On the Liapounoff limit of error in the theory of probability. *Ark. Math. Astr. och Fysik* **28A**, 1–19. [81]

Ewens, W. J. (1969). "*Population Genetics.*" Methuen, London. [235]

Fabian, V. (1968). On asymptotic normality in stochastic approximation. *Ann. Math. Statist.* **39**, 1327–1332. [238]

Feigin, P. D. (1978). The efficiency criteria problem for stochastic processes. *Stochastic Processes Appl.* **6**, 115–127. [165]

Feller, W. (1971). "*An Introduction to Probability Theory and Its Applications,*" Vol. II, 2nd ed., Wiley, New York. [9, 72, 89, 201, 205, 270]

Foldes, L. (1978). Martingale conditions for optimal saving—discrete time. *J. Math. Econ.* **5**, 83–96. [P]

Foster, D., and Williams, D. (1978). The Hawkins sieve and Brownian motion. *Compositio Math.* **37**, 277–289. [229]

Fraser, D. (1973). The rate of convergence of a random walk to Brownian motion. *Ann. Probability* **1**, 699–701. [111]

Freedman, D. A. (1971). "*Brownian Motion and Diffusion.*" Holden-Day, San Francisco, California. [270]

Freedman, D. A. (1975). On tail probabilities for martingales. *Ann. Probability* **3**, 100–118. [54, 116]

Furstenberg, H., and Kesten, H. (1960). Products of random matrices. *Ann. Math. Statist.* **31**, 457–469. [225, 226]

Galambos, J. (1978). "*The Asymptotic Theory of Extreme Order Statistics.*" Wiley, New York. [201, 202]

Gänssler, P., and Häusler, E. (1980). Remarks on the functional central limit theorem for martingales. *Z. Wahrsch. Verw. Gebiete* **50**, 237–243. [52, 98]

Gänssler, P., Strobel, J., and Stute, W. (1978). On central limit theorems for martingale triangular arrays. *Acta. Math. Acad. Sci. Hungar.* **31**, 205–216. [52]

Gaposhkin, V. F. (1972). Convergence and limit theorems for sequences of random variables. *Theor. Probability Appl.* **17**, 379–400. [208]

Garsia, A. M. (1965). A simple proof of Eberhard Hopf's maximal ergodic theorem. *J. Math. and Mech.* **14**, 318–382. [281]

Garsia, A. M. (1973). "*Martingale Inequalities: Seminar Notes on Recent Progress.*" Benjamin, New York. [11, 282]

Gebelein, H. (1941). Das statistische Problem der Karelation als Variations und Eigenwertproblem und sein Zusammenhang mit der Ausgleichungsrechnung. *Z. Angew. Math. Mech.* **21**, 364–379. [147]

Gikhman, I. I., and Skorokhod, A. V. (1969). "*Introduction to the Theory of Random Processes.*" Saunders, Philadelphia, Pennsylvania. [183]

Gilat, D. (1977). Every nonnegative submartingale is the absolute value of a martingale. *Ann. Probab.* **5**, 475–481. [8]

Gnedenko, B. V., and Kolmogorov, A. N. (1954). "*Limit Distributions for Sums of Independent Random Variables.*" Addison-Wesley, Reading, Massachusetts. [55, 82]

Gordin, M. I. (1969). The central limit theorem for stationary processes. *Soviet Math. Dokl.* **10**, 1174–1176. [127, 128]

Grams, W. F. (1972). *Rates of Convergence in the Central Limit Theorem for Dependent Variables.* Ph.D. dissertation, Florida State Univ. [89]

Granger, C. W., and Andersen, A. (1978a). "*An Introduction to Bilinear Time Series Models.*" Vandenhoek and Ruprecht, Gottingen. [196, 197]

Granger, C. W., and Andersen, A. (1978b). On the invertibility of time series models. *Stochastic Processes. Appl.* **8**, 87–92. [196, 197]

Hall, P. (1977). Martingale invariance principles. *Ann. Probab.* **5**, 875–887. [52, 59, 67]

Hall, P. (1978a). On the duality between the behaviour of sums of independent random variables and the sums of their squares. *Math. Proc. Cambridge Philos. Soc.* **84**, 117–121. [70]

Hall, P. (1978b). The convergence of moments in the martingale central limit theorem. *Z. Wahrsch. Verw. Gebiete* **44**, 253–260. [70]

Hall, P. (1979a). On the Skorokhod representation approach to martingale invariance principles. *Ann. Probab.* **7**, 371–376. [100]

Hall, P. (1979b). On the invariance principle for U-statistics. *Stochastic Processes. Appl.* **9**, 163–174. [P]

Hall, P., and Heyde, C. C. (1976). On a unified approach to the law of the iterated logarithm for martingales. *Bull. Austral. Math. Soc.* **14**, 435–447. [116]

Hall, W. J. and Loynes, R. M. (1977). Weak convergence of processes related to likelihood ratios. *Ann. Statist.* **5**, 330–341. [P]

Hammersley, J. M. (1974). Postulates for subadditive processes. *Ann. Probab.* **2**, 652–680. [214]

Hammersley, J. M., and Welsh, J. A. D. (1965). First passage percolation, sub-additive processes, stochastic networks and generalized renewal theory. *In* "*Bernoulli-Bayes-Laplace Anniversary Volume*" (J. Neyman and L. M. Le Cam, eds.). Springer-Verlag, Berlin and New York. [213, 222]

Hannan, E. J. (1970). "*Multiple Time Series.*" Wiley, New York. [135, 146, 177, 180, 182, 187, 191, 192, 194]

Hannan, E. J. (1973). The asymptotic theory of linear time-series models. *J. Appl. Probab.* **10**, 130–145. [194]

Hannan, E. J. (1976). The asymptotic distribution of serial covariances. *Ann. Statist.* **4**, 396–399. [194]

Hannan, E. J., and Heyde, C. C. (1972). On limit theorems for quadratic functions of discrete time series. *Ann. Math. Statist.* **43**, 2058–2066. [136, 184, 187]

Hannan, E. J., and Quinn, B. G. (1979). The determination of the order of an autoregression. *J. Roy. Statist. Soc. Ser. B* **41**, 190–195. [P]

Hardy, G. H. and Littlewood, J. E. (1914). Some problems of Diophantine approximation. *Acta Math.* **37**, 155–239. [115]

Harris, T. E. (1963). "*The Theory of Branching Processes.*" Springer-Verlag, Berlin and New York. [5, 165, 195]

Hartman, P., and Wintner, A. (1941). On the law of the iterated logarithm. *Amer. J. Math.* **63**, 169–176. [10, 116]

Hausdorff, F. (1913). "*Grundzuge der Mengenlehre.*" Veit, Leipzig. [115]

Hawkins, D. (1958). The random sieve. *Math. Mag.* **31**, 1–3. [226]

Heyde, C. C. (1969). On extended rates of convergence for the invariance principle. *Ann. Math. Statist.* **40**, 2178–2179. [111]

Heyde, C. C. (1971). Some central limit analogues for supercritical Galton-Watson processes. *J. Appl. Probab.* **8**, 52–59. [6]

Heyde, C. C. (1973). An iterated logarithm result for martingales and its application in estimation theory for autoregressive processes. *J. Appl. Probab.* **10**, 146–157. [194]

Heyde, C. C. (1974a). On the central limit theorem for stationary processes. *Z. Wahrsch. Verw. Gebiete* **30**, 315–320. [133]

Heyde, C. C. (1974b). On martingale limit theory and strong convergence results for stochastic approximation procedures. *Stochastic Processes Appl.* **2**, 359–370. [238]

Heyde, C. C. (1974c). On estimating the variance of the offspring distribution in a simple branching process. *Adv. Appl. Probab.* **6**, 421–433. [196]

Heyde, C. C. (1974d). An iterated logarithm result for autocorrelations of a stationary linear process, *Ann. Probab.* **2**, 328–332. [194]

Heyde, C. C. (1975a). On the central limit theorem and iterated logarithm law for stationary processes. *Bull. Austral. Math. Soc.* **12**, 1–8. [140]

Heyde, C. C. (1975b). Remarks on efficiency in estimation for branching processes. *Biometrika* **62**, 49–55. [157, 165, 171]

Heyde, C. C. (1977a). On central limit and iterated logarithm supplements to the martingale convergence theorem. *J. Appl. Probab.* **14**, 758–775. [11, 80, 81, 125]

Heyde, C. C. (1977b). The effect of selection on genetic balance when the population size is varying. *Theoret. Population Biol.* **11**, 249–251. [236]

Heyde, C. C. (1977c). An optimal property of maximum likelihood with application to branching process estimation. *Bull. Int. Statist. Inst.* **47**(2), 407–417. [157]

Heyde, C. C. (1978a). A log log improvement to the Riemann hypothesis for the Hawkins random sieve. *Ann. Probab.* **6**, 870–875. [229]

Heyde, C. C. (1978b). On an optimal asympotic property of the maximum likelihood estimator of a parameter from a stochastic process. *Stochastic Processes Appl.* **8**, 1–9. [161]

Heyde, C. C. and Brown, B. M. (1970). On the departure from normality of a certain class of martingales. *Ann. Math. Statist.* **41**, 2161–2165. [85]

Heyde, C. C., and Feigin, P. D. (1975). On efficiency and exponential families in stochastic process estimation. *In* "*Statistical Distributions in Scientific Work*" (G. P. Patil, S. Kotz, and J. K. Ord, eds.), Vol. 1. pp. 227–240. Reidel, Dordrecht. [157, 165, 167, 171]

Heyde, C. C., and Leslie, J. R. (1971). Improved classical limit analogues for Galton-Watson processes with or without immigration. *Bull Austral. Math. Soc.* **5**, 145–155. [6, 195]

Heyde, C. C., and Scott, D. J. (1973). Invariance principles for the law of the iterated logarithm for martingales and processes with stationary increments. *Ann. Probab.* **1**, 428–436. [116, 140, 145]

Heyde, C. C., and Seneta, E. (1972). Estimation theory for growth and immigration rates in a multiplicative process, *J. Appl. Probab.* **9**, 235–256. [178, 181]

Heyde, C. C., and Seneta, E. (1974). Notes on "Estimation theory for growth and immigration rates in a multiplicative process," *J. Appl. Probab.* **11**, 572–577. [178]

Heyde, C. C., and Seneta, E. (1975). The genetic balance between random sampling and random population size. *J. Math. Biol.* **1**, 317–320. [237]

Heyde, C. C., and Seneta, E. (1977). "*I. J. Bienaymé. Statistical Theory Anticipated.*" Springer-Verlag, Berlin and New York. [P]

Hinderer, K. (1975). Neuere Resultate in der Stochastischen Dynamischen Optimierung. *Z. Angew. Math. Mech.* **55**, T16–T26. [253]

Hirschfeld, O. (1935). A connection between correlation and contingency. *Math. Proc. Cambridge Philos. Soc.* **31**, 520–524. [147]

Ho, Y. C. (1963). On the stochastic approximation method and optimal filtering theory. *J. Math. Anal. Appl.* **6**, 152–154. [238]

Holmes, P. T., and Strawderman, W. E. (1969). A note on the waiting times between record observations. *J. Appl. Probab.* **6**, 711–714. [5]

Holmes, P. T., and Strawderman, W. E. (1970). On the law of the iterated logarithm for inter-record times. *J. Appl. Probab.* **7**, 432–439. [5]

Hopf, E. (1937). "*Ergodentheorie*," Ergebnisse Math., 5, No. 2. Springer, Berlin (Reprinted by Chelsea, New York, 1948). [281]

Ibragimov, I. A. (1962). Some limit theorems for stationary processes. *Theory Probab. Appl.* **7**, 349–382. [128, 133, 277]

Ibragimov, I. A. (1963). A central limit theorem for a class of dependent random variables. *Theory Probab. Appl.* **8**, 83–89. [51, 84]

Ibragimov, I. A. (1975). A note on the central limit theorem for dependent random variables. *Theory Probab. Appl.* **20**, 135–141. [128, 148, 151]

Ibragimov, I. A., and Linnik, Yu. V. (1971). "*Independent and Stationary Sequences of Random Variables.*" Wolters-Noordhoff, Groningen. [7, 82, 132, 133, 135, 142, 146, 151, 277, 280]

Ibragimov, I. A., and Rozanov, Yu. A. (1978). "*Gaussian Random Processes.*" Springer-Verlag, Berlin and New York. [147]

Iglehart, D. L. (1971). Multiple channel queues in heavy traffic: law of the iterated logarithm. *Z. Wahrsch. Verw. Gebiete* **17**, 168–180. [4]

Ingham, A. E. (1932). "*The Distribution of Prime Numbers.*" Cambridge Univ. Press, London and New York. [227]

Iosifescu, M., and Tăutu, P. (1973). "*Stochastic Processes and Applications in Biology and Medicine*," Vol. 2. Springer-Verlag, Berlin and New York. [235]

Ishitani, J. (1977). A central limit theorem for the subadditive process and its application to products of random matrices. *Publ. Res. Inst. Math. Sci. Kyoto Univ.* **12**, 565–575. [223, 225, 226]

Jain, N. C., Jogdeo, K., and Stout, W. F. (1975). Upper and lower functions for martingales and mixing processes. *Ann. Probab.* **3**, 119–145. [116]

Johnstone, I. M. (1978). *Problems in Limit Theory for Martingales and Posterior Distributions from Stochastic Processes.* M.Sc. Thesis, The Australian National Univ., Canberra. [52]

Junco, A. del (1977). On the decomposition of a subadditive stochastic process. *Ann. Probab.* **5**, 298–302. [217]

Kato, Y. (1978). Rates of convergence in central limit theorem for martingale differences. *Bull. Math. Statist.* **18**, 1–8. [84]

Katznelson, Y. (1968). "*Harmonic Analysis.*" Wiley, New York. [191]

Keilson, J. and Wishart, D. G. M. (1964). A central limit theorem for processes defined on a finite Markov chain. *Math. Proc. Cambridge Philos. Soc.* **60**, 547–567. [7]

Kendall, D. G. (1967). On finite and infinite sequences of exchangeable events. *Studia Sci. Math. Hungar.* **2**, 319–327. [202, 204]

Kersting, G. (1977). Almost sure approximation of the Robbins-Monro procedure by sums of independent random variables. *Ann. Probab.* **5**, 954–965. [240]

Khintchine, A. (1924). Uber einen Satz der Wahrscheinlichkeitsrechnung. *Fund. Math.* **6**, 9–20. [116]

Khintchine, A. (1933). Zu Birkhoffs Lösung des Ergodenproblems. *Math. Ann.* **107**, 485–488. [281]

Kingman, J. F. C. (1968). The ergodic theory of subadditive processes. *J. Roy. Statist. Soc. Ser. B* **30**, 499–510. [213]

Kingman, J. F. C. (1973). Subadditive ergodic theory. *Ann. Probab.* **1**, 883–909. [214, 217]

Kingman, J. F. C. (1976). Subadditive processes. *In* "*École d' Eté de Probabilités de Saint Flour V-1975.*" Springer Lecture Notes in Mathematics No. 539. [214]

Kingman, J. F. C. (1978). Uses of exchangeability. *Ann. Probab.* **6**, 183–197. [202]

Klimko, L. A., and Nelson, P. I. (1978). On conditional least squares estimation for stochastic processes. *Ann. Statist.* **6**, 629–642. [173, 174, 176, 178]

Klopotowski, A. (1977a). Limit theorems for sums of dependent random vectors in $\mathbb{R}^d$. *Dissertationes Math.* **156**. [52]

Kĺopotowski, A. (1977b). Mixtures of infinitely divisible distributions as limit laws for sums of dependent random variables. *Preprint No. 4/77, Institute of Mathematics, Nicholas Copernicus Univ. Torun, Poland.* [52]

Kolmogorov, A. N. (1929). Das Gesetz des iterierten Logarithmus. *Math. Ann.* **101**, 126–135. [116]

Komlós, J. (1967). A generalization of a problem of Steinhaus. *Acta Math. Acad. Sci. Hungar.* **18**, 217–229. [205, 208]

Landers, D., and Rogge, L. (1976). On the rate of convergence in the central limit theorem for Markov chains. *Z. Wahrsch. Verw. Gebiete* **35**, 57–63. [89]

Lévy, P. (1935a). Propriétés asymptotiques des sommes de variables aléatoires enchainées. *Bull. Sci. Math.* **59** (ser. 2), 84–96, 109–128. [2, 51]

Lévy, P. (1935b). Propriétés asymptotiques des sommes de variables aléatoires indépendantes ou enchainées. *J. Math. Pures Appl.* **14**, (ser. 9), 347–402. [2, 51]

Lévy, P. (1937). "*Théorie de l'addition des variables aléatoires.*" Gauthier-Villars, Paris. [2, 51]

Lévy, P. (1954). "*Théorie de l'addition des variables aléatoires,*" 2nd ed. Gauthier-Villars, Paris. [116]

Ljung, L. (1978). Strong convergence of a stochastic approximation algorithm. *Ann. Statist.* **6**, 680–696. [242]

Loève, M. (1977). "*Probability Theory I,*" 4th ed. Springer-Verlag, Berlin and New York. [7, 9, 10, 29, 55, 175, 281]

Loève, M. (1978). "*Probability Theory II,*" 4th ed. Springer-Verlag, Berlin and New York. [22, 281]

Loynes, R. M. (1969). The central limit theorem for backwards martingales. *Z. Wahrsch. Verw. Gebiete* **13**, 1–8. [51, 78]

Loynes, R. M. (1970). An invariance principle for reversed martingales. *Proc. Amer. Math. Soc.* **25**, 56–64. [51]

Loynes, R. M. (1976). A criterion for tightness for a sequence of martingales. *Ann. Probab.* **4**, 859–862. [100]

Loynes, R. M. (1978). On the weak convergence of U-statistics, and of the empirical process. *Math. Proc. Cambridge Philos. Soc.* **83**, 269–272. [P]

Major, P., and Révész, P. (1973). A limit theorem for the Robbins-Monro approximation. *Z. Wahrsch. Verw. Gebiete* **27**, 79–86. [240]

Mandl, P. (1974). The use of optimal stationary policies in the adaptive control of linear systems. *In Proc. Symp. to Honour J. Neyman* pp. 223–242. [252]

Mandl, P. (1977). Some results in the adaptive control of linear systems. *In* "*Information Theory, Statistical Decision Functions, Random Processes*" (J. Kozesnik, ed.), Vol. A, pp. 399–410. Reidel, Dordrecht. [267]

Matthews, J. P. (1970). A central limit theorem for absorbing Markov chains. *Biometrika* **57**, 129–139. [7]

McLeish, D. L. (1974). Dependent central limit theorems and invariance principles. *Ann. Probab.* **2**, 620–628. [44, 51–53, 57, 59]

McLeish, D. L. (1975a). An invariance principle for strongly multiplicative systems. *Acta Math. Acad. Sci. Hungar.* **26**, 81–85. [7]

McLeish, D. L. (1975b). A maximal inequality and dependent strong laws. *Ann. Probab.* **3**, 829–839. [19, 20, 280]

McLeish, D. L. (1975c). Invariance principles for dependent variables. *Z. Wahrsch. Verw. Gebiete* **32**, 165–178. [128]

McLeish, D. L. (1976). Functional and random central limit theorems for the Robbins-Monro process, *J. Appl. Probab.* **13**, 148–154. [240]

McLeish, D. L. (1977). On the invariance principle for nonstationary mixingales. *Ann. Probab.* **5**, 616–621. [128]

McQueen, J. B. (1973). A linear extension of the martingale convergence theorem, *Ann. Probab.* **1**, 263–271. [P]

Mihoc, G., and Craiu, M. (1972). "*Statistical Inference for Dependent Variables.*" Ed. Acad. Rep. Soc. Romania, Bucarest (in Romanian). [155]

Monroe, I. (1972). On embedding right continous martingales in Brownian motion. *Ann. Math. Statist.* **43**, 1293–1311. [269]

Monroe, I. (1978). Processes that can be embedded in Brownian motion. *Ann. Probab.* **6**, 42–56. [269]

Moran, P. A. P. (1962). "*The Statistical Processes of Evolution Theory.*" Oxford Univ. Press, London and New York. [235]

Nakata, T. (1976). On the rate of convergence in mean central limit theorem for martingale differences. *Rep. Statist. Appl. Res. Univ. Japan Sci. Eng.* **23**, 10–15. [84]

Nelson, P. I. (1970). A class of orthogonal series related to martingales. *Ann. Math. Statist.* **41**, 1684–1694. [P]

Neudecker, W., and Williams, D. (1974). The "Reimann hypothesis" for the Hawkins random sieve. *Compositio Math.* **29**, 197–200. [229]

Neumann, J. von (1932). Proof of the quasi-ergodic hypothesis. *Proc. Nat. Acad. Sci. U.S.A.* **18**, 70–82. [280]

Neuts, M. F. (1967). Waiting times between record observations. *J. Appl. Probab.* **4**, 206–208. [5]

Nevel'son, M. B., and Has'minskii, R. Z. (1973/76). "*Stochastic Approximation and Recursive Estimation,*" Translations of Math. Monographs 47, Amer. Math. Soc., Providence, Rhode Island. [238]

Neveu. J. (1965). "*Mathematical Foundations of the Calculus of Probability.*" Holden-Day, San Francisco, California. [31, 56]

Neveu, J. (1975). "*Discrete-Parameter Martingales.*" North-Holland Publ., Amsterdam. [31, 118]

Olshen, R. (1974). A note on exchangeable sequences. *Z. Wahrsch. Verw. Gebiete.* **28**, 317–321. [204]

Oodaira, H., and Yoshihara, K. (1971a). The law of the iterated logarithm for stationary processes satisfying mixing conditions. *Kōdai Math. Sem. Rep.* **23**, 311–334. [7, 128]

Oodaira, H., and Yoshihara, K. (1971b). Note on the law of the iterated logarithm for stationary processes satisfying mixing conditions. *Kōdai Math. Sem. Rep.* **23**, 335–342. [7, 128]

Oodaira H., and Yoshihara, K. (1972). Functional central limit theorems for strictly stationary processes satisfying the strong mixing conditions. *Kōdai Math. Sem. Rep.* **24**, 259–269. [7, 128]

Pagan, A. R. (1980). Some identification and estimation results for regression models with stochastically varying coefficients. *J. Econometrics* (to appear). [P]

Pakes, A. G. (1980). Some remarks on a one-dimensional random walk with repulsion. *J. Austral. Math. Soc. Ser. A* (to appear). [P]

Pakshirajan, R. P., and Sreehari, M. (1970). The law of the iterated logarithm for a Markov process. *Ann. Math. Statist.* **41**, 945–955. [7]

Patil, G. P. (1962). Certain properties of generalized power series distribution. *Ann. Inst. Statist. Math.* **41**, 179–182. [171]

Petrov, V. V. (1975). "*Sums of Independent Random Variables.*" Springer-Verlag, Berlin and New York. [82, 90]

Philipp, W. (1969a). The law of the iterated logarithm for mixing stochastic processes. *Ann. Math. Statist.* **40**, 1985–1991. [7]

Philipp, W. (1969b). The central limit problem for mixing sequences of random variables. *Z. Wahrsch. Verw. Gebiete.* **12**, 155–171. [7]

Philipp, W. (1969c). The remainder in the central limit theorem for mixing stochastic processes. *Ann. Math. Statist.* **40**, 601–609. [89]

Philipp, W., and Stout, W. F. (1975). Almost sure invariance principles for partial sums of weakly dependent random variables. *Mem. Amer. Math. Soc.* No 161. [6–8, 128]

Plosser, C. I., Schwert, G. W., and White, H. L. (1979). Differencing as a Test of Specification. Working paper no. GPB79-7. Dept. Econ. Univ. Rochester. [P]

Prabhu, N. U. (1965). "*Stochastic Processes.*" Macmillan, New York. [4]

Pratt, J. W. (1960). On interchanging limits and integrals. *Ann. Math. Statist.* **31**, 74–77. [281]

Quine, M. P. (1976). Asymptotic results for estimators in a subcritical branching process with immigration. *Ann. Probab.* **4**, 319–325. [178]

Quine, M. P. (1977). Correction to "Asymptotic results for estimators in a subcritical branching process with immigration". *Ann Probab.* **5**, 318. [178]

Raikov, D. A. (1938). On the decomposition of the Gaussian and Poisson laws. *Izv. Akad. Nauk. SSSR. Ser. Mat.* 91–124. [55, 69]

Rao, C. R. (1963). Criteria of estimation in large samples. *Sankhyā Ser. A* **25**, 189–206. [156, 166]

Rao, C. R. (1973), "*Linear Statistical Inference and its Applications,*" 2nd ed. Wiley, New York. [155, 165]

Rao, M. M. (1961). Consistency and limit distributions of estimators of parameters in explosive stochastic difference equations. *Ann. Math. Statist.* **32**, 195–218. [172]

Rao, M. M. (1966). Inference in stochastic processes II. *Z. Wahrsch. Verw. Gebiete.* **6**, 317–335. [158]

Rényi, A. (1963). On stable sequences of events. *Sankhyā Ser. A* **25**, 293–302. [56]

Rényi, A. (1970). "*Foundations of Probability.*" Holden-Day, San Francisco, California. [205]

Rényi, A., and Révész, P. (1963). A study of sequences of equivalent events as special stable sequences. *Publ. Math. Debrecen* **10**, 319–325. [204]

Révész, P. (1968). "*The Laws of Large Numbers.*" Academic Press, New York. [22, 180]

Riesz, F. (1945). Sur la théorie ergodique. *Comment. Math. Helv.* **17**, 221–239. [281]

Robinson, P. M. (1977). The estimation of a nonlinear moving average model. *Stochastic Processes Appl.* **5**, 81–90. [197]

Rootzén, H. (1976). Fluctuations of sequences which converge in distribution. *Ann. Probab.* **4**, 456–463. [57, 135]

Rootzén, H. (1977a). On the functional central limit theorem for martingales. *Z. Wahrsch. Verw. Gebiete.* **38**, 199–210. [52, 98]

Rootzén, H. (1977b). A note on convergence to mixtures of normal distributions. *Z. Wahrsch. Verw. Gebiete.* **38**, 211–216. [52, 59]

Rootzén, H. (1980). On the Functional Central Limit Theorem for Martingales, Part II. Preprint. [52, 98]

Rosén, B. (1967a). On the central limit theorem for sums of dependent random variables. *Z. Wahrsch. Verw. Gebiete.* **7**, 48–52. [51, 97]

Rosén, B. (1967b). On asymptotic normality of sums of dependent random variables. *Z. Wahrsch. Verw. Gebiete.* **7**, 95–102. [51]

Rosén, B. (1967c). On the central limit theorem for a class of sampling procedures. *Z. Wahrsch. Verw. Gebiete.* **7**, 103–115. [51]

Rosenblatt, M. (1956). A central limit theorem and a strong mixing condition. *Proc. Nat. Acad. Sci. U.S.A.* **42**, 43–47. [128, 132]

Rosenblatt, M. (1971). "*Markov Processes. Structure and Asymptotic Behavior.*" Springer-Verlag, Berlin and New York. [20]

Rosenkrantz, W. A. (1967). On rates of convergence for the invariance principle. *Trans. Amer. Math. Soc.* **129**, 542–552. [111]

Roussas, G. G. (1972). "*Contiguity of Probability Measures.*" Cambridge Univ. Press, London and New York. [155, 159, 162, 163, 167]

Sacks, J. (1958). Asymptotic distribution of stochastic approximation procedures. *Ann. Math. Statist.* **29**, 373–405. [238, 242]

Salem, R., and Zygmund, A. (1947). On lacunary trigonometric series, I. *Proc. Nat. Acad. Sci. U.S.A.* **33**, 333–338. [7]

Salem, R., and Zygmund, A. (1948). On lacunary trigonometric series II. *Proc. Nat. Acad. Sci. U.S.A.* **34**, 54–62. [7]

Salem, R., and Zygmund, A. (1950). La loi du logarithme itéré pour les séries trigonométriques lacunaires. *Bull. Sci. Math.* (*2*) **74**, 209–224. [7]

Saridis, G. N. (1974). Stochastic approximation methods for identification and control—a survey. *IEEE Trans. Autom. Control* **19**, 798–809. [238]

Saridis, G. N. (1977). "*Self-Organizing Control of Stochastic Systems.*" Dekker, New York. [238]

Sawyer, S. (1967). A uniform rate of convergence for the maximum absolute value of partial sums in probability. *Commun. Pure Appl. Math.* **20**, 647–658. [11, 270]

Sawyer, S. (1968). Uniform limit theorems for the maximum cumulative sum in probability. *Trans. Amer. Math. Soc.* **132**, 363–367. [111]

Sawyer, S. (1972). Rates of convergence for some functionals in probability. *Ann. Math. Statist.* **43**, 273–284. [89, 111]

Sen, P. K. (1979). Invariance principles for the coupon collectors problem: a martingale approach. *Ann. Statist.* **7**, 372–380. [P]

Schmetterer, L. (1966). On the asymptotic efficiency of estimates. *In* "*Research Papers in Statistics. Festschrift for J. Neyman*" (F. N. David, ed.), pp. 301–317. Wiley, New York. [156]

Scott, D. J. (1971). An invariance principle for reversed martingales. *Z. Wahrsch. Verw. Gebiete.* **20**, 9–27. [53]

Scott, D. J. (1973). Central limit theorems for martingales and for processes with stationary increments using a Skorokhod representation approach. *Adv. Appl. Probab.* **5**, 119–137. [44, 52, 104, 140]

Scott, D. J. (1977). A central limit theorem for martingales and an application to branching processes. *Stochastic Processes Appl.* **6**, 241–252. [52]

Scott, D. J. (1979). On a Connection Between Martingales and Reversed Martingales. Preprint No. 3, Mathematisches Institut, Univ. Muenchen. [78]

Seneta, E. (1974). A note on the balance between random sampling and population size. *Genetics* **77**, 607–610. [237]

Serfling, R. J. (1968). Contributions to central limit theory for dependent variables. *Ann. Math. Statist.* **39**, 1158–1175. [7, 280]

Shorack, G. R., and Smythe, R. T. (1976). Inequalities for max $|S_k|/b_k$ where $k \in N^r$. *Proc. Amer. Math. Soc.* **54**, 331–336. [16]

Silvey, S. D. (1961). A note on maximum likelihood in the case of dependent random variables. *J. Roy. Statist. Soc. Ser. B* **23**, 444–452. [155]

Simmons, G. F. (1963). "*Introduction to Topology and Modern Analysis.*" McGraw-Hill, New York. [128]

Shorokhod, A. V. (1965). "*Studies in the Theory of Random Processes.*" Addison-Wesley, Reading, Massachusetts. [269, 270]

Smith, J. C. (1945). Asymptotic distribution of sums of Rademacher functions. *Bull. Amer. Math. Soc.* **51**, 941–944. [64]

Smythe, R. T. and Wierman, J. C. (1978). "*First Passage Percolation on the Square Lattice.*" Springer Lecture Notes in Mathematics No. 671. [213]

Solo, V. (1978). *Time Series Recursions and Stochastic Approximations.* Ph.D. thesis, The Australian National Univ., Canberra. [240, 242]

Stout, W. F. (1970a). A martingale analogue of Kolmogorov's law of the iterated logarithm. *Z. Wahrsch. Verw. Gebiete* **15**, 279–290. [116]

Stout, W. F. (1970b). The Hartman-Wintner law of the iterated logarithm for martingales. *Ann. Math. Statist.* **41**, 2158–2160. [10, 116]

Stout, W. F. (1974). "*Almost Sure Convergence.*" Academic Press, New York. [22, 31, 280]

Strassen, V. (1964). An invariance principle for the law of the iterated logarithm. *Z. Wahrsch. Verw. Gebiete.* **3**, 211–226. [116, 119]

Strassen, V. (1967). Almost sure behaviour of sums of independent random variables and martingales. *Proc. 5th Berkeley Symp. Math. Statist. Probab.*, **2**, 315–343. [52, 269]

Sunklodas, J. (1977). An estimation of the convergence rate in the central limit theorem for weakly dependent random variables. *Litovsk. Mat. Sb.* **17**, 41–51, 211–212. (Russian. Lithuanian and English summaries.) [89]

Sweeting, T. J. (1978). On efficient tests for branching processes. *Biometrika* **65**, 123–127. [165]

Takahashi, S. (1951a). On the central limit theorem. *Tohoku Math. J.* (2) **3**, 316–321. [64]

Takahashi, S. (1951b). On the asymptotic distribution of the sum of independent random variables. *Proc. Japan Acad.* **27**, 393–400. [64]

Takahashi, S. (1975). Probability limit theorems for trigonometric series. *In* "*Limit Theorems of Probability*" (P. Révész, ed.), pp. 381–397. North-Holland Publ, Amsterdam. [7]

Tsypkin, Ya. Z. (1971). "*Adaptation and Learning in Automatic Systems.*" Academic Press, New York. [238]

Venter, J. H. (1967). An extension of the Robbins-Monro procedure. *Ann. Math. Statist.* **38**, 181–190. [238]

Villalobos, A. G. (1974). Some functional limit theorems for dependent random variables. *Ann. Probab.* **2**, 1090–1107. [7]

Ville, J. (1939). "*Etude Critique de la Notion de Collectif.*" Gauthier-Villars, Paris. [1]

Volkonskii, V. A., and Rozanov, Yu. A. (1959). Some limit theorems for random functions. I. *Theory Probab. Appl.* **4**, 178–197. [277]

Wald, A. (1948). Asymptotic properties of the maximum likelihood estimate of an unknown parameter of a discrete stochastic process. *Ann. Math. Statist.* **19**, 40–46. [158]

Wasan, M. T. (1969). "*Stochastic Approximation.*" Cambridge Univ. Press, London and New York. [238]

Watson, R. K. (1980a). A useful random time-scale transformation for the standard epidemic model. *J. Appl. Probab.* (to appear). [P]

Watson, R. K. (1980b). An application of a martingale central limit theorem to the standard epidemic model. *Stochastic Processes Appl.* (to appear). [P]

Weber, N. C. (1979). Uniform Bounds for the Departure from Normality of Reversed Martingales. Preprint. [96]

Weiss, L. (1971). Asymptotic properties of maximum likelihood estimators in some nonstandard cases. *J. Amer. Statist. Assoc.* **66**, 345–350. [158]

Weiss, L. (1973). Asymptotic properties of maximum likelihood estimators in some nonstandard cases II. *J. Amer. Statist. Assoc.* **68**, 428–430. [158]

Weiss, L., and Wolfowitz, J. (1966). Generalized maximum likelihood estimators. *Theory Probab. Appl.* **11**, 58–81. [157, 160]

Weiss, L., and Wolfowitz, J. (1974). "*Maximum Probability Estimators and Related Topics*." Springer Lecture Notes in Mathematics No. 424. [155, 156, 164]

Wetherill, G. B. (1966). "*Sequential Methods in Statistics*." Methuen, London. [238]

Whittaker, E. T., and Watson, G. N. (1927). "*Modern Analysis*," 4th ed. Cambridge Univ. Press, London and New York. [232]

Wunderlich, M. C. (1974). A probabilistic setting for prime number theory. *Acta Arith.* **26**, 59–81. [227]

Wunderlich, M. C. (1976). The prime number theorem for random sequences. *J. Number Theory* **8**, 369–371. [227]

Zacks, S. (1971). "*The Theory of Statistical Inference*." Wiley, New York. [155]

Zolotarev, V. M. (1967). A generalization of the Lindeberg-Feller Theorem. *Theory Probab. Appl.* **12**, 608–618. [52]

Index to Theorems, Corollaries, and Examples

Index*

*P refers to the Preface.